MEDICAL RADIOLOGY

Diagnostic Imaging

E. Neri · M. Cosottini · D. Caramella (Eds.)

MR Angiography of the Body

Technique and Clinical Applications

With Contributions by

M. Anzidei · V. Battaglia · E. Belloni · F. Bianchi · P. Boraschi · G. Caproni
C. Catalano · A. Carriero · F. de Cobelli · M. Cosottini · F. Donati · L. Faggioni
R. Fossaceca · M. Francone · I. Iacucci · A. Laghi · M. Lombardi · P. Lucchesi
A. del Maschio · M. Milanesi · E. Neri · P. Paolantonio · I. Pesaresi · M. Rengo
A. Roberta · S. Salemi · M. di Terlizzi · F. Turini · V. Vallini · A. Vanzulli · C. Vignali
F. Vullo · V. Zampa

Foreword by

A. L. Baert

With 91 Figures in 182 Separate Illustrations, 35 in Color and 5 Tables

Springer

Emanuele Neri, MD
Diagnostic and Interventional Radiology
Department of Oncology, Transplants,
and Advanced Technologies in Medicine
University of Pisa
Radiodiagnostica 1 Universitaria
Ospedale Cisanello. Via Paradisa 2,
56100 Pisa
Italy

Mirco Cosottini, MD
Unit of Neuroradiology
Department of Neurosciences
University of Pisa
Via Roma 67
56126 Pisa,
Italy

Davide Caramella, MD
Diagnostic and Interventional Radiology
Department of Oncology, Transplants,
and Advanced Technologies in Medicine
University of Pisa
Via Roma 67
56126 Pisa
Italy

Medical Radiology · Diagnostic Imaging and Radiation Oncology
Series Editors:
A. L. Baert · L. W. Brady · H.-P. Heilmann · H. Hricak
M. Knauth · M. Molls · C. Nieder · M. F. Reiser

Continuation of Handbuch der medizinischen Radiologie
Encyclopedia of Medical Radiology

ISBN: 978-3-540-79716-6 e-ISBN: 978-3-540-79717-3

DOI: 10.1007/978-3-540-79717-3

Springer Heidelberg Dordrecht London New York

Medical Radiology · Diagnostic Imaging and Radiation Oncology ISSN 0942-5373

Library of Congress Control Number: 2009930364

Cover design: Publishing Services Teichmann, 69256 Mauer, Germany

Printed on acid-free paper

9 8 7 6 5 4 3 2 1

Springer is part of Springer Science+Business Media (www.springer.com)

Foreword

Indirect angiography following intravenous bolus injection of iodinated contrast media entered clinical practice as an alternative less-invasive method to direct (arterial) angiography when digital substraction X-ray techniques became available.

Because of its technical shortcomings, IVDSA was later replaced by CT angiography as another less-invasive modality for vascular imaging. The introduction of multislice CT scanners and rapid progress in electronic image reconstruction allowed CT angiography to become the actually well-known, highly performing, and reliable radiological modality.

MR angiography offers better patient safety than CT angiography because of the absence of ionizing irradiation. As compared with ultrasound vascular imaging, it offers and superior accuracy as well as better contrast resolution.

This book offers a comprehensive overview of the latest technological advances in MR angiography, including up to date image acquisition techniques as well as optimal sequences. Special attention is given to flow-based angiography, optimal and safe use of MR contrast media, rapid and reliable image procession, as well as to the correct interpretation of common artifacts.

The main section of this book is devoted to the current clinical applications of MR angiography in the different body organs and anatomic regions as well as to the important role of this method in the managing of transplant patients. The book is exquisitely illustrated with multiple high-quality demonstrative cases and as always presented in technically impeccable printing and lay out

The editors are internationally well known for their superior and longstanding knowledge in MR angiography and image procession. They are supported by a group of distinguished collaborating authors, who are also experts in the field.

I am very much indebted to the editors and especially to E. Neri for preparing this outstanding volume in a record short time period, which enabled them to include the latest technical advances in this rapidly evolving important radiological method.

It is highly recommended to general and organ specialized radiologists as a most welcome update of their knowledge and as a practical guide in their daily practice. I am convinced that this volume, which nicely completes the list of topics dealt with in our series, will meet appropriate interest and success with our readers.

Leuven, Belgium

Albert L. Baert

Foreword

Indirect angiography following intravenous bolus injection of iodinated contrast media entered clinical practice as an alternative less-invasive method to direct (arterial) angiography when digital substraction X-ray techniques became available.

Because of its technical shortcomings, IVDSA was later replaced by CT angiography as another less-invasive modality for vascular imaging. The introduction of multislice CT scanners and rapid progress in electronic image reconstruction allowed CT angiography to become the actually well-known, highly performing, and reliable radiological modality.

MR angiography offers better patient safety than CT angiography because of the absence of ionizing irradiation. As compared with ultrasound vascular imaging, it offers and superior accuracy as well as better contrast resolution.

This book offers a comprehensive overview of the latest technological advances in MR angiography, including up to date image acquisition techniques as well as optimal sequences. Special attention is given to flow-based angiography, optimal and safe use of MR contrast media, rapid and reliable image procession, as well as to the correct interpretation of common artifacts.

The main section of this book is devoted to the current clinical applications of MR angiography in the different body organs and anatomic regions as well as to the important role of this method in the managing of transplant patients. The book is exquisitely illustrated with multiple high-quality demonstrative cases and as always presented in technically impeccable printing and lay out

The editors are internationally well known for their superior and longstanding knowledge in MR angiography and image procession. They are supported by a group of distinguished collaborating authors, who are also experts in the field.

I am very much indebted to the editors and especially to E. Neri for preparing this outstanding volume in a record short time period, which enabled them to include the latest technical advances in this rapidly evolving important radiological method.

It is highly recommended to general and organ specialized radiologists as a most welcome update of their knowledge and as a practical guide in their daily practice. I am convinced that this volume, which nicely completes the list of topics dealt with in our series, will meet appropriate interest and success with our readers.

Leuven, Belgium Albert L. Baert

Preface

Magnetic resonance angiography, in short MRA, is a fast growing imaging procedure that fills many sessions of our busy MR suites.

However, at early introduction, when only unenhanced sequences could be acquired by using phase contrast and time-of-flight, MRA was considered just an optional tool that could be used to complement a standard MRI examination.

Nowadays, the technique is a well-established noninvasive angiographic examination, substitutive of conventional diagnostic angiography and of paramount importance in many clinical situations.

The use of contrast-enhanced sequences has improved both contrast and spatial resolution, making MRA a key imaging tool in the study of abdominal aorta and peripheral vessels, where it plays a significant role in the planning of interventional procedures.

Moreover, MRA is increasingly moving toward the study of more complex clinical situations, as in the case of transplantation surgery.

This book was made possible because of the valuable contributions of many outstanding experts in this field, and it covers all the relevant clinical applications of MRA in the study of the whole body. Starting from the basic imaging techniques, it moves to a detailed description of the vascular anatomy and deals with the most appropriate strategies for the diagnosis of vascular pathology.

We hope that this book may help practicing radiologists in making the best use of this exciting methodology, and encourage referring physicians to include MRA in the diagnostic work-up of their patients.

Pisa, Italy

Emanuele Neri
Mirco Cosottini
Davide Caramella

Contents

Image Acquisition Technique and Sequences 1

1 **Flow-Based MRA** 3
Angelo Vanzulli

2 **MR Angiography Contrast Agents** 7
Ilaria Pesaresi and Mirco Cosottini

3 **Image Acquisition Technique and Sequences Contrast-Enhanced MRA** 17
Marco Francone, Michele Anzidei, Ilaria Iacucci, Francesco Vullo, and Carlo Catalano

4 **Artifacts in MR-Angiography** 27
Andrea Laghi, Pasquale Paolantonio, Marco Rengo, and Paola Lucchesi

5 **Image Processing** 35
Lorenzo Faggioni and Emanuele Neri

Clinical Applications 45

6 **Radiologic Vascular Anatomy** 47
Claudio Vignali, Valentina Vallini, and Francesca Bianchi

7 **Neck Vessels** 73
Mirco Cosottini and Ilaria Pesaresi

8 **Thoracic Aorta and Pulmonary Vessels** 87
Marco Di Terlizzi, Ambrosini Roberta, Rita Fossaceca, and Alessandro Carriero

9 **Heart and Coronary Arteries** 105
Massimo Lombardi and Matteo Milanesi

10 **Abdominal Aorta and Renal Arteries** 115
Virna Zampa, Francesca Turini, Lorenzo Faggioni, and Gabriele Caproni

11 **Peripheral Vessels** 127
Francesco De Cobelli, Elena Belloni,
and Alessandro Del Maschio

Special Topics: Transplants 143

12 **MRA in Liver and in Orthotopic Liver Transplants** 145
Piero Boraschi, Francescamaria Donati,
Simonetta Salemi, and Francesca Turini

13 **MRA in Transplanted Pancreas and Kidneys** 157
Piero Boraschi, Francescamaria Donati,
and Valentina Battaglia

Subject Index 169

List of Contributors 175

Image Acquisition Technique and Sequences

Flow-Based MRA

1

Angelo Vanzulli

CONTENTS

1.1 Time of Flight (TOF) MR Angiography *4*

1.2 Phase Contrast (PC) MR Angiography *4*

1.3 T2-Based MR Angiography *5*

References *5*

ABSTRACT

MR imaging is particularly suited to depict vessels and this can be obtained using sequences that highlight flow. In the era of nephrogenic sistemic fibrosis these unenhanced sequences are used more frequently than in the past.

These sequences are based on three principles: time of flight, phase contrast, and fresh blood acquisitions. The physical principles of these acquisitions are explained in order to understand the basis of MR angiography.

In the evaluation of vessel lumen, MR sequences can be applied in order to detect flow. These sequences are based mainly on three principles: (a) using the inflow into the slab of fresh spins, which are not subjected to all the radiofrequency pulses, unlike the stationary tissue: time-of-flight phenomenon; (b) detecting the phase changes of moving spins subjected to a gradient: phase contrast; (c) optimizing an ultrafast spin echo or gradient echo sequence, in order to obtain bright signal from blood due to its T2 properties.

In all three techniques, the signal and resulting image are produced by physiological flow, and these methods will be degraded by artifacts in case of complex or turbulent flows [1].

In general, the MR angiography techniques use sequences derived from T1-weighted gradient echo sequences. In fact vessels generally appear as hyposignals in spin echo sequences because of the outflow effect. The spins in the blood are excited during the slice selection pulse. At time TE/2, the flowing spins move out of the slice and are not subjected to the 180° pulse: therefore they do not give any signal in the original voxel (signal void). When the flow is turbulent and not laminar, some spins are back in the

Angelo Vanzulli, MD
Niguarda Cà Granda Hospital, Piazza Ospedale Maggiore 3, 20162 Milano, Italy

original voxel at the time of the echo refocusing and give some signal (artifact) [2].

1.1 Time of Flight (TOF) MR Angiography

In time-of-flight MR angiography [3], gradient-echo sequences are optimized to favor the vascular signal over that of the surrounding tissues by saturating the stationary tissue signal with very short TR (TR largely inferior to the tissues T1). The longitudinal magnetization of these tissues does not have time to regrow after some 90° pulses and after their signal weakens. The flowing spins coming into the slice volume have not been saturated yet, giving high signal intensity. The strength of the vascular signal is proportional to the flow velocity (faster flow gives higher signal intensity) and depends on the length and orientation of the vessel (the vascular signal will be higher if the slice is perpendicular to the axis of the vessel) due to shorter travel of the spins into the slice volume. In the case of a long vessel tract traveling into the slice, flowing spins will receive many saturating pulses, lowering the signal they can return from them. The shorter the TR and the higher the flip angle, the stronger is the stationary tissues suppression, but slow flowing spins can also be suppressed. A longer TE causes dephasing of the flowing spins, lowering vessel signal intensity. An increased slice thickness also causes a longer travel of the flowing spins into the imaging plane with some degree of vessel signal loss due to the progressive saturation of the spins.

The main limitations of time-of-flight MRA are signal loss linked to spin dephasing when the flow is complex or turbulent (stenosis), when the flow is too slow or oriented parallel to the slice plane and poor signal suppression of the stationary tissues when substances with very short T1 relaxation time are present (fat, blood degradation products) [4]. Vascular contrast can be improved by suppressing the signal coming from static tissues, by means of a magnetization transfer preparation pulse or by selective excitation of water, or by fat saturation [5].

To selectively visualize arterial or venous flow, presaturation bands can be applied upstream to the selected slice (arterial suppression to detect venous flow only) or downstream to the selected slice (venous suppression).

Time of Flight MR angiography can be obtained in 2D or 3D mode.

In 2D acquisition, single thin slices are obtained in sequential order while in 3D acquisition, a thicker slice is excited and many single partitions are reconstructed from this thick slab by different phase encoding steps. The main advantage of the 2D technique is better sensitivity to slow flows (shorter travel of the moving spins into the slice), with the possibility of using higher flip angles (giving better stationary tissue saturation). The main drawback of 2D acquisition is poor through plane spatial resolution due to the thickness of the slices.

Contrary to the 2D TOF and 3D TOF, volumetric imaging gives good spatial resolution in all spatial directions, with a better signal-to-noise ratio, but the flowing spins travel through a thick volume, with progressive signal loss due to in-plane saturation. The slowest flows are more sensitive to this artifact. Flow saturation can be reduced by (a) dividing 3D acquisition into thinner slabs (MOTSA: multiple overlapping thin slab acquisition), (b) by using a variable excitation angle that is lower as the flow enters the volume and higher as it leaves the volume (TONE: tilted optimized nonsaturating excitation), thus compensating relaxation of short T1 tissues.

1.2 Phase Contrast (PC) MR Angiography

Phase contrast angiography relies on dephasing the moving spins submitted to a bipolar gradient in gradient echo acquisitions. In the presence of a bipolar gradient of a given intensity and time, the moving spins will dephase in proportion to their velocity while stationary spins will be dephased and rephased by the opposite gradients, returning to their original status. So the flow velocity is proportional to the phase shift and the stationary tissues phase shift is null.

Similar to spatial encoding in the phase direction, the possible phase values range from $-\pi$ to $+\pi$. Beyond this range of values, aliasing occurs, causing wrong velocity encoding [6].

The encoding gradient characteristics are thus defined in order to encode flows within a certain velocity range from −venc to +venc; this range has to be determined by the user before acquiring the data. Any velocity outside this range will be poorly encoded (similar to what happens in pulsed and color Doppler with pulse repetition frequency-PRF). The bipolar gradients are applied in a particular axis (x, y, or z) and the dephasing of the spins moving along the gradient axis is proportionate to their velocity, gradient intensity and the application time of a gradient lobe. Magnetic field heterogeneities can cause stationary spin dephasing. To compensate for these dephasings, a second acquisition is obtained, reversing the order of the encoding gradient lobes, and subtracting the two acquisitions: the moving spins will cumulate dephasings in the opposite direction, while dephasing of stationary spins due to field heterogeneities will be identical for both acquisitions, disappearing in subtraction.

To depict the movements in all the directions of space, the acquisitions are repeated with flow-encoding gradients in each of the three spatial directions. An additional acquisition with no flow-encoding gradient will serve as a reference.

The images acquired with flow-encoding gradient are summed to depict flow in all directions and subtracted from the reference image without encoding gradient. With this subtraction only vessels are depicted, as well as those that run parallel to the imaging plane.

Phase data allow the measurement of flow velocity and direction [7].

The limitations of this technique relate to intra-voxel dephasing in case of complex or turbulent flow with loss of flow signal in the presence of vascular loops, bifurcations, or stenoses.

Phase contrast MRA can be obtained from a thick slab 2D acquisition, allowing synchronization with ECG pulse (cine MR phase contrast angiography) or allowing to test different flow velocities for further 3D longer acquisitions.

To obtain a quantitative evaluation of flows, the slice plane has to be perpendicular to flow direction. Thus a flow velocity curve can be obtained as a function of time, which, when coupled with the area of the vessel section, will permit one to calculate flow rate.

In phase contrast 3D MR angiography, each partition is encoded in three directions, resulting in long acquisition time.

The 3D acquisition gives thinner partitions with better image quality than in 2D single-slice technique.

Because Phase Contrast MR Angiography is more sensitive than time of flight to slow flow, its main application is cerebral venous imaging. The scan time of these sequences can be greatly reduced by parallel imaging [8].

1.3 T2-Based MR Angiography

These MR angiography techniques are based on two different sequences: a very short TR-TE gradient echo acquisition called balanced fast gradient echo (or tru-FISP) and a 3D half-fourier fast spin echo sequence prospectively synchronized with ECG pulse.

The first technique (balanced fast gradient echo) relies on an ultrafast gradient echo sequence in which the TR value is exactly equal to 2TE. This causes the production of a stimulated spin echo signal every three gradient echo read-out which gives a particular high signal for fluids and blood (proportional to T2/T1).

This sequence allows the detection of the vessel, but fluids also have the same high signal intensity, preventing one from obtaining MIP images.

The other MRA technique based on T2 relaxation of blood is based on a 3D ultrafast half Fourier spin echo acquisition with prospective ECG synchronization and fat suppression with an inversion pulse (STIR) [9].

The half Fourier spin echo sequence is fast (less than 1 s per slice) and its movements are insensitive. The center of k-space is filled by precocious echo (short effective TE) in which the flow-induced dephasing has not reduced the vascular signal yet. The acquisition plane is parallel to the axes of the vessels. The phase encoding direction is chosen parallel to the axes of the vessels in order to prevent the read-out gradient from weakening the vascular signal.

Prospective ECG synchronization triggers acquisitions at the same moment in the cardiac cycle. The optimal interval between R wave of the ECG and acquisition is determined by a prior calibration sequence and optimized for the vessel of interest. This interval is chosen with different 2D images obtained with different time delays.

A 3D acquisition with a time delay adjusted to obtain signals from slow flowing venous vessels can be used as a reference and be subtracted from the arterial acquisition to depict only arteries.

STIR preparation reduces the signal from the surrounding fat tissues.

References

1. Ersoy H, Zhang H (2006) Peripheral MR angiography. J Cardiovasc Magn Reson 8(3):517–528
2. Foo TK, Polzin JA (2005) MR angiography physics: An update. Magn Reson Imaging Clin N Am 13(1):1–22

3. Anderson CM, Lee RE (1993) Time-of-flight techniques: Pulse sequences and clinical protocols. Magn Reson Imaging Clin N Am 1:217–227
4. McCauley TR, Monib A, Dickey KW, et al. (1994) Peripheral vascular occlusive disease: Accuracy and reliability of time-of-flight MR angiography. Radiology 192:351–357
5. Atkinson D, Brant-Zawadzki M, Gillan G, et al. (1994) Improved MR angiography: Magnetization transfer suppression with variable flip angle excitation and increased resolution. Radiology 190(3):890–894
6. O'Donnell M (1985) NMR blood flow imaging using multiecho, phase contrast sequences. Med Phys 12:59–64
7. Steinberg FL, Yucel EK, Dumoulin CL, et al. (1990) Peripheral vascular and abdominal applications of MR flow imaging techniques. Magn Reson Med 14:315–320
8. Wilson GJ, Hoogeveen RM (2004) Parallel imaging in MR angiography. Top Magn Reson Imaging 15(3):169–185
9. Ho KY, de Haan MW, Oei TK, et al. (1997) MR angiography of the iliac and upper femoral arteries using four different inflow techniques. AJR Am J Roentgenol 169:45–53

Miyazaki M, Sugiura S (2000) Non-contrast-enhanced MR angiography using 3D ECG-synchronized half-Fourier fast spin echo. J Magn Reson Imaging 12(5):776–783

MR Angiography Contrast Agents

2

ILARIA PESARESI and MIRCO COSOTTINI

CONTENTS

2.1 Magnetic Properties *7*

2.2 Relaxivity *8*
2.2.1 Paramagnetic Contrast Agents *8*
2.2.2 Superparamagnetic and Ferromagnetic Contrast Agents *8*

2.3 Susceptibility Effect *9*

2.4 Contrast Agents for Vascular Imaging *9*
2.4.1 Paramagnetic Gadolinium Agents *9*
2.4.1.1 Extracellular Fluid Agents *9*
2.4.1.2 Blood-Pool Agents *11*
2.4.2 Superparamagnetic Ultrasmall Iron Oxide Particles *13*

2.5 Safety *14*

2.6 Future Perspectives *15*
2.6.1 Contrast Agent Use at High Field *15*
2.6.2 Other Contrast Agents *15*
2.6.2.1 Gadolinium-Based Particulate Agents *15*
2.6.2.2 Hyperpolarized Contrast Agents *15*
2.6.2.3 Chemical Exchange Saturation Transfer (CEST) *15*

References *16*

ABSTRACT

Gadolinium-based agents are the most widely employed agents for CE-MR angiography. They act as positive paramagnetic substances, thus they increase the vascular signal by strongly reducing the *T1* relaxation time of blood. Several Gd-based agents have been approved for clinical application to date. On the basis of their distribution, Gd-agents are subdivided into extra-cellular fluid (ECF) and blood-pool agents.

Conventional ECF agents have been employed for long time in vascular imaging. Given their brief plasma half-life, they require a careful timing of the MRA acquisition in order to properly sample the contrast bolus first pass.

Blood-pool agent have been recently introduced. They are characterized by high relaxivity constants. Since they persist in the vascular compartment for a quite long time, blood-pool agents allow the steady state acquisition of angiograms with higher spatial resolution.

Contrast agents for magnetic resonance imaging reached clinical validation in 1988 and were initially employed for the visualization of diseased tissues. MR angiography imaging was related to flow-based phenomenon until the introduction of MRA with contrast agents in early 1990s (CREASY et al. 1990).

2.1 Magnetic Properties

MR contrast agents are represented by molecules endowed with not-null magnetic properties, which act by affecting the relaxation times of the surrounding water protons.

ILARIA PESARESI, MD
Diagnostic and Interventional Radiology, Department of Oncology, Transplants, and Advanced Technologies in Medicine, University of Pisa, Via Roma 67, 56100 Pisa, Italy
MIRCO COSOTTINI, MD
Unit of Neuroradiology, Department of Neurosciences, University of Pisa, Via Roma 67, 56100 Pisa, Italy

The magnetic property of a substance derives from the existence of unpaired electrons within the externals orbitals, a typical characteristic of transitional metal (e.g., iron, manganese) and lanthanide ions (e.g., gadolinium). The number of unpaired electrons influences the magnetic moment of a substance, which reflects the efficiency of that substance to modify the environmental magnetic field (magnetic susceptibility).

Substances with not-null magnetic properties are divided into paramagnetic, ferromagnetic, and superparamagnetic. Paramagnetic substances acquire a net magnetism only when exposed to an external magnetic field. They are made of multiple ions which act as isolated magnetic dipoles without reciprocal magnetic interaction, so that the overall net magnetization is given by the sum of single ion magnetizations.

Ferromagnetic and superparamagnetic materials are characterized by a cooperative interaction among the constituting magnetic elements so that the resulting magnetic field is much higher than the sum of the single dipoles. While superparamagnetic materials acquire a net magnetic field only when submitted to an external magnetic drift, ferromagnetic ones maintain a net magnetism even under neutral external conditions. Iron oxides may exhibit either superparamagnetism (when arranged in small particles) or ferromagnetism (when organized in large crystals).

2.2 Relaxivity

All contrast agents shorten both T1 and T2 relaxation times of surrounding protons. This effect on proton relaxation is ruled by the relaxivity constant, which indicates the contrast agent's ability to decrease the T1 (longitudinal relaxivity, r1) and T2 (transversal relaxivity, r2) relaxation times of the water protons per unit (mM) concentration of metal ion.

The relaxation effect differs for paramagnetic, superparamagnetic, and ferromagnetic contrast agents.

2.2.1 Paramagnetic Contrast Agents

The relaxation effect of paramagnetic substances in solution is mainly realized through dipolar interactions between the paramagnetic ions and the hydrogen molecules of water belonging to the so-called inner sphere. Indeed, paramagnetic contrasts can be modelled as isolated ions surrounded by an inner and an outer sphere of interacting water molecules. While the energetic transfer between the metal ion and inner sphere water molecules is explained by dipolar interactions, the relaxation effect on outer-sphere molecules is explained by the Curie-spin relaxation theory (Gueron 1975; Muller et al. 2001; Caravan and Randall 2006).

The relaxation efficiency (i.e., relaxivity) of paramagnetic materials is governed by many factors such as the magnetic moment of the ion, the number of water molecules within the inner-sphere and the rate of exchange among the inner sphere and the bulk solvent. Moreover, the local magnetic field generated by the ion must fluctuate at a rate (ω_i) close to the hydrogen Larmor frequence (ω_H) in order to stimulate hydrogen relaxation.

As free metal and lanthanide ions are embedded into chelates to avoid in vivo toxicity, contrast agent relaxivity is influenced by both paramagnetic ion and molecular ligands properties.

In particular, ligands influence the number of water molecules within the inner sphere and the fluctuation rate of the ionic magnetic field ω_{i}.

Paramagnetic agents modify proton relaxation according to the following equations:

$$1/T1 = 1/T1_0 + r1 \times C \quad (2.1)$$

$$1/T2 = 1/T2_0 + r2 \times C \quad (2.2)$$

where T1 and T2 identify the final relaxation times in the presence of the contrast agent, $T1_0$ e $T2_0$ the initial relaxation times, *C* the molar concentration of the contrast agent, r1 and r2 the longitudinal and transversal relaxivity respectively of the contrast agent.

For all medically used paramagnetic contrast agents r2 is higher than r1. However, as $1/T1_0$ is lower than $1/T2_0$, a dominant effect on T1 relaxation is observed.

2.2.2 Superparamagnetic and Ferromagnetic Contrast Agents

The relaxation effect of superparamagnetic and ferromagnetic agents is not based on direct dipolar interactions. These agents are constituted by particles of variable dimension which form a stable suspension. As no water molecules are admitted within the particles, direct interactions between ions and water molecules are prevented. The relaxation effect is exerted on water molecules diffusing near the particles (the outer sphere) through the Curie-spin relaxation (Muller et al. 2001; Gueron 1975).

Most superparamagnetic and ferromagnetic agents are made of iron oxide crystals enveloped by a dextran or siloxan coating which prevents agglomeration. Each crystal is made of several thousands of magnetic ions which interact cooperatively so that the magnetic moment of the entire crystal tends to align with the external magnetic field.

Ultrasmall particles of iron oxides (USPIOs) contain a single iron oxides crystal within the particle core and show a superparamagnetic behavior. Small particles of iron oxide (SPIOs) contain multiple iron oxide crystals within the particle core and behave as ferromagnetic agents.

Relaxivity constants are strictly influenced by the particles core dimension. In particular, r2/r1 ratio increases with increasing particle size; thus smaller particles are much better T1-shortening agents than larger ones (Roch et al. 1999).

USPIOs show high r1 and r2 values, so they can be employed to either increase signal intensity on T1-weighted images or decrease signal on T2-weighted images.

SPIOs are characterized by the highest r2 constant, and then they cause a dramatic signal drop on T2-weighted images.

To sum up, paramagnetic substances are classified as *T1 agents* or *positive agents* because they relevantly increase signal intensity on T1-weighted images. Ferromagnetic agents (SPIOs) are classified as *T2 agents* or *negative agents* as they cause a signal drop on T2-weighted images. Superparamagnetic agents (USPIOs) can be classified as T1 or T2 agents as they can either increase signal intensity on T1-weighted images or decrease signal on T2-weighted images.

2.3 Susceptibility Effect

Even if contrast agents are generally exploited for their effect on proton relaxation, they can also exert a relevant susceptibility effect (*T2* agents*).

All contrast agents locally increase the static magnetic field (effective magnetic field). If field inhomogeneities reach a sufficient strength, they can significantly affect signal intensity by broadening the Larmor frequency and fastening the protons dephasing within a voxel. This effect is most evident with long TE (on T2-weighetd images) as intravoxel dephasing and diffusion of water molecules through regions of variable magnetic field become more evident (Caravan and Randall 2006).

The existence of strong field inhomogeneities subsequent to contrast agent administration can be due to crystals aggregation, as observed with SPIOs, or due to compartmentalization.

Ferromagnetic, superparamagnetic, and paramagnetic agents can all undergo compartmentalization in tissues or vessels and manifest a medically relevant susceptibility effect.

2.4 Contrast Agents for Vascular Imaging

Vascular MR contrast agents for clinical application are mainly represented by positive paramagnetic agents. Few studies have dealt with superparamagnetic iron oxides for vascular imaging, however, to date these agents have generally been restricted to the investigation of the reticulo-endothelial-system (RES).

2.4.1 Paramagnetic Gadolinium Agents

Most contrast agents approved for human applications are gadolinium-based. Gadolinium is a lanthanide ion endowed with a high magnetic moment and a proper magnetic field fluctuation rate which result in a high relaxivity constant. Given its high toxicity, gadolinium needs to be embedded into chelates to be administered in vivo. Different gadolinium chelates show different relaxation properties and different biodistribution and must be subdivided into extracellular fluid (ECF) and blood-pool agents.

2.4.1.1 Extracellular Fluid Agents

ECF agents represent the first generation of clinically approved gadolinium contrast-agents and their role in contrast-enhanced magnetic resonance imaging is well established. They are characterized by a brief vascular phase, then they rapidly equilibrate in the extracellular space (distribution half-life of about 5 min) reaching roughly the same concentration in the vascular and interstitial compartments.

Given their brief vascular phase, the adoption of ECF agents for MR angiography implies a careful timing and a fast imaging in order to capture the contrast bolus first pass in the arterial district.

The clearance of ECF agents is mainly mediated by the renal system, with an elimination half-life of about 80 min.

All gadolinium ECF agents are characterized by an eight-coordinate ligand binding to gadolinium (III) to prevent the release of ions in solution. Gadolinium ligands are represented by small molecules (polyaminocarboxylate/phosphonate derivatives) which differ in their electrical charge, either neutral or negative, and in their chemical structure, either cyclic or linear (Table 2.1 and Fig. 2.1).

Neutral chelates allow lower osmolarity in solution with respect to the ionic ones and can be formulated at high-concentration.

Linear chelates have been reported to present a slightly lower thermodynamic stability with respect to cyclic complexes and they might theoretically facilitate the release of gadolinium ions because of biochemical competitions for the binding site (Wiginton et al. 2008). However, no evidence of relevant effects in vivo has been reported.

Despite different chemical structures, ECF agents are characterized by almost the same relaxation properties, biodistribution, and plasma half-life.

ECF agents are employed in contrast-enhanced MRA for bright blood imaging. Their effect on T1 relaxation is ruled by (2.1) and thereafter, it linearly depends on the concentration of contrast agents. While performing CEMRA examination, the concentration of contrast media within the vessel depends

Table 2.1. Gadolinium-based extracellular fluid contrast agents approved for marketing

Type	Generic name	Trade name	Chemical abbreviation	Chemical structure	R1 (0.5 T, 37°C) (1/mM/s)	R2 (0.5 T, 37°C) (1/mM/s)	Osmolarity[a] (Osm/Kg)
Ionic	Gadopentate dimeglumine	Magnevist (Bayer Schering)	Gd-DTPA	Linear	3.8	3.8[b]	1.96
	Gadoterate meglumine	Dotarem (Guerbet)	Gd-DOTA	Cyclic	3.6	4.8	1.35
Neutral	Gadodiamide	Omniscan (GE Healthcare)	Gd-DTPA-BMA	Linear	3.9	4.3[b]	0.79 – (1.90)
	Gadoteridol	Prohance (Bracco)	Gd-HP-DO3A	Cyclic	3.7	4.8[b]	0.63 – (1.91)
	Gadobutrol	Gadovist (Bayer Schering)	Gd-BT-DO3A	Cyclic	3.6		0.57 – (1.39)
	Gadoversetamide	OptiMARK (Mallinckrodt)	Gd-DTPA-BMEA	Linear	4.7		1.11

[a]Osmolarity values at 0.5 M concentration, except those in parentheses (1 M concentration)
[b]Relaxivity at 0.5 T

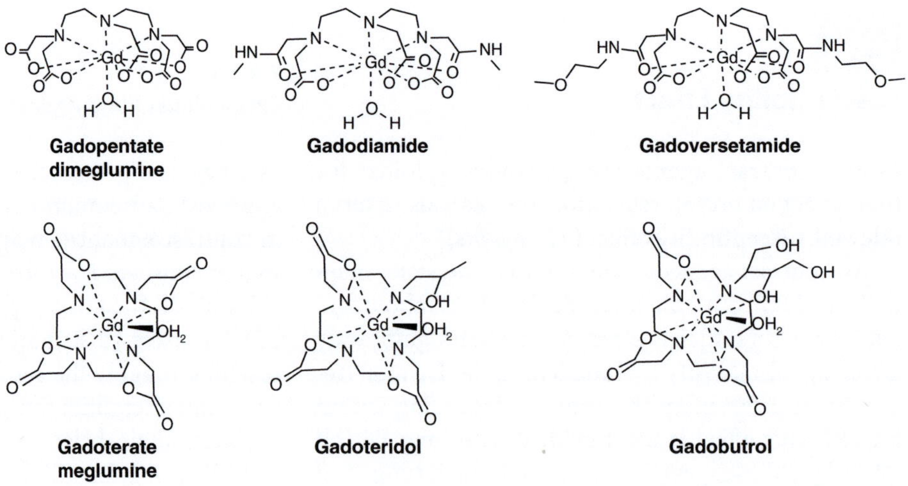

Fig. 2.1. Chemical structure of gadolinium-based extracellular fluid (ECF) agents

not only on contrast media formulation but also on physiological parameters as well as the injection rate and the contrast amount, as detailed in Chap. 7.

Contrast agents formulated at high concentration potentially allow higher relaxation effects. However, they modify the bolus geometry by reducing the bolus length and are thus particularly suited for fast angiographic imaging.

2.4.1.2 Blood-Pool Agents

While ECF agents rapidly leak out from the vascular compartment, the blood-pool agents (BPAs) are confined within vessels for a quite long time.

In order to keep gadolinium contrast complexes within the vascular compartment, two main approaches have been adopted. The first approach consists in increasing the size of gadolinium ligands (large-size BPAs) and the second one in utilizing small gadolinium ligands reversibly bound to plasma proteins (small-size BPAs).

Large-size BPAs were initially obtained through covalent binding of gadolinium to macromolecules such as polylysine, dextran, or modified bovine serum albumin. These compounds showed a negligible leakage in the extravascular compartment and provided optimal vascular imaging. Moreover, they were associated with very high gadolinium relaxivity because large-size ligands lessen the gadolinium magnetic field fluctuation rate and favor energetic exchange between the protons and gadolinium. Despite such promising features, large-size BPAs did not achieve clinical approval as they were limited by a very-slow blood clearance and by potentially dangerous immunologic effects.

To overcome these problems, a second generation of large-size BPAs has been developed. This new class of BPAs agents (Table 2.2 and Fig. 2.2) is made of gadolinium complexes large enough to persist within vessels for long time but small enough to be excreted

Table 2.2 Gadolinium-based blood pool agents approved for marketing or under clinical investigation

Type	Generic name	Trade name	Chemical abbreviation	Affinity for albumin	Bound-fraction (%)	R1 (0.5 T/37°C) (1/mM/s)	R2 (0.5 T/37°C) (1/mM/s)
Large-size BPAs	Gadomer 17,24	Gadomer-17 (Bayer Schering)	Gd-DTPA-17 cascade polymer	–	–	11.9[a]	16.5[a]
	Gadomelitol	Vistarem (Guerbet)	P792	–	–	45	50
Small-size BPAs	Gadobenate	Multihance (Bracco)	Gd-BOPTA	Weak	10	Buffer: 4.4 Plasma: 9.7	5.6
	Gadofosveset trisodium	Vasovist[b] (Bayer Schering)	MS-325	Strong	91	Buffer: 6.6 Plasma: 50	
	Gadocoletic acid	B22956 (Bracco)	B22956	Strong	95	Buffer: 39 Plasma: 44.5	

[a]Relaxivity at 1 T
[b]Approved for marketing in Europe and USA

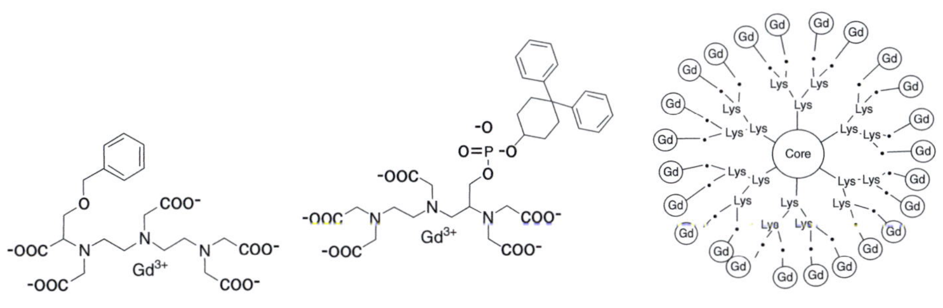
Gadobenate Gadofosveset Gadomer 17,24

Fig. 2.2. Chemical structure of some gadolinium-based blood-pool agents through glomerular filtration. They are still characterized by high relaxivity. An example is represented by Gd-DTPA-17 cascade polymer (*Gadomer-17*, Bayer Schering), a macromolecular complex with a dentritic architecture which includes multiple gadolinium ions. *Gadomer-17* does not show significant affinity for plasma proteins and undergoes renal clearance.

Biodegradable macromolecular complexes which decompose through disulfide-thiol exchange, facilitating the excretion of gadolinium chelates, are currently under development (Lu et al. 2004).

Small-size BPAs (Table 2.2) are designed to reversibly bind to plasma proteins, in particular to albumin which represents the protein with the highest concentration in plasma (about 0.67 mM). The not-covalent binding to albumin is mediated by hydrophobic moieties attached to the chelating agents.

This class of contrast agents can be further subdivided on the basis of the level of affinity for plasma proteins. Low-affinity BPAs (e.g., Gd-BOPTA, *Multihance*) are characterized by bound-fractions of about 10% while high-affinity BPAs (e.g., Gadofosveset trisodium, *Vasovist* or B22956) reach bound-fractions of 90–95%.

Similar to large-size BPAs, small-size BPAs have r1 relaxivity that is much higher than ECF agents, because of the lower fluctuation rate of gadolinium magnetic field secondary to the albumin binding.

It should be noted that neither large nor small-size BPAs can be considered pure blood-pool compounds, as they show a minimal diffusion into the interstitial space (Bremerich et al. 2007).

Contrast enhanced MR angiography may take advantage from BPAs. Given the extended imaging window (concentration of BPAs in plasma remains stable for over 1 h), BPAs minimize the problem of sequence temporization and temporal resolution. Furthermore, they potentially allow higher spatial resolution with acquisitions at the steady-state (Fig. 2.3)

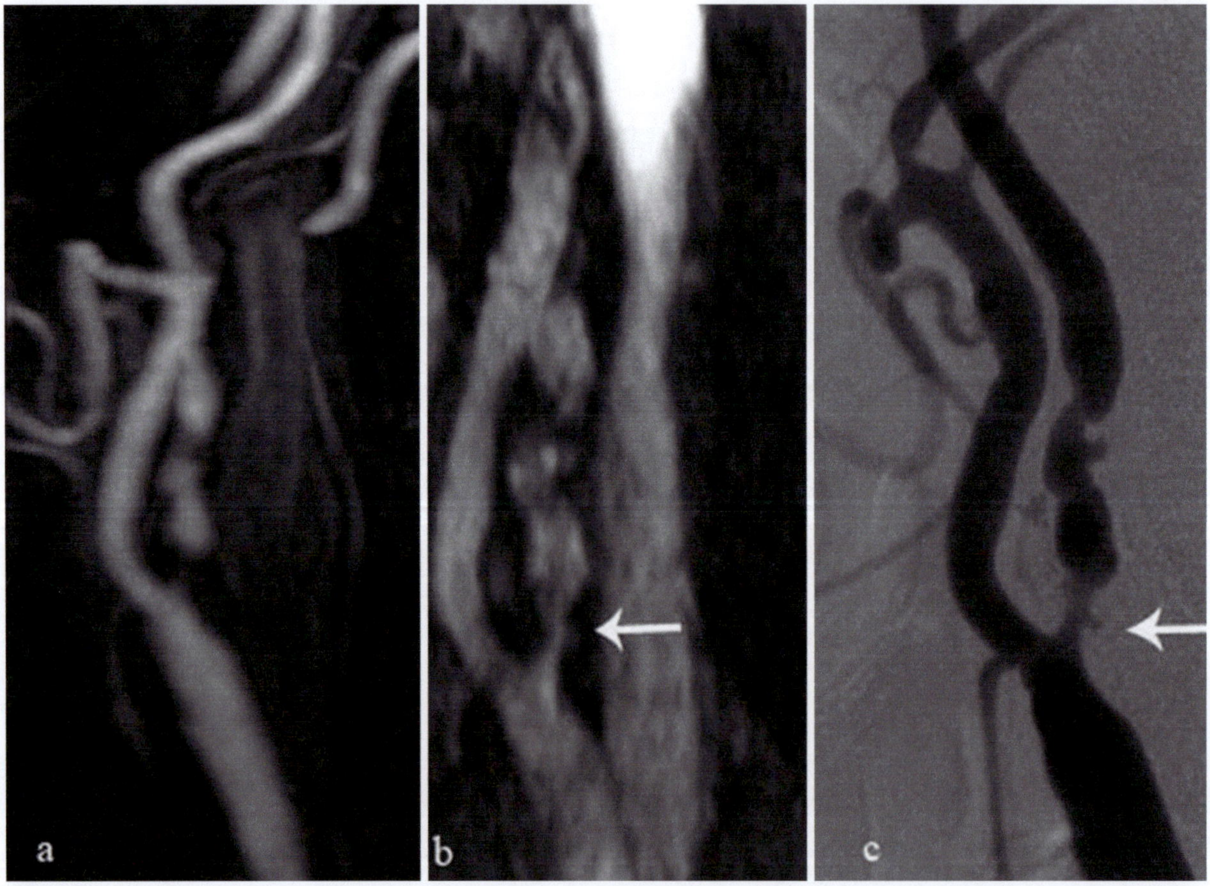

Fig. 2.3. Contrast-enhanced MR angiography with a gadolinium blood-pool agent (Gadofosveset trisodium, Vasovist, Bayer Schering). First pass contrast-enhanced angiograms (**a**) correctly classify the patient as having severe stenosis (>70%). Nevertheless, the stenosis is overestimated with respect to the gold standard DSA (**c**) and the residual patent lumen is poorly defined. The high spatial resolution MRA at steady state (**b**) increases the vessels' sharpness and allows a more clear depiction of the stenosis morphology (ulcerated plaque)

and extend the investigation from the arterial to the venous district. All gadolinium-based BPAs, due to their high r1 relaxivities, provide potentially higher vessel-to-background signal ratio than ECF agents allowing the utilization of lower concentrations and doses. Nevertheless, the effective signal gain on first pass magnetic resonance angiograms with BPAs, with respect to ECF agents, is controversial. Indeed, according to both literature (Klessen et al. 2007) and personal data (Fig. 2.4), BPAs may show a reduced vessels-to-background relative contrast. This finding could be explained by the nonlinear relationship between the T1 relaxation effect and the concentration of small-size BPAs. Indeed, the relaxivity of small-size BPAs is strongly influenced by the albumin-bound fraction. The reversible reaction between contrast complexes and albumin reaches equilibrium after a time interval, which depends on the reciprocal affinity constant. During the contrast agent's first pass in the arterial district, the equilibrium between bound and unbound fractions may not be achieved yet. Therefore, the first pass contrast relaxivity may be lower with respect to the steady-state one (Caravan and Randall 2006).

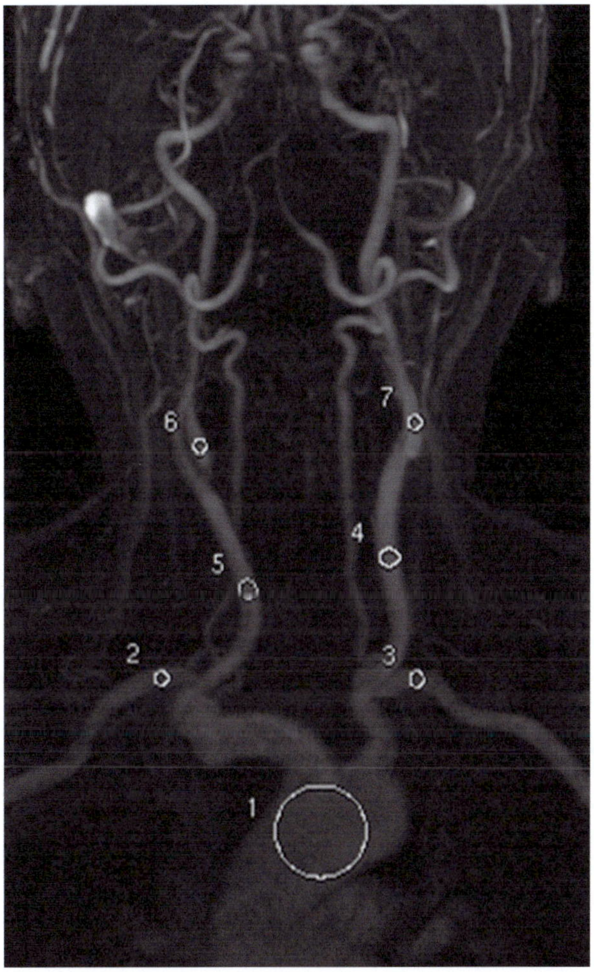

Fig. 2.4. Vessels-to-background relative contrast on first pass CEMRA with gadolinium blood-pool and ECF agents. The signal enhancement of a blood-pool agent (Gadofosveset trisodium, Vasovist, Bayer Schering) is compared with that of ECF agents on first-pass MRA. Vessels' signal is measured by placing regions of interest in the aortic arch, in the subclavian and in the carotid arteries. The relative vessels-to-background contrast (RC) is calculated by the ratio (SV − SF)/(SV + SF), where SV is the signal into the vessels and SF the signal of the background. Gadofosveset shows a significantly lower signal enhancement in two regions

To date the only gadolinium blood-pool agent approved for marketing is Gadofosveset trisodium (*Vasovist*, Schering).

2.4.2 Superparamagnetic Ultrasmall Iron Oxide Particles

USPIOs (Table 2.3) represent true BPAs, as they are too large to leak out from vessels. While SPIOs (Table 2.3) are rapidly removed from the blood stream by endocytosis in liver, spleen, or lymp nodes through the reticulo-endothelial system (vascular half-life of about 10 min), USPIOs persist within vessels for quite a long time so that they can be exploited to obtain high-resolution steady-state angiograms. Subsequent to RES clearance, iron oxide agents are incorporated in the body iron pool.

To date, SH U555 C (*Supravist*, Bayer Schering) is the only USPIO agent which has completed the clinical trials, even though it has not achieved FDA approval. It is a derivative of ferucarbutran with smaller dimensions with respect to *Resovist*, Bayer Schering (particle diameter of about 20 nm vs. 50 nm).

While SPIO agents and some USPIOs (e.g., Sinerem, Guerbet) can be administered only by slow intravenous infusion, SH U555 C admits bolus injection, so it can be employed for both first pass and high-resolution steady state angiography (up to 42 min postinjection) (Bremerich et al. 2007).

Given their superparamagnetic properties, USPIOs can be adopted as T1 agents to perform bright blood contrast-enhanced angiography. To avoid signal loss due to rapid intravoxel dephasing (high r2 values and susceptibility effect), angiographic sequences with very low echo times are required (Bremerich et al. 2007).

Table 2.3. Iron oxides-based contrast agents approved for marketing or under clinical investigation

Particle type	Generic name	Trade name	Chemical abbreviation	Crystals diameter (nm)	Particle diameter (nm)	R1 (0.5 T, 37°C) (1/mM/s)	R2 (0.5 T, 37°C) (1/mM/s)
USPIOs	Ferumoxtran	Sinerem[a] (Guerbet) Combidex (AMAG Pharmaceuticals)	AMI-227	4.3–4.9	50	22.7	53.1
	Feruglose	Clariscan (GE Healthcare)	NC 100150	4–7	20	21.8	35.3
	Ferucarbotran	Supravist (Bayer Schering)	SH U555 C	3–5	20	22	45
SPIOs	Ferumoxide	Endorem[b] (Guerbet) Feridex[b] (AMAG Pharmaceuticals)	AMI-25	4.3–4.8	200	24	107
	Ferucarbotran	Resovist[b] (Bayer Schering)	SH U555 A	4.2	62	20	190

[a]Withdrawn from market
[b]Approved for marketing in Europe and/or USA

2.5 Safety

ECF gadolinium agents have been widely employed in clinical setting and their safety profiles have been deeply investigated. For example, Gd-DTPA (*Magnevist*) has been used in more than 45 million magnetic resonance imaging procedures since 1988 and is currently used globally in more than 5 million applications annually. The broadest category of spontaneously reported adverse events, that is subjective symptoms, occurs in less than 0.01% of procedures. Within the total number of adverse events reported, the distribution of serious and non-serious reports was 9.3 and 90.7% respectively (Knopp et al. 2006).

With regard to blood-pool gadolinium agents, in particular Gadofosveset trisodium (Vasovist), phase II and phase III clinical trials report a safety profile comparable to that of conventional ECF gadolinium agents (Shamshi et al. 2006).

As gadolinium agents require very low doses to ensure adequate contrast enhancement compared to iodinate contrast agents, for many years they were considered safe compounds to be used in patients with renal function deficiency. In the last years, an increasing alert on gadolinium-related nephrogenic systemic fibrosis (NFS) has arisen.

NFS is a sclerosing disorder clinically characterized by indurated dermal plaques, mainly distributed in the lower extremities, and by fibrosis in other tissues such as striated muscles, myocardium, lungs, and dura mater (Wiginton et al. 2008).

The first case of NSF was reported by Cowper et al. in 2000.The association between NSF and gadolinium contrast agents relies on the identification of gadolinium deposits in skin biopsies (Wiginton et al. 2008). Up to now, the number of reported cases in the USA (about 130) is extremely small when compared with the millions of patients who have been exposed to gadolinium-based agents, which makes the overall risk of developing NSF very low (Geraldes and Laurent 2009). NFS has been exclusively identified in patients with severely impaired renal function. In these patients, the longer half-life of the contrast agents potentially increases the risk of trans-metallation with the release of free gadolinium ions in tissues (Bongartz 2007).

Although there is currently evidence of a strong association between gadolinium exposure and the development of NSF in patients with severe renal failure, the causative role of gadolinium in the pathogenesis of the disease can not be stated with absolute certainty (Kurtkoty et al. 2008).

On 23 May 2007 the U.S Food and Drug Administration has asked manufacturers to include a new boxed warning on all gadolinium-based contrast agents, stating that patients with severe kidney insufficiency are at risk of developing NSF. FDA has assessed that all gadolinium-based contrast agents marketed in the United States should be equally treated in this regard as it is not possible to know the extent of risk associated with each agent (Kanal et al. 2008).

Given the great concern on gadolinium-related NSF in patients with renal function impairment, an increasing interest has been addressed to iron oxide agents. Indeed, USPIOs clearance is mediated by the reticulo-endothelial system and their safety profile is not altered in patients with kidney insufficiency (Neuwelt et al. 2009).

2.6 Future Perspectives

2.6.1 Contrast Agent Use at High Field

Even though most contrast-enhanced MR angiography studies are currently performed with clinical 1.5 T scanners, higher magnetic fields (from 3 up to 7 T) are getting more and more widespread. By increasing the external static magnetic field (B_0) both the relaxation properties of hydrogen protons and of contrast agents get modified.

Indeed, as B_0 increases, water protons lessen their longitudinal relaxation and fasten their transversal relaxation. The dependence of contrast agent relaxivity from B_0 strength varies on the basis of the magnetic and chemical properties of the compounds.

Gadolinium ECF compounds manifest a minimal relaxivity field-dependence, while blood-pool gadolinium agents show relevant r1 variations at different B_0. Such discrepancy between ECF and BPAs agents is mainly determined by the different fluctuation rates of gadolinium magnetic field (ω_g). Indeed, BPAs ligands (both large-size and small-size ligands) slow down molecule tumbling and reduce ω_g. This effect may strongly enhance r1 with a variable effectiveness depending on the external magnetic field B_0.

Iron oxide agents dramatically increase their transversal relaxivity (r2) together with their susceptibility effect at higher magnetic fields. As a consequence, a confounding $T2^*$ shortening may affect the signal intensity on bright-blood angiograms so that the signal-to-noise ratio from blood is similar at 1.5 and 3 T (Bremerich et al. 2007).

2.6.2 Other Contrast Agents

A huge variety of new contrast agents are currently under development. In some cases they represent a further evolution of traditional paramagnetic and superparamagnetic agents, while in other cases they exploit completely different principles.

2.6.2.1 Gadolinium-Based Particulate Agents

Gadolinium (III) can be employed in numerous particulate contrast agents in order to improve their relaxation efficiency (Geraldes and Laurent 2009). For example, Gd^{3+} ions can be bound to amphiphylic chelates in micelles or liposomes, or to porous materials like zeolites. Gadolinium oxide nanoparticles can also be formed. Although promising, none of these agents has passed the preclinical status to date.

2.6.2.2 Hyperpolarized Contrast Agents

The signal intensity on magnetic resonance imaging is related to the number of protons which align to the external magnetic field. At 1.5 T approximately 0.0006% of protons are polarized.

In order to increase signal intensity, contrast agents made of hyperpolarized nuclei have been designed. Hyperpolarization is obtained by using a high-power laser on noble gases (^{3}He, ^{129}Xe) or ^{13}C-enriched molecules. In order to be exploited as contrast agents, these compounds require T1 relaxation times long enough to allow imaging before the recovery of the equilibrium status. ^{13}C-labeled water-soluble compounds have been employed for contrast-enhanced magnetic resonance angiograms in rats (Caravan and Randall 2006).

2.6.2.3 Chemical Exchange Saturation Transfer (CEST)

Chemical exchange saturation transfer (CEST) agents contain exchangeable hydrogen atoms (-NH, -OH, etc) which resonate at a Larmor frequency different from that of the bulk water. CEST protons can be saturated by an off-set RF pulse and can transfer their magnetization to water molecules. This effect results in water signal loss and negative image contrast. By utilizing proper pulse sequences a positive-contrast effect can also be obtained (Caravan and Randall 2006; Geraldes and Laurent 2009). CEST agents currently require a very high dose to obtain a relevant effect on signal intensity, so further development is needed for in vivo applications.

References

Bongartz G (2007) Imaging in the time of NFD/NFS: Do we have to change our routines concerning renal insufficiency? MAGMA 20:57–62

Bremerich J, Bilecen D, Reimer P (2007) MR angiography with blood pool contrast agents. Eur Radiol 17:3017–3024

Caravan P, Randall B (2006) Contrast agents: Basic principles. In: Edelman RR, Zlatkin MB, Hesselink JR (eds) Clinical magnetic resonance imaging. Saunders Elsevier, Philadelphia, PA, pp 361–366

Cowper SE, Robin HS, Steinberg SM, et al. (2000) Scleromyxoedema-like cutaneous disease in renal-dialysis patients. Lancet 356:1000–1001

Creasy JL, Price RR, Presbrey T, et al. (1990) Gadolinium-enhanced MR angiography. Radiology 175:280–283

Geraldes CFGC, Laurent S (2009) Classification and basic properties of contrast agents for magnetic resonance imaging. Contrast Media Mol Imaging 4:1–23

Gueron M (1975) Nuclear relaxation in macromolecules by paramagnetic ions: A novel mechanism. J Magn Reson 19:58–66

Kanal E, Broome DR, Martin DR, et al. (2008) Response to the FDA's May 23 2007, nephrogenic systemic fibrosis update. Radiology 246:11–14

Klessen C, Hein PA, Huppertz A, et al. (2007) First-pass whole body magnetic resonance angiography (MRA) using the blood pool contrast medium gadofosveset trisodium. Invest Radiol 42(9):659–664

Knopp MV, Balzer T, Esser M, et al. (2006) Assessment of utilization and pharmacovigilance based on spontaneous adverse event reporting of gadopentetate dimeglumine as a magnetic resonance contrast agent after 45 million administrations and 15 years of clinical use. Invest Radiol 41(6): 491–499

Kurtkoti J, Snow T, Hiremagalur B (2008) Gadolinium and nephrogenic systemic sclerosis: Association or causation. Nephrology 13:235–241

Lu ZR, Parker DL, Goodrich KC, et al. (2004) Extracellular biodegradable macromolecular gadolinium (III) complexes for MRI. Magn Reson Med 51:27–34

Muller RN, Roch A, Colet JM, et al. (2001) Particulate magnetic contrast agents. In: Merbach AE, Tçth E (eds) Chemistry of contrast agents in medical magetic resonance imaging. Wiley, Chichester, UK, pp 417–435

Neuwelt EA, Hamilton BE, Varallyay CG, et al. (2009) Ultrasmall superparamagnetic iron oxides (USPIOs): A future alternative magnetic resonance (MR) contrast agent for patients at risk for nephrogenic systemic fibrosis (NSF)? Kidney Int 75(5):465–474

Roch A, Muller RN, Gillis P (1999) Theory of proton relaxation induced by superparamagnetic particles. J Chem Phys 110:5403–5411

Shamshi K, Yucel EK, Chamberlin P (2006) A summary of safety of gadofosveset (MS-325) at 0.03 mmol/kg body weight dose: Phase II and phase III clinical trials data. Invest Radiol 41(11):822–830

Wiginton CD, Kelly B, Oto A, et al. (2008) Gadolinium-based contrast exposure, nephrogenic systemic fibrosis, and gadolinium detection in tissue. AJR 190:1060–1068

Image Acquisition Technique and Sequences Contrast-Enhanced MRA

3

Marco Francone, Michele Anzidei, Ilaria Iacucci, Francesco Vullo, and Carlo Catalano

CONTENTS

3.1 Introduction *18*

3.2 Basic Principle of CE-MRA *18*

3.3 Contrast Administration Strategy *19*

3.4 Imaging and Acquisition Technique *19*
3.4.1 Fast 3D Gradient Echo Sequences and Acquisition Technique *19*
3.4.2 Parallel Imaging *21*
3.4.3 Time-Resolved CE-MRA *21*
3.4.4 Steady-State Imaging *21*

3.5 k-Space Filling Strategies *23*

3.6 Conclusions *25*

References *25*

ABSTRACT

Although MR angiography (MRA) can be performed using either black-blood or bright blood, wich have the disvantage of high sensitive to flow related-artifacts, in-plane saturation, and field inhomogeneity, and are also limited by the long acquisition times. Currently, the use of contrast medium for imaging of most vascular districts, agrees to obtain images of high diagnostic accuracy, that can be acquired in seconds rather than minutes with few flow-related artifacts. Such introduction explains the reason, Currently, MR angiography has proven to be the best approach for imaging of most vascular districts is the use of contrast-enhanced (CE-MRA), offering the opportunity to detect vascular disease rapidly and early in the course of the disease.

CE-MRA is based on the principle of shortening the *T1* relaxation of blood by intravenously injecting Gd-chelate contrast agents, wich results in a significant difference in signal intensity between flowing blood and stationary tissue, in T1 arterial or venous phase. Further advantage is shorter scan duration.

Contrast administration should be optimized in order to concentrate the highest amount of gadolinium in the acquisition temporal window within district of interest, and imaging should be ideally performed at the peak of vascular enhancement, when a maximum difference exists between signal intensity of the target vessel and the surrounding overlapping structures.

Three different techniques are currently available for this purpose: the test bolus scan, the automated bolus detection, and the MR fluoroscopic trigger.

Marco Francone, MD, PhD
Michele Anzidei, MD
Ilaria Iacucci, MD
Francesco Vullo, MD
Carlo Catalano, MD
Department of Radiology, University of Rome "La Sapienza," Policlinico Umberto I, viale Regina Elena 324, 00161 Roma, Italy

Obviously, to deliver the best contrast, k-space center-filling must correspond to the moment of peak intravascular contrast; for this purpose, the acquisition of central k-space data is timed to the contrast bolus arrival within the target vessels

The sequences best suited to CE-MRA imaging are:

- Fast 3D-GRE sequences (e.g., spoiled gradient echo SPGR-General Electric Medical Systems, Waukesha, WI); fast low-angle shot FLASH-Siemens Medical), used to obtain high spatial resolution and a good signal-to-noise.
- Parallel imaging that results in accelerated image acquisition while maintainingfull spatial resolution and image contrast.
- Time-resolved CE-MRA acquisition techniques, providing dynamic information along with high-resolution, extended anatomical coverage, and without contamination by the venous signal.
- Steady-state imaging, which enables one to take advantage of the long temporal window during the equilibrium phase of the contrast material administration.

3.1 Introduction

Although MR angiography (MRA) can be performed using either black-blood (i.e., dark signal of blood-flowing spins) or bright blood sequences (time-of-flight, phase contrast and steady-state free precession), these techniques are highly sensitive to flow related-artifacts, in-plane saturation, and field inhomogeneity, and also limited by the long acquisition times (i.e., minutes).

As a result, unenhanced MRA have progressively shown limited clinical impact for the assessment of many vascular territories (Prince et al. 2003).

Currently, the best approach for imaging of most vascular districts is the use of contrast-enhanced MRA (CE-MRA) techniques, which have rapidly evolved over the past decade as a consequence of the continuous hardware development (i.e., scanners and gradients), and implementation of new sequences that brought significant improvements in contrast-to-noise-ratio, temporal resolution, and reliability of the examinations.

Compared with unenhanced techniques, CE-MRA has proven to be highly accurate, robust, and reproducible, and can be performed in seconds rather than minutes with few flow-related artifacts (Zhang et al. 2007).

This chapter will specifically focus on the basic principles, and different techniques of CE-MRA with particular emphasis on three-dimensional gradient echo (3D-GRE) imaging, time-resolved MRA, and the role of parallel imaging. Some practical aspects of contrast media administration optimization and k-space filling strategies will also be reviewed.

3.2 Basic Principle of CE-MRA

CE-MRA is based on the principle of shortening the T_1 relaxation of blood by intravenously injecting Gd-chelate contrast agents. This results in a significant difference in signal intensity between flowing blood and stationary tissue at heavily T1-weighted arterial-phase imaging, leading to the high signal intensity of blood on post-Gd T1-weighted sequences.

Signal intensity (i.e., T1-weighted shortening) within the target vessel obviously depends on the amount of gadolinium concentrated within the vascular bed at the moment of acquisition; therefore, imaging should be ideally performed at the peak of vascular enhancement, when a maximum difference exists between signal intensity of the target vessel and the surrounding overlapping structures.

Unlike time-of-flight (TOF) or phase contrast (PC) imaging, signal of vessels in CE-MRA is based on the intrinsic T_1 signal of blood, and rather less on flow effects; therefore, as already mentioned above, this technique is less flow sensitive, which means that vascular signal is not hampered by the numerous flow-related artefacts that may occur during acquisition such as signal loss from spin saturation, or slow-flow phenomena potentially mimicking a significant stenosis or vascular occlusion particularly in small vessels caliber-like carotid arteries (Prince et al. 2003).

A further relevant advantage of CE-MRA is that the usual loss of signal-to-noise ratio (SNR) from faster scanning with most MR pulse sequences can be compensated for by injecting the same dose of contrast faster over a shorter scan duration. In this way, faster scanning can attain higher-quality images with less motion artifacts (Prince et al. 2003; Zhang et al. 2007).

With CE-MRA, the pause between arterial enhancement and the onset of venous enhancement (i.e., arteriovenous window) offers an opportunity for arterial phase imaging that shows only the arteries. When this selective arterial phase is too short, simple additional maneuvers such as thigh compression with tourniquets inflated to a pressure intermediate between arterial and venous (e.g., 60 mmHg) delays onset of venous enhancement in the last station of a single-injection, multistation peripheral or whole-body MRA. This can compensate for the fact that acquisition speed is still too slow to keep up with the bolus, and eliminates venous superimposition within the legs, one of the last remaining hurdles to routine implementation of peripheral and whole-body CE-MRA into clinical practice.

3.3 Contrast Administration Strategy

Contrast administration should be optimized in order to concentrate the highest amount of gadolinium in the acquisition temporal window within district of interest. This simple concept emphasizes the importance of contrast administration in CE-MRA as vascular visualization relies heavily on timing – i.e., the moment in which imaging is actually performed.

It is obvious that delay between contrast administration and image acquisition is highly variable, and depends on various factors including patient's individual characteristics (i.e., cardiac output and vascular anatomy), injection parameters (i.e., flow rate, monophasic vs. biphasic injection, use of saline), and intrinsic characteristics of the Gd-chelate administered (i.e., concentration, relaxivity, viscosity).

Optimal timing for image acquisition should be therefore individually determined, and three different techniques are currently available for this purpose, i.e., the test bolus scan, the automated bolus detection, and the MR fluoroscopic trigger; all these methods permit reliable and time-efficient CE-MRA acquisitions (Huston et al. 1999). Briefly, optimal delay in test bolus scan is usually calculated by administering a 1–2 mL dose of Gd-chelate contrast agent prior to main MRA acquisition, using the same injection rate of the main bolus, and measuring its arrival time in the vessel of interest; the automated bolus detection is based on the concept that contrast bolus arrival and initiation of the MRA data acquisition are automated and integrated into a single pulse sequence (Ho and Foo 1998) whereas real time CE-MRA timing utilizes a fluoroscopic trigger (Riederer et al. 2000) in which monitoring is performed by the operator visually using a continuous fast 2D spoiled gradient echo pulse sequence in with imaging centered over the vascular bed.

The rate of injection influences the peak of Gd-concentration, and thus the achievable arterial SNR. Generally, a faster injection rate will result in higher arterial SNR, but shorter bolus duration and earlier venous enhancement (Kopka et al. 1998). Injection rates of 2 mL/s are ideal for most CE-MRA applications, with little benefit shown for higher rates. Slower injection rates result in lower overall Gd concentrations, but extend the plateau phase of arterial enhancement and delay venous enhancement. Slower injection rates (e.g., 0.8–1.0 mL/s) have generally been used for multistation CE-MRA in order to prolong the arterial phase, and delay venous enhancement, both of which are desirable to ensure adequate arterial signal for multistation coverage (Meaney et al. 1999).

3.4 Imaging and Acquisition Technique

3.4.1 Fast 3D Gradient Echo Sequences and Acquisition Technique

The sequences best suited to CE-MRA imaging are fast 3D-GRE sequences (e.g., spoiled gradient echo (SPGR), General Electric Medical Systems, Waukesha, WI; fast low-angle shot (FLASH), Siemens Medical Solutions, Erlangen, Germany; T1-fast field echo (T1-FFE), Philips Medical Systems, Best, The Netherlands), with short TR (≤5 ms) to maximize T1-weighting of images and short TE (1–2 ms) to minimize T2* effects (Table 3.1). They are strongly T1-weighted sequences with destruction of residual transverse magnetization. Given that echo time is very short, the gradients are not flow compensated.

Three-dimensional acquisition is used to obtain high spatial resolution and a good signal-to-noise; a fat signal suppression method is also recommended to weaken the background signal.

It is important to note that flip angle significantly affects gradient echo signal intensity and should be of 30–45° when performing CE-MRA at 1.5 T. Also, flip angle influences the desired degree of contrast

Table 3.1. Schematic representation of different manufacture's 3D GRE sequences with corresponding scanning parameters

Imaging parameters	General electric	Philips	Siemens
Pulse sequence 3D FSPGR Imaging options Repetition time (TR)	3D FSPGR (Fast, GX) – (e.g., 4–6 ms)	3D FFE Contrast enhancment - T1 Minimum (e.g., 4–6 ms)	3D FLASH Minimum (e.g., 3–5 ms)
Echo time (TE)	Minimum (e.g., 1–2 ms)	Minimum (e.g., 1–2 ms)	Minimum (e.g., 1–2 ms)
Flip angle (FA) Bandwidth	45° ±32.25 kHz (option: ±62.5 kHz)	40° WFS = 0.9 (at 448 matrix = +57 kHz)	25° Variable (± 590 Hz/pixel)
Field of view (FOV)	30–40 cm (option: 0.8 FOV)	400 mm, RFOV = 0.75	400 mm
Matrix Number of partitions Partition thickness k-space	256 or 512 × 192–256 40–60 1.0–2.5 mm Elliptical centric Centric Reverse sequential (with partial Fourier or 0.5 NEX)	448 × 258 40–60 1.0–2.5 mm CENTRA low_high linear half scan (=partial Fourier)	256–512 × 192–384 60–80 1.0–2.5 mm Elliptical centric ± partial Fourier
Number of excitations (NEX or NSA) Timing	1 (option: 0.5) SMARTPREP, Flouro trigger or test bolus	1 (option: 0.5) Bolus trak or test bolus	1 (option: 0.5) Care bolus or test bolus
Misc. options	ZIP × 2 ZIP 512 1024 ASSET	Overcontiguous slices Reconstruct 256, 512, 1024 SENSE	Reconstruct 256, 512, 1024 SENSE, GRAPPA

between vessels and background tissue; the higher the flip angle the greater the background suppression although signal from gadolinium contrast may also become attenuated.

Using state-of-the-art MRI scanners, high resolution images could be obtained using isotropic or almost isotropic voxels up to 0.8 mm^3 with an image matrix of 384 × 384 mm; as already stated above, acquisition time should be ideally tailored according to patient's individual hemodynamic characteristic, contrast administration parameters adopted and obviously the length of the vascular district assessed.

To obtain high-quality images, it is important to have specialized coils overlying the patients to ensure a high signal to noise ratio. Image acquisition should be ideally performed 3 times.

A precontrast 3D MRA acquisition is recommended to determine proper anatomic coverage, positioning of the 3D volume, and also to potential unwanted aliasing artefacts. It is also recommended that precontrast 3D MRA acquisition should be performed in an identical fashion to the actual CE-MRA acquisition, as this will familiarize the patient to both the breath holding procedure and the expected length for the breath hold. In addition, the precontrast scan can serve as a mask for subsequent image subtraction; subtraction techniques allow to subtract background signal, and maximize the SNR of the images (Yuji Watanabe and Akira 2000).

For the CE-MRA, acquisitions are typically performed in both the arterial and delayed phases. The addition of the second delayed phase acquisition will ensure proper visualization of vessels with slow flow or depiction of a late filling false channel in the case of an aortic dissection. Additionally, the second acquisition can often distinguish artefacts from true vascular findings. Finally, the inclusion of the second postcontrast 3D MRA acquisition may provide additional venous depiction that may be of clinical importance in certain specific clinical situations.

Images can be further postprocessed with maximum and minimum intensity projections and volume rendering to generate more visually appealing images, but it should be noted that these images should not typically be used for image interpretation as manipulation of the original data can lead to over- or underestimation of critical dimensions (e.g., in the context of stenosis, extravascular pathology may also be obscured or overlooked).

3.4.2 Parallel Imaging

With the gain in speed that it entails, parallel imaging is particularly well suited to CE-MRA. Parallel imaging techniques (e.g., sensitivity encoding or SENSE (Van den Brink et al. 2003) and simultaneous acquisition of spatial harmonics or SMASH (Maki et al. 2002; Sodickson et al. 2000)) use the spatial-encoding proprieties of multiphase array coils elements to reduce the number of spatial-encoding steps required to generate an image, thereby overcoming the gradient hardware performance limits imposed by on conventional MRA techniques. In this way only a fraction of phase-encoding steps have to be acquired directly which result in accelerated image acquisition while maintaining full spatial resolution and image contrast.

Besides increased temporal resolution at a given spatial resolution, the time savings due to parallel imaging can also be used to improve the spatial resolution in a given imaging time.

3.4.3 Time-Resolved CE-MRA

Conventional static 3D GRE MRA techniques create high-spatial-resolution structural studies, but fail to provide physiological information inherent in the delivery of blood or contrast; as a consequence, dynamic assessment of contrast media bolus progression through a vascular territory, or vascular lesion has long been an essential, and exclusive feature of conventional X-ray angiography, especially X-ray digital subtraction angiography (DSA).

However, the recent development of faster MR scanners with improved gradient performances and innovative software has enabled the performance of analogous dynamic viewing of Gd-chelate contrast media using fast 2D and/or 3D MRA acquisitions (Zou et al. 2008; Du et al. 2007; Brauck et al. 2007).

The clinical availability of temporally enhanced time-resolved acquisition techniques such as TRICKS (time-resolved imaging of contrast kinetics) revolutionizes the way that MRA is performed in the clinical setting, providing dynamic information along with high-resolution, extended anatomical coverage, and without contamination by the venous signal (Fig. 3.1) (Du and Bydder 2007).

Conversion of an MRA acquisition from single-phase high-resolution to time-resolved generally involves serial scan repetition as the contrast agent is delivered to the target object or organ. Protocol modifications typically include reduction in spatial resolution and anatomical coverage (fewer and/or thicker slices) in the interest of faster cycling of the image capture. The trade-off in quality inherent in this brute force approach with conventional imaging techniques is sufficiently severe to limit applicability in clinical practice.

Modern time-resolved techniques reduce these negative trade-off aspects by creatively altering the way in which image k-space is captured. With TRICKS and its variants, the center of k-space is sampled more often than the periphery (see next paragraphs). This produces multiple physiologic snapshots per each full pass through k-space, and typically delivers a four-fold increase in temporal resolution without a reduction in the SNR. The addition of parallel imaging to a time-resolved protocol increases by twofold or more the temporal resolution, although with the typical reduction in SNR.

3.4.4 Steady-State Imaging

The use of blood-pool contrast agents, in addition to conventional first-pass arterial phase imaging, also enables the use of steady-state acquisitions (Parmelee et al. 1997; Caravan et al. 2002; Rohrer et al. 2005; Grist et al. 1998). With use of this technique, which enables one to take advantage of the long temporal window during the equilibrium phase of the contrast material administration the imaging protocols are adjusted virtually without time limitations owing to the use of preparation pulses, smaller fields of view, an increased in-plane resolution, and extra-thin three-dimensional partitions (Anzidei et al. 2009). The technique usually applied for steady state imaging is a modified three-dimensional high-spatial-resolution (up to 0.7 mm section thickness, 512 × 512 matrix, voxel size of 0.7 × 0.7 × 0.7 mm) spoiled gradient-echo sequence. Using this technique, two approaches can be chosen in order to compensate for reduced SNR, which is associated with smaller voxel size: either the use of a short TR (4–7 ms), low flip angle, and multiple excitations or a longer TR (8–15 ms), higher flip angle, and using lower bandwidths (Figs. 3.2 and 3.3). The initial observations suggest that the second strategy provides higher SNR.

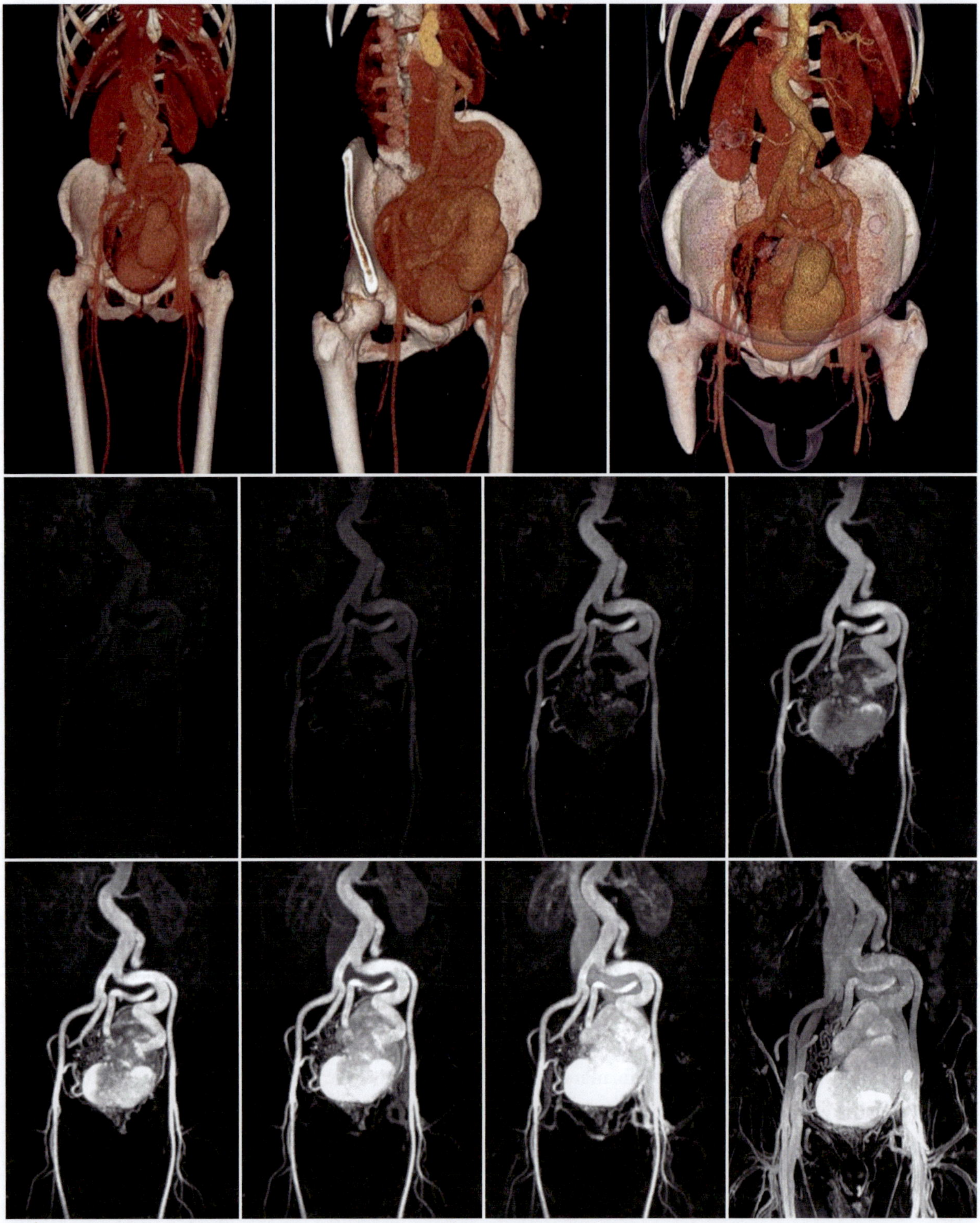

Fig. 3.1. Time-resolved MRA of a large artero-venous malformation (AVM) of the pelvis in a 50-year-old male with abdominal discomfort and a palpable mass. Dynamic MR imaging shows early and progressive filling of the AVM nidus, with evidence of an aneurismatic left internal iliac artery and dilated draining veins

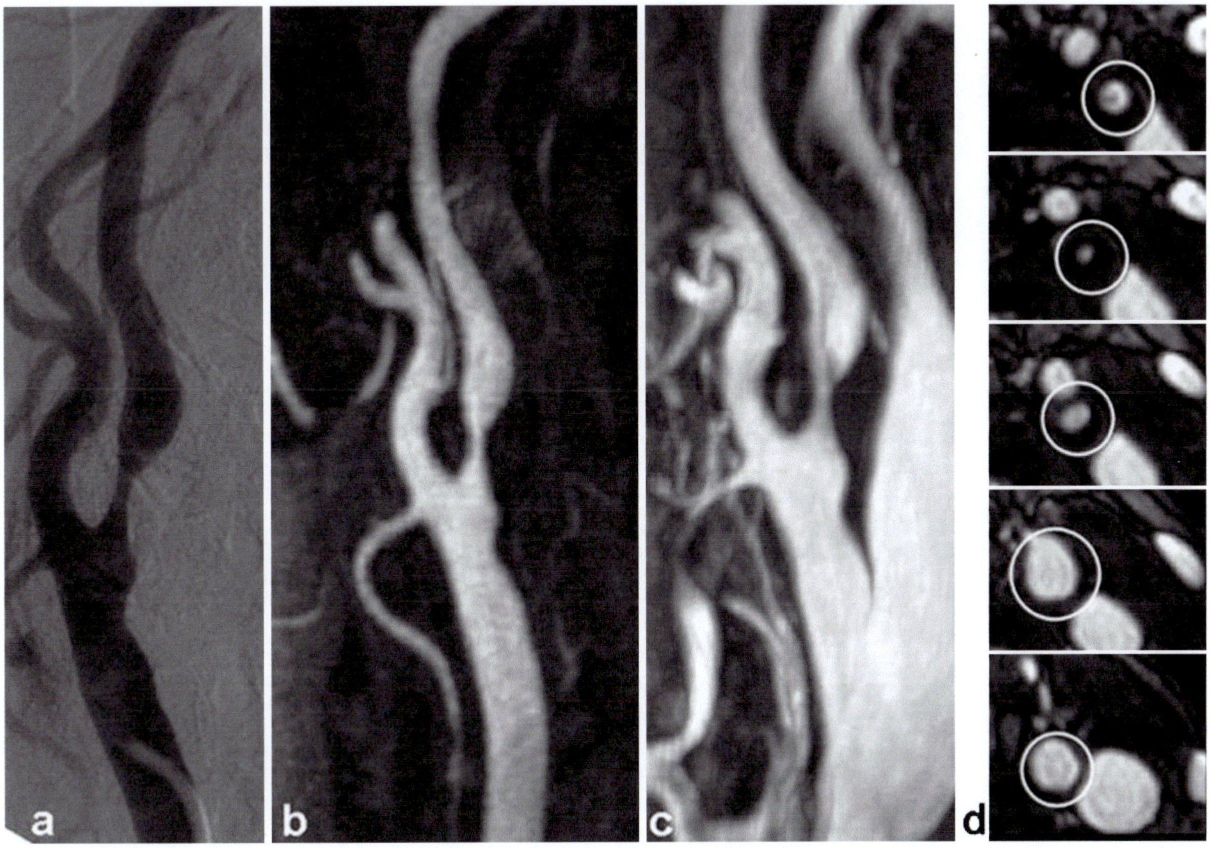

Fig. 3.2. Eighty percent stenosis (NASCET grade IV) of the left internal carotid artery (ICA) due to a long and slightly irregulae plaque. Images show good correlation between DSA findings and thin-MIP (**a**) reconstruction from FP (**b**) and steady state (**c**) CE-MRA. Multiple partition from coaxial reconstruction of steady state dataset offer a high resolution view of the plaque surface and residual lumen, confirming stenosis grading (**d**)

3.5 k-Space Filling Strategies

To deliver the best contrast, k-space center filling must correspond to the moment of peak intravascular contrast as the center of k-space contains the signal-to-noise, and contrast information for the image (also called low spatial frequency data), so it represents the major structures of the image, and thus most of the gross image form and contrast. Also, a high rate of change of the contrast agent concentration during the acquisition of central k-space must be avoided to prevent ringing artifacts, arising in the Fourier transform (Zhang et al. 2007).

k-Space filling should be therefore optimized (spiral or elliptical centrifugal trajectory, partial refilling, shared data on the k-space periphery) in CE-MRA to rapidly acquire its center (image contrast), and meet the constraints of the transient passage of the contrast agent.

Traditional k-space acquisition schemes are linear in that k-space data is filled sequentially from top to bottom (sequential k-space filling), and the central portion of the k-space is filled in the middle of the acquisition period.

Contrast agent arrival and imaging however are best synchronized if the acquisition of central k-space data is timed to the contrast bolus arrival within the target vessels. For this purpose, centric and more recently elliptical centric phase ordering schemes have been introduced for CE-MRA (Flamm and Muthupillai 2004). In centric phase ordering, the central lines of k-space are filled in the first half of the acquisition period. With elliptical centric phase ordering the most central points (i.e., radial distance from center) are acquired first, resulting in a shortened and more compact imaging period for the acquisition of central k-space data (Fig. 3.4).

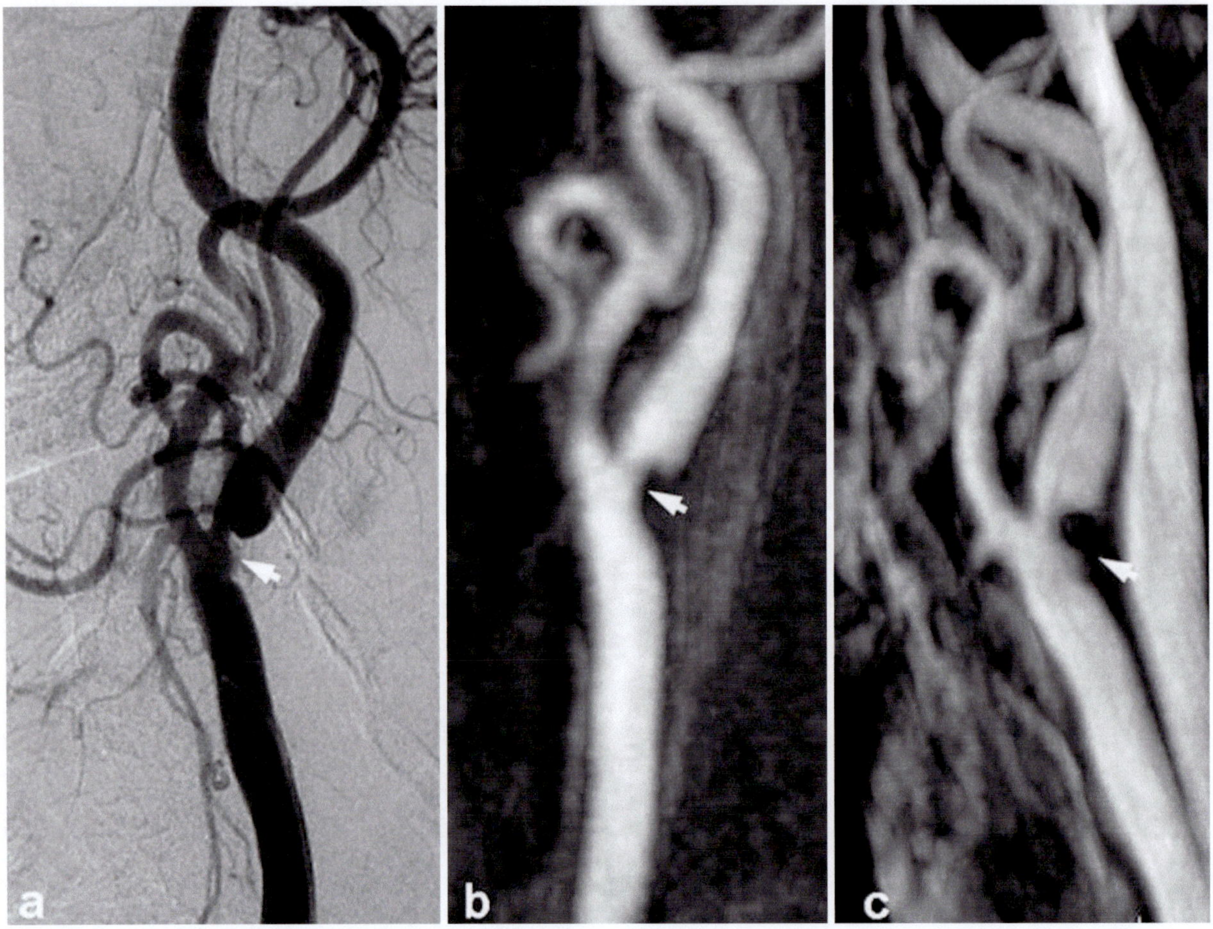

Fig. 3.3. Discrepancy of grading ICA stenosis (*arrow*) between the different CE-MRA reading modalities. DSA shows a 75% stenosis (NASCET grade IV) of the left ICA; on thin-MIP (**a**) reconstruction from first pass dataset (**b**) stenosis was underestimated at 66% (NASCET grade III), while on steady state images it was correctly measured as a NASCET grade IV (**c**)

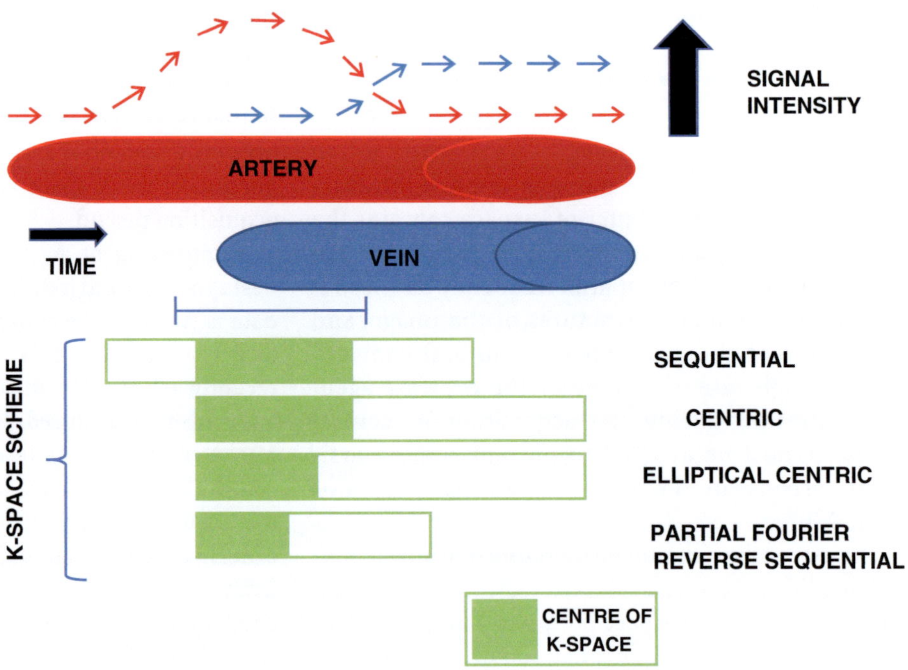

Fig. 3.4. Figure shows alignment of preferential arterial-phase enhancement for different k-space filling schemes used for CE-MRA. It is important to remark that central k-space data (i.e., low spatial frequency data) should be acquired during the plateau phase of arterial enhancement. In the conventional sequential k-space scheme, the central k-space data is acquired during the middle of the data acquisition period whereas in both conventional centric and elliptical centric acquisition schemes, central k-space filling occurs at the beginning of imaging

3.6 Conclusions

MRA has evolved rapidly since the early studies by Prince et al. (1993), becoming a reference technique for noninvasive assessment of vascular anatomy, and hemodynamic significance of stenoses, particularly suited in pediatric patients.

Currently, the optimal quality for MRA image can only be achieved administering contrast agents, which allow to overcome many of the limitation of unenhanced MRA techniques; the recent development of blood-pool agent has further improved image quality enabling acquisition of high resolution submillimetric images during steady state imaging.

Furthermore, faster imaging speed can also be used to improve the temporal resolution of CE MRA, and time-resolved MRA has emerged an interesting application in selected diagnostic situations allowing the possibility to dynamically assess progression of contrast bolus through the vascular bed.

Safe, accurate mapping of vascular anatomy with MRA offers the opportunity to detect vascular disease rapidly and early in the course of the disease, while there is time to salvage end-organ function. It gives both anatomical and functional information, and can image multiple fields of view for a more comprehensive assessment of systemic vascular disease.

References

Anzidei M, Napoli A, Marincola BC, et al. (2009) Gadofosveset-enhanced MR angiography of carotid arteries: does steady-state imaging improve accuracy of first-pass imaging? Comparison with selective digital subtraction angiography. Radiology 251:457–466

Brauck K, Maderwald S, Vogt FM, et al. (2007) Time-resolved contrast-enhanced magnetic resonance angiography of the hand with parallel imaging and view sharing: initial experience. Eur Radiol 17:183–192

Caravan P, Cloutier NJ, Greenfield MT, et al. (2002) The interaction of MS-325 with human serum albumin and its effect on proton relaxation rates. J Am Chem Soc 124:3152–3162

Du J, Bydder M (2007) High-resolution time-resolved contrast-enhanced MR abdominal and pulmonary angiography using a spiral-TRICKS sequence. Magn Reson Med 58:631–635

Du J, Fain SB, Korosec FR, et al. (2007) Time-resolvedcontrast-enhanced carotid imaging using undersampled projection reconstruction acquisition. J Magn Reson Imaging 25: 1093–1099

Flamm SD, Muthupillai R (2004) Coronary artery magnetic resonance angiography. J Magn Reson Imaging 19: 686–709

Grist TM, Korosec FR, Peters DC, et al. (1998) Steady-state and dynamic MR angiography with MS-325: initial experience in humans. Radiology 207:539–544

Ho VB, Foo TK (1998) Optimization of gadolinium-enhanced magnetic resonance angiography using an automated bolus-detection algorithm (MR SmartPrep). Original investigation. Invest Radiol 33:515–523

Huston J 3rd, Fain SB, Riederer SJ, et al. (1999) Carotid arteries: maximizing arterial to venous contrast in fluoroscopically triggered contrast-enhanced MR angiography with elliptic centric view ordering. Radiology 211:265–273

Kopka L, Vosshenrich R, Rodenwaldt J, et al. (1998) Differences in injection rates on contrast-enhanced breath-hold three-dimensional MR angiography. AJR Am J Roentgenol 170:345–348

Maki JH, Wilson GJ, Eubank WB, et al. (2002) Utilizing SENSE to achieve lower station sub-millimeter isotropic resolution and minimal venous enhancement in peripheral MR angiography. J Magn Reson Imaging 15(4):484–491

Meaney JF, Ridgway JP, Chakraverty S, et al. (1999) Stepping-table gadolinium-enhanced digital subtraction MR angiography of the aorta and lower extremity arteries: preliminary experience. Radiology 211:59–67

Parmelee DJ, Walovitch RC, Ouellet HS, et al. (1997) Preclinical evaluation of the pharmacokinetics, biodistribution, and elimination of MS-325, a blood pool agent for magnetic resonance imaging. Invest Radiol 32:741–747

Prince MR, Grist TM, Debatin JF (2003) 3D contrast MR angiography, 3rd ed. Springer, Berlin

Prince MR, Yucel EK, Kaufman JA, et al. (1993) Dynamic gadolinium-enhanced three dimensional abdominal MR arteriography. J Magn Reson Imaging 3:877–881

Riederer SJ, Bernstein MA, Breen JF, et al. (2000) Three-dimensional contrast-enhanced MR angiography with real-time fluoroscopic triggering: design specifications and technical reliability in 330 patient studies. Radiology 215:584–593

Rohrer M, Bauer H, Mintorovitch J, et al. (2005) Comparison of magnetic properties of MRI contrast media solutions at different magnetic field strengths. Invest Radiol 40: 715–724

Sodickson DK, McKenzie CA, Li W, et al. (2000) Contrast-enhanced 3D MR angiography with simultaneous acquisition of spatial harmonics: A pilot study. Radiology 217:284–289

Van den Brink JS, Watanabe Y, Kuhl CK, et al. (2003) Implications of SENSE MR in routine clinical practice. Eur J Radiol 46:3–27

Watanabe Y, Dohke M, Okumura A, et al. (2000) Dynamic subtraction contrast-enhanced MR angiography: technique, clinical applications, and pitfalls. Radiographics 20:135–152

Zhang H, Maki JH, Prince MR (2007) 3D contrast-enhanced MR angiography. J Magn Reson Imaging 25:13–25

Zou Z, Ma L, Cheng L, et al. (2008) Time-resolved contrast-enhanced MR angiography of intracranial lesions. J Magn Reson Imaging 27(4):692–699

Artifacts in MR-Angiography

4

Andrea Laghi, Pasquale Paolantonio, Marco Rengo, and Paola Lucchesi

CONTENTS

4.1 Introduction *27*

4.2 Classification *28*

4.3 Radiofrequency Artifacts *28*

4.4 Flow Artifacts *28*
4.4.1 Turbulence's Artifact *28*
4.4.2 The Artifact Due to Saturation *29*

4.5 Hinge Artifact *29*

4.6 Geometric Artifacts *30*
4.6.1 The Hypointensity Linear Horizontal Artifact *30*
4.6.2 The Artifact Due to Noninclusion of the Vase in the Excited Volume *30*

4.7 Magnetic Susceptibility Artifacts *30*

4.8 Maki Artifact *31*

4.9 Vascular Blurring *31*

4.10 Patient Artifacts *31*

4.11 Artifacts from Postprocessing *32*
4.11.1 The Artifact Due to Projection of Background Noise *32*
4.11.2 Step Artifact *33*

References *33*

ABSTRACT

Artifacts represent potential pitfalls in MR-angiography: radiologists should be aware of them to prevent their appearance. They can be classified according to the mechanism responsible for their genesis. Flow Artifacts are strictly related to modification of signal intensity induced by dephasing proton mechanism when the blood flow loses the characteristic laminar pattern; they are turbulence'artifact and artefact due to saturation. Hinge artefact, typical of 3DTOF sequence, is characterized by alternating hyper-hypointense punctiform images. Geometrical artefacts are caused by wrong geometric positioning of angiographic sequence or saturation bands; they are called hypointensity linear horizontal artefact and the artefact due to non-inclusion of the vase in the excited volume. Artifacts related to wrong acquisition time with respect to contrast medium bolus arrival are maki artefact and vascular blurring. Voluntary and involuntary movement are the primary cause of patient artifacts like the gost artifact. The main artifacts from post-processing consist of the artifact from projection of the background noise and the step artifact. Magnetic susceptibility artifacts are caused by paramagnetic external substances. Finally we remember radiofrequency artifacts, due to the instability of radiofrequency or nonhomogeneity of magnetic field or instability of the the system receiver-trasmitter.

Andrea Laghi, MD
Pasquale Paolantonio, MD
Marco Rengo, MD
Paola Lucchesi
Department of Radiological Sciences, "Sapienza" – University of Rome, Polo Pontino, I.C.O.T. Hospital, Viale Franco Fagiana 34, 04100 Latina, Italy

4.1 Introduction

Artifacts often corrupt MRI images and can be generated using basic MR sequences as well as more complex MR applications like MR-angiography (Arena 1995). Artifacts represent false image without any relationship with the patient's anatomy. They have many causes and consequences on image appearance due to complex technical reasons. Artifacts may represent potential pitfalls in MR-angiography. Radiologist and radiographers should be aware of them to prevent their appearance and recognize the diagnostic pitfalls they can mimic (Patton 1994; Carriero et al. 1994, 1995; Pusey et al. 1986; Glockner et al. 2005; Foo et al. 2005; Bradley and Stark 1992; Colagrande and Pazzaglini 2007). The better the understanding of MR images built, the better the dealing with artifacts. Some artifacts that usually affect basic MR acquisition such as aliasing, chemical shift, cross-talk, and ghost are less important in Angiographic application of MRI. In this chapter we will focus our attention on artifacts in MR-Angiography.

4.2 Classification

Artifacts in MR-Angiography can be classified according to the mechanism responsible for their genesis in:

- Radiofrequency artifacts (instability of radiofrequency, inhomogeneous magnetic field, instability of the system receiver-transmitter)
- Flow artifacts
- Hinge artifacts
- Geometrical artifacts
- Magnetic susceptibility artifacts
- Artifacts related to wrong acquisition time with respect to contrast medium bolus arrival (Maki artifact, blurring)
- Artifacts related to the patient (voluntary and involuntary movement)
- Postprocessing artifacts

4.3 Radiofrequency Artifacts

Instability of radio frequency or nonhomogeneity of magnetic field, or instability of the system receiver–transmitter may determine artifacts in basic MR sequences as well as in MR-angiography (Glockner et al. 2005; Foo et al. 2005). Radiofrequency Artifacts may, however, be eliminated and prevented, thanks to strict and constant technical maintenance of the various components constituting the hardware since they are operator and patient independent. The most common radiofrequency artifact is due to the contamination of magnetic bore by radiofrequency waves originating outside the faraday cage that lead to multiple lineary artifacts, each one of them involving very few number of pixels.

4.4 Flow Artifacts

Sequence Artifacts are related to the adopted technique and the choice of angiographic sequence (Foo et al. 2005; Penman et al. 1986). These artifacts especially affect non-contrast-enhanced-MRA sequence and are strictly related to modification of signal intensity induced by dephasing protons mechanism when the blood flow loses the characteristic laminar pattern.

4.4.1 Turbulence's Artifact

Turbulence's artifact generates an absence of signal of the spins of blood because of their whirling and turbulent motion (arterial flow), which is accentuated near physiological shrinkage or pathological stenosis (Penman et al. 1986).

The distribution speed of spins in the turbulent areas changes at every excitement, causing, in this way, dependence from the time of intravoxel coherence that leads to a further loss of signal. The result is an overstatement of stenosis or a pronounced reduction in the caliber of the vessel.

For example, in MR-A of carotid arteries we can find diagnostic traps that lead to the diagnostic suspect of occlusion in patients with carotid stenosis >70% (overstatement) and the diagnostic suspect of stenosis of 50% in patients with stenosis of 70% (understatement) (Fig. 4.1).

The negative influence of this artifact can be reduced by the use of "inflow-effect"; for this reason MR-A of carotid is actually performed using a multislab technique acquired on the perpendicular plane of the vessel in relation to the pattern of blood flow. In addition, the signal of turbulent spins can be fully recovered using a short TE variable between 4 and 6 ms.

In relation to MR-A of vertebral arterials, the same phenomenon of turbulence can determine the

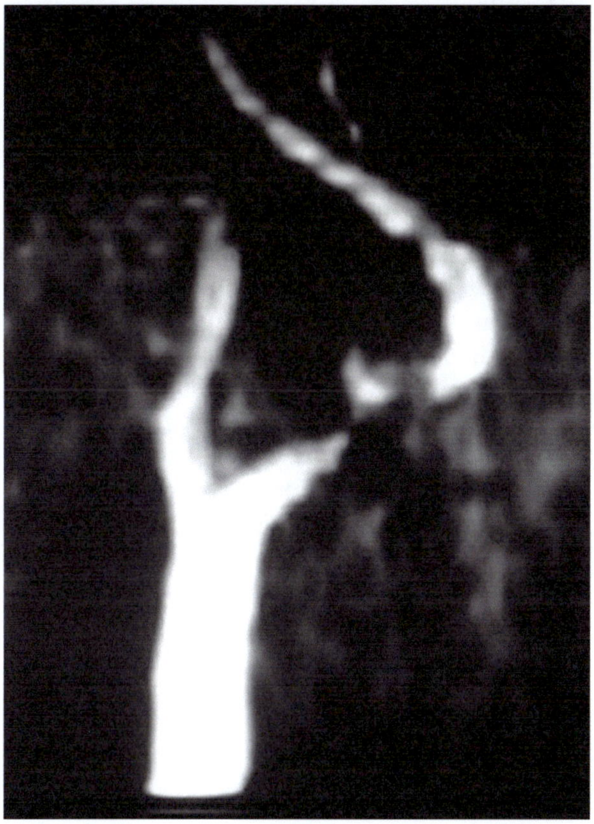

Fig. 4.1. MIP reconstruction based on time of flight MR angiography of carotid artery showing an overestimation of degree of stenosis due to turbulence's artifact. The artifact is generated by the turbulent motion of flowing blood at the level of the stenosis

suspicion of an occlusion of one of the two arteries in patients with spinal stenosis of 70% (overstatement). Also in this case, the modalities to avoid diagnostic pitfalls are similar to those already suggested for the carotid vessels and consist of applying multislab techniques with short TE acquired in a perpendicular plane in relation to the vessel's course that allow visualization of turbulent spin with high speed.

4.4.2 The Artifact Due to Saturation

The artifact due to saturation is more common with the time of flight 3D technique and occurs with reduction and/or disappearance of the signal intensity from the mobile spins; this artifact is due to the accumulation of RF pulses by the spins that move with the excited volume (Foo et al. 2005). It has been estimated that the total loss of signal of the mobile spins occurs in this sequence after about 10–20 pulses of radio frequency. In relation to MR-A of vertebral arterials, this pitfall can determine mistaken diagnosis of occlusion of one of the two arteries. Therefore the mis-visualization of a vertebral artery with acquisition of volume according to the coronal plane cannot lead to a diagnosis of occlusion but should suggest the presence of the "theft of succlavia" syndrome. This suspicion can be confirmed by integrating 3D TOF sequence with 2D TOF acquisition: the visualization of vertebral vessel in 2D TOF and its absence in 3D TOF is suggestive of the diagnosis of "theft."

In the study of renal arteries, the diagnostic trap that determines suspect occlusion in patients with normal renal arteries is the artifact due to saturation: there is loss of signal of mobile spins which accumulate 15–20 RF impulse during 3D TOF acquisition; this phenomenon is responsible for nonvisualization of the distal part of renal arteries (Carriero et al. 1994, 1995).

To optimize sequences, the reduction of flip angle with increase of TR, mount up the depth to which the mobile spins can penetrate the selected volume without being unduly affected by saturation artifact for accumulation of radio frequency.

4.5 Hinge Artifact

Another artifact typical of 3D TOF sequences, is the "hinge artifact" that occurs along the direction of the coding frequency; It is characterized by alternating hyper-hypointense punctiform images. The genesis of the "hinge artifact" is caused by the presence of "stimulated echoes" which occur when sequences with multiple impulses of radio frequency are used (Arena 1995; Glockner et al. 2005; Foo et al. 2005).

For example in the study of intracranial circle, this diagnostic trap could lead to a wrong diagnosis of venous angioma in patients with small intraparenchymal spots of methemoglobin; the high signal of methemoglobin is not discriminated from the high signal of mobile spin by post processing algorithms. Spin echo sequences and the careful analysis of MR-A during the acquisition process allow to correctly identify methemoglobin.

This kind of artifact is common when a head coil with linear polarization is used. In this situation hyperintense pixels near the vessel can be incorrectly confused with small aneurysm of a cerebral artery; this is because of a sub-optimal signal noise to ratio.

In angio-thoracic-MR, this pitfall could cause wrong diagnosis like thoracic aneurysm of the aorta. In cases of suspected diagnosis of aneurysm, it is possible to find structures that are already hyperintense before the acquisition (thrombus or chesty process with chronic bronchiectasis), which, being in contact with major thoracic vessels, may simulate an aneurysm.

These diagnostic traps can be avoided using Phase Contrast or Magnitude Contrast sequences; the subtraction allows canceling this hyperintensity in both moments of acquisition.

4.6 Geometric Artifacts

Sometimes MRA may be affected by artifacts because of wrong geometric positioning of angiographic sequence or saturation bands. These artifacts are the so called Hypointensity linear horizontal artifact and the artifact due to noninclusion of the vase in the excited volume (Glockner et al. 2005; Foo et al. 2005; Bradley and Stark 1992).

4.6.1 The Hypointensity Linear Horizontal Artifact

The Hypointensity linear horizontal artifact consists of a linear signal void present in 2D or 3D acquisitions acquired using a presaturation impulse. The genesis of this artifact is due to a wrong position of the saturation band positioned too close to the excited stratum. The additional radiofrequency can saturate even the spins around the volume excited.

This diagnostic trap can lead to interpretation of a signal vacuum as thrombus especially during MR-A of peripheral vessels.

One or more dark longitudinal stripes can be seen within the aortic lumen simulating the presence of a dissection.

4.6.2 The Artifact Due to Noninclusion of the Vase in the Excited Volume

The artifact due to noninclusion of the vase in the excited volume is frequent in anatomical districts where there is an abnormal course of the vessels (coiling, kinking); it is an exclusive prerogative of 3D technique and it is the result of wrong positioning of the acquisition volume compared to the course of the vessel.

All the sequence artifacts described can be removed by optimizing the use of MR-A sequences, and in particular their intrinsic parameters. Echo times as short as possible (4–6 ms) and variable flip angle allow to recover artifacts from turbulence and saturation. Elimination of the layer, which has the hinge artifact, before postprocessing allows to delete it completely without degrading the final quality of the image. A proper positioning between presaturation impulse (FRODO technique) and the layer or volume excited and a wide safety margin in place acquisition volumes, help respectively to prevent linear and horizontal hypointensity artifact and the artifact owing to complete noninclusion of the vase in the excited volume.

4.7 Magnetic Susceptibility Artifacts

Magnetic susceptibility artifacts are not an absolute prerogative of MR-A sequences and can be caused by paramagnetic external substances (dental prostheses, clips, etc..) which alter the geometry of the main magnetic field or by interfaces between different tissue sensitivity for the magnetic field (i.e., air-vessel) (Glockner et al. 2005; Foo et al. 2005; Bradley and Stark 1992).

The first ones determine a distortion of the image bounded by a small hyperintense edge, while interfaces artifacts are responsible for the presence of a hypointense board of a few millimetres thick. These artifacts are minimized, but not completely eliminated using short TE sequences or 3D contrast MR-A. To get the shortest possible TE, it should be used as the widest band-width available on the imaging system used. Intravascular stents can be particularly problematic depending on the stent composition. The most MR compatible stents are nitinol or platinum, which are totally nonmagnetic. Even nonmagnetic stents may obscure the artery lumen by attenuating the radiofrequency signal. This can be at least partially overcome in platinum and nitinol stents using a higher flip angle.

Artifacts caused by external paramagnetic substances may be partially removed by a thorough medical history of the patient.

4.8 Maki Artifact

Maki artifacts may occur using contrast-enhanced – MR-angiography. Maki artifact appears as a double contour of the vessel wall with low signal intensity of a central zone within the vessel.

This artifact depends on a too early acquisition of MR sequence with respect to the arrival of contrast medium bolus in the target vessel (Zhang et al. 2007; Maki et al. 1996).

Therefore contrast medium reaches the ideal concentration into the target vessel during the later part of sequence acquisition. In other terms, central lines of k-space are sampled before the arrival of contrast medium into the vascular district, while the good concentration of contrast medium is reached during peripheral k-space sampling where information of image contour is collected.

This mechanism explains the low grade of vascular enhancement as well as the effect on vessel wall contours (Fig. 4.2).

To prevent this artifact a strict monitoring of the contrast medium arrival into the target vessel using bolus-tracking system is essential. Moreover it is essential to start image acquisition immediately when contrast medium has reached the entire vascular district, especially if a central sampling of the k-space is used to prevent the maki artifact in the distal portion of the vascular district while obtaining an excellent enhancement in the proximal portion.

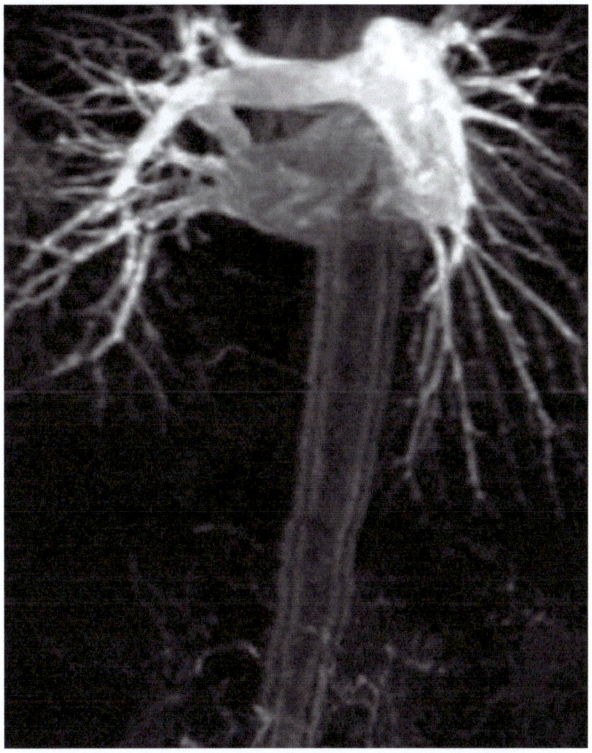

Fig. 4.2. Maki artifact: MIP reconstruction based on CE-MRA acquisition. Image shows a good contrast enhancement of pulmonary arteries while abdominal aorta presents double contour with low signal intensity of the central vessel area. This artifact is due to mis-registration of central k-space line with respect to contrast medium bolus arrival. (Reprint with permission by manuale di RM addominale, Poletto editore)

4.9 Vascular Blurring

This artifact may affect CE-MRA and is easily recognizable like a blurring of vessel wall (Carriero et al. 1994, 1995; Pusey e al. 1986; Glockner et al. 2005; Foo et al. 2005; Penman et al. 1986).

Similar to Maki artifact, the vascular blurring is also due to a suboptimal synchronization between acquisition of central k-space lines and the arrival of contrast medium bolus into the target vessel. This artifact depends on the variation of signal intensity of target vessel during the central k-space sampling due to the usage of a too short bolus of contrast medium. Usually this artifact may occur when a limited dose of contrast medium is used or when injection flow is high as well as when the acquisition of CE-MRE sequence starts before the vascular peak enhancement of the target vessel.

In other terms if the signal intensity of the vessel lumen shows relevant variation during the central k-space sampling with increasing signal intensity over time, the central portion of k-space will finally reflect inhomogeneity. This k-space inhomogeneity will lead to signal blurring (Fig. 4.3).

The vascular blurring can be prevented by avoiding high flow rate injection for contrast medium and an underdosage of contrast medium as well by optimizing the image acquisition using bolus tracking system.

4.10 Patient Artifacts

Patient artifacts are common in Magnetic Resonance and particularly frequent in MR-A; the voluntary and involuntary movement (hyperperistalsis, throbbing

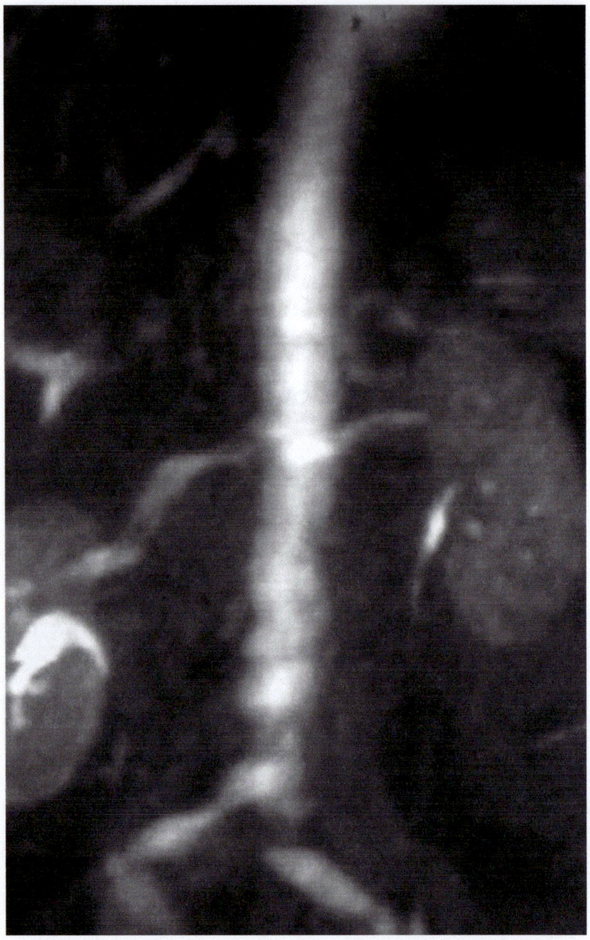

Fig. 4.3. MIP reconstruction based on CE-MRA acquisition. Note the blurring artifact at the level of abdominal aorta. The artifact is due to the wrong synchronization between the contrast arrival and the k-space sampling. During the central sampling of the k-space, the contrast medium reaches a good concentration into the vessel lumen while a poor concentration is achieved during peripheral sampling of the k-space leading to a loss of spatial resolution

of the arterial blood vessels) are the primary cause of the ghost artifact. Ghost artifact manifests itself with the presence of hyper or hypointense bands with monotonous repetition of anatomical region considered along the phase encoding. This causes a quality decay of the image both in the layer and volume acquisition. Motion artifacts are particularly frequent in the study of thorax-abdominal vessels where it is not possible to maintain the apnoea (Arena 1995; Foo et al. 2005).

The change of position following a voluntary movement can lead completely to off-axis blood vessels so during the reconstruction for example, if it is a 2D acquisition, the image shows a typical segmented profile of vascular lumen especially evident in coronal reconstruction of axial acquisitions.

In 3D instead, you get an out-of-focus image in general because it establishes a causal dependence on the time between the various phases concerning the various repetitions of the sequence. We must bear in mind that there is an interval time between the encoding phase and the frequency and also in recording of x and y positions which helps to prevent the artifact from moving, leading to an erroneous recording space.

The tortuosity inherent to the thoracic aorta and its branches may result in the inadvertent exclusion of important portions of the arterial anatomy from the 3D imaging volume. To prevent this artifact, a 3D data set can be collected initially without contrast to assure that the relevant anatomy is encompassed within the imaging volume.

4.11 Artifacts from Postprocessing

The main artifacts from postprocessing consist of the artifact from projection of the background noise and the step artifact (Foo et al. 2005; Bradley and Stark 1992; Maki et al. 1996).

4.11.1 The Artifact Due to Projection of Background Noise

The artifact due to projection of background noise has its rational in a limited or wrong contrast to noise; spins in motion, which have a signal intensity equal to or even lower than that of stationary tissues are not in fact found by postprocessing algorithm and will not be represented in the angiogram after reconstruction.

In addition, MIP reduces the size of the vessel because of different phase precession of spins with different speed; spins at the centre of the vase have the same speed, so will have a homogeneously hyperintense signal; spins that flow behind the wall have on the contrary, a different phase of precession that leads to a decay of the signal, when it approximates to 180.

On the other hand basic signal of hyperintense tissues such as methemoglobin is not differentiated from hyperintensity of mobile spins by postprocessing algorithm; it follows that tissues of nonvascular origin can be represented on definitive angiogram simultaneously to those of vascular.

4.11.2 Step Artifact

This artifact is exclusive of 2D sequences acquired according to the axial plan it is characterized by an irregularity of vascular profile evident after postprocessing. Its genesis is to be searched, especially, in nonperfect alignment of the profile of vases. It is frequent in MR-A of peripheral arterial blood vessels in which the axial acquisition without cardiac synchronization can happen at any stage of the cardiac cycle, resulting in signal acquisition during the top of the sphygmic wave that conditions the axial diameter of vessel and the nonperfect alignment of their profile.

In order to prevent the artifact from postprocessing (MIP) which causes the disappearance of the vases with intensity signal like that of background noise, or the representation of stationary tissues with signal intensity overlap or greater than that of the vases, it is necessary in the first case to improve the contrast to noise, and in the second case, use MR-A techniques to subtract as the technical phase contrast. Acquisitions synchronization with the cardiac circle for peripheral circle and correction factors in the development of postprocessing, permit finally, to prevent and correct the artifact by step.

References

Arena L, et al. (1995) MR imaging artifacts that simulate disease: how to recognize and eliminate them. Radiographics 15:1373–1394

Bradley WG, Jr, Stark DD (1992) Magnetic resonance imaging. Mosby-Year Book, St Louis

Carriero A, et al. (1994) Artifacts in magnetic resonance angiography. Radiol Med 88:765–770

Carriero A, et al. (1995) Diagnostic pitfalls in magnetic resonance angiography. Radiol Med 90:719–725

Colagrande S, Pazzaglini S (2007) Manuale di RM addominale. Poletto, Milano

Foo TK, et al. (2005) MR angiography physics: an update. Magn Reson Imaging Clin N Am 13:1–22

Glockner JF, et al. (2005) MR angiography interpretation: techniques and pitfalls. Magn Reson Imaging Clin N Am 13: 23–40

Maki JH, et al. (1996) Three-dimensional contrast-enhanced MR angiography. Top Magn Reson Imaging 8:322–344

Patton JA (1994) MR imaging instrumentation and image artifacts. Radiographics 14:1083–1096

Penman WH, et al. (1986) Artifacts from pulsatile flow in MR imaging. J Comput Assist Tomogr 10:473–483

Pusey E, et al. (1986) Magnetic resonance imaging artifacts: Mechanism and clinical significance. Radiographics 6: 891–911

Zhang H, et al. (2007) 3D contrast-enhanced MR angiography. J Magn Reson Imaging 25:13–25

Image Processing

5

Lorenzo Faggioni and Emanuele Neri

CONTENTS

5.1 Introduction 35
5.2 Multiplanar Reformation (MPR) 35
5.3 Maximum Intensity Projection (MIP) 37
5.4 Shaded Surface Display (SSD) 39
5.5 Volume Rendering (VR) 40
5.6 Virtual Endoscopy (VE) 41
References 43

Lorenzo Faggioni, MD
Emanuele Neri, MD
Diagnostic and Interventional Radiology, Department of Oncology, Transplants, and Advanced Technologies in Medicine,University of Pisa, Radiodiagnostica 1 Universitaria, Ospedale Cisanello. Via Paradisa 2, 56100 Pisa, Italy

5.1 Introduction

Magnetic resonance angiography (MRA) sequences generate an amount of data that represents the signal intensity of voxels from the image volume, distributed on consecutive slices and oriented on a user-defined plane. In particular, the availability of wide anatomic coverage coils and high-field scanners, together with fast sequences and high relaxivity contrast media, allows to easily obtain MRA series formed by a high number of thin slices, usually partially overlapped on each other. This poses the problem to achieve a synthetic view of such a large quantity of analytical information, as derived from each single partition, through either a panoramic or a targeted visualization of a given vascular territory. The demonstration of a particular vascular anatomy may also raise the issue of separating or eliminating other structures contained in the same image volume (such as veins in an MR arteriogram, superimposed vessels or stationary tissues with contrast enhancement).

In this chapter, the following main techniques for the postprocessing of MRA images will be illustrated:

- Multiplanar reformation (MPR)
- Maximum intensity projection (MIP)
- Shaded surface display (SSD)
- Volume rendering (VR)
- Virtual endoscopy (VE)

5.2 Multiplanar Reformation (MPR)

The MPR technique allows creating a bidimensional image on an arbitrary plane from a dataset of complanar images. More precisely, the information about position and signal intensity of voxels belonging to a stack of images acquired on parallel planes with a given spatial orientation (for example, a series of axial images) is used to produce an output image where voxels are geometrically projected on a user-defined plane with a different inclination (e.g., a coronal, sagittal, or oblique plane), as a function of position and signal intensity of voxels from source images (Lell et al. 2006). In this way, it is possible to obtain an integrated representation of structures located on planes different from the native one, which is usually chosen so to optimize technical parameters such as image acquisition time or to minimize artifacts.

A typical vascular imaging application of the MPR algorithm is the continuous reconstruction of vessels which are not entirely included in single sections of the image volume, in particular of those with an anatomical inclination different from that of the acquisition plane. In this sense, it is noteworthy that the majority of workstations and image processing software applications allow to automatically perform MPR on standard planes (axial, coronal, sagittal: orthogonal MPR) by generating additional stacks of reformatted images spanning from a given initial to a given final limit of the source dataset that can be saved in DICOM format as independent image series. The thickness and spacing of reformatted images are user-definable; in particular, increasing slice thickness can be useful in order to increase signal-to-noise ratio (SNR) and/or to reduce total image number, and is a good practice if the dataset does not contain findings that do not require the highest available spatial resolution in order to be recognized. More generally, the user is free to define oblique reconstruction planes arbitrarily by orienting and moving an electronic caliper on the three standard orthogonal planes, on which vessels are identified based on their anatomical course. In this way, it is possible, for example, to create images on planes perpendicular to the longitudinal axis of the vessel in order to measure its diameter or section area, thereby avoiding the measurement error due to the geometric distortion that occurs whenever such measures are taken on native images from vessels coursing nonperpendicularly to the acquisition plane. This is of particular importance, for example, for the evaluation of conditions associated with widening or narrowing of the vessel lumen, such as aneurysms or vascular stenoses, for which the exact measurement of dimensional parameters may affect treatment planning.

It is intuitive that the quality of MPR reconstructions is optimal when the dataset is composed of isotropic voxels. Yet, this condition, which is easily fulfilled on modern MDCT scanners, is still seldom accomplished in MRA owing to limitations related to prolongation of imaging time and worsening of SNR and contrast-to-noise ratio (CNR). In fact, the need for low section thickness in phase direction poses a penalty on acquisition time (proportional to the number of phase encoding steps and image averages), which forces the user to reduce matrix size in phase direction and use single averaging or k-space undersampling techniques in order to keep acquisition time down to a reasonable level, usually resulting in lower spatial resolution in phase direction and/or increased image noise (Prince et al. 2003). A practical method for increasing spatial resolution in phase direction is represented by acquiring partially overlapped slices (typically, with 50% overlap), which also leads to reduction of partial volume effects with preserved SNR as compared with nonoverlapped sections.

A variant of the MPR algorithm is CPR (curved planar reformation), that allows to project native images from a succession of planes lying on a user-defined polygonal trajectory. CPR is particularly useful to display vessels with a tortuous course in the three spatial directions, such as coronary arteries or the arteries of the celiac district. This operation can be carried out by selecting a series of points aligned along the center of the lumen (centerline) on one or more MPR projections, thus obtaining a continuous rectified representation of the vessel course (Fig. 5.1). With CPR, it may also be easier to perform measurements of vessel size, as the selected vessel is visualized continuously without distortions of its lumen contour.

MPR and CPR have the advantage of preserving the whole spatial and contrast information of the dataset with relative computational simplicity, as well as being relatively operator-independent, because it is not necessary to perform any segmentation of source images prior to reconstruction. On the other hand, the 2D nature of MPR and CPR images and their projective character do not allow to achieve a panoramic depiction of multiple vessels and, in particular, of branch vessels oriented on multiple planes. For these reasons, as illustrated below, "raw" MPR and CPR tools are usually associated to the MIP algorithm, which can yield a simultaneous representation of multiple vessels located inside a user-defined spatially oriented slab.

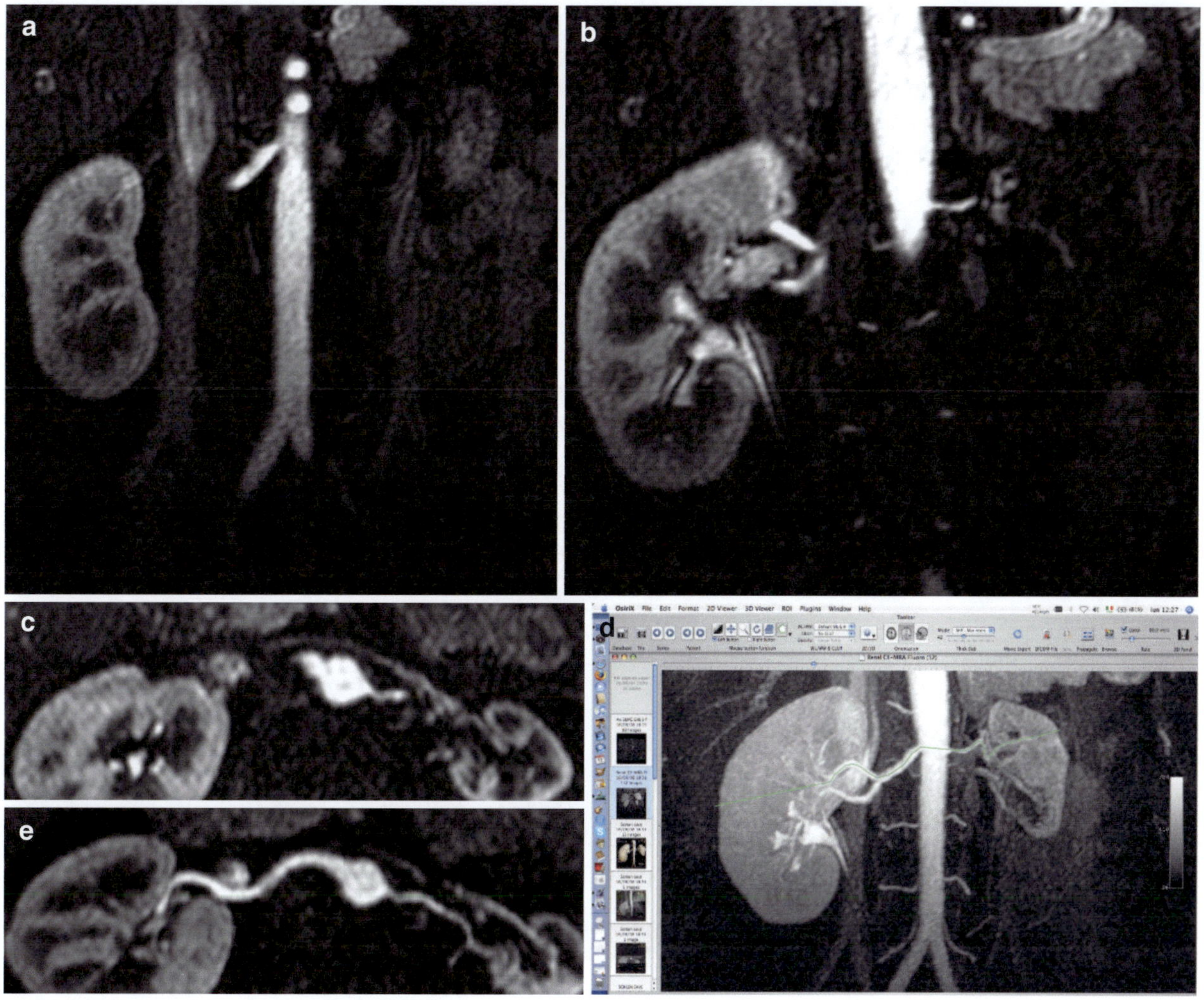

Fig. 5.1. CEMRA of normal renal arteries. (**a**, **b**) Single native images acquired on the coronal plane do not allow to evaluate the renal arteries over their entire course, nor to yield reliable information about the presence of stenosis. (**c**) Axial MPR image allows to exclude stenosis only within the proximal third of both renal arteries included in the reformat plane. (**d**, **e**) By tracing a curve (*green line*) that follows the course of the renal arteries as displayed on a coronal MIP image, it is possible to visualize the entire course of the vessels, which appear to be patent

5.3 Maximum Intensity Projection (MIP)

With this algorithm, a slab with a defined thickness and a given spatial orientation is selected, and for each set of voxels aligned along the chosen direction, only that with the maximal intensity is represented on the resulting reconstructed image. For better clarity, let us suppose to draw through the slab an array of parallel lines (ray casting), located at a distance from each other equal to the size of the voxel in the direction perpendicular to that of the oriented lines. Among all the voxels crossed by each line, the MIP algorithm extracts that with the highest signal intensity and adds it to the output image, discarding the information relative to the other voxels. As a consequence:

- The MIP image is exclusively formed by the maximum intensity voxels included in each oriented slice in which the slab is subdivided.
- Because a small fraction (about 10%) of the whole dataset information is used, the MIP algorithm requires little computational resources, resulting in fast postprocessing times even on low-power workstations.
- Since the MIP algorithm requires only two parameters to be run (slab inclination and thickness), it is only moderately operator-dependent as well as conceptually simple, thus having the potential to reduce the time needed for reconstruction.

- Eliminating nonmaximal intensity voxels leads to loss of information related to image depth and contrast resolution among structures with different signal intensity inside the slab (Calhoun et al. 1999; Fishman et al. 2006; Lell et al. 2006).

The MIP technique plays a fundamental role for the postprocessing of MRA data, as it allows displaying vessels (which have high signal intensity due to either flow phenomena in TOF and PC sequences, or T1 shortening of intravascular blood in CEMRA acquisitions) to their whole extent on user-defined planes, thereby achieving angiography-like reconstructions (Fig. 5.2). Moreover, in contrast with pure MPR techniques, it is possible to visualize the course of multiple vessels inside the slab, such as branching or collateral vessels.

A great advantage of MIP on MRA over CT angiography images is that there is no need to remove bones, whose density is near that of contrast-enhanced vessels, resulting in vessel obscuration especially on thick-slab MIP reconstructions of CT angiograms (Kabul and Hagspeil 2006). However, for a similar reason, it is necessary to ensure that source images do not contain structures with signal intensity similar to that of vessels (such as methemoglobin deposits or fat tissue). This can be accomplished by using fat suppression techniques and/or by subtracting to CEMRA images a mask series acquired with the same imaging parameters before administration of contrast material.

The selection of slab thickness is relevant. In fact, on one hand, the choice of a thick slab guarantees good panoramicity, but on the other hand it can lead to underestimate the diameter of small vessels or to overestimate stenoses because of partial volume averaging phenomena, especially on images with suboptimal stationary tissue suppression or low SNR (Lell et al. 2006, 2007). This is due to the fact that the image contrast of a vessel parallel to the imaging plane varies as a Gaussian function with its maximum (proportional to the vessel diameter)

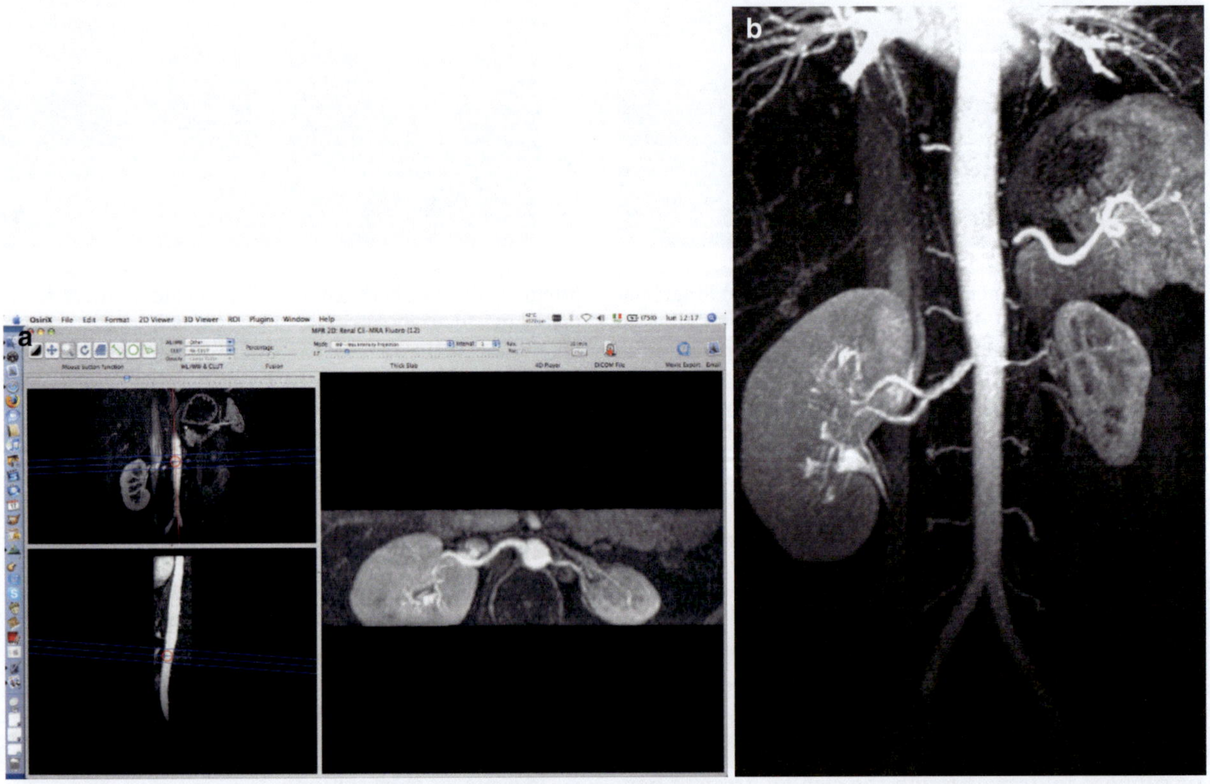

Fig. 5.2. Case of Fig. 5.1. (**a**) By selecting an oriented thin slab volume centered at a level corresponding to the origin of the renal arteries (*left side*), it is possible to display the course of the renal arteries on a para-axial plane (*right side*). Notice the partial superimposition of the right renal vein on the ipsilateral renal artery at the hilum and the visualization of a pair of lumbar arteries arising from the abdominal aorta, that were not present on MPR and CPR reconstructions. (**b**) Thick-slab MIP on the coronal plane, allowing a paronamic view of the renal arteries and the aorto-iliac vessels. The mid and distal third of the splenic artery, the proximal tract of the lumbar arteries and part of the cardiopulmonary circulation included in the field of view are also displayed

corresponding to the vessel centerline, symmetrically tending to zero with increasing distance from this latter (Prokop et al. 1997; Sun and Parker 1999). For this reason, CNR of vessels on MIP images decreases with increasing background signal intensity and reduced vessel size, and selection of the thinnest possible slab for MIP can be beneficial, as it leads to minimize inclusion of background tissue into the slab. Consequently, thin-slab MIP will be advisable if stenosis grading is to be performed, so as to include as few voxels as possible in the slab. MIP with minimal slab thickness is required for representation of thrombi, as those are hypointense to blood on MRA images and would therefore be obscured on thicker slab MIP, as well as for depiction of aortic dissection (Glockner 2001; Lell et al. 2006; Prokop et al. 1997; Westenberg et al. 2000).

Furthermore, the choice of a moderate slab thickness permits to avoid or reduce the problem of artero-venous overlap in territories (such as iliac, thoracic, and cervical vessels) where arteries and veins are close to each other with a parallel course. In particular, MIP slab thickness should be kept to a minimum when concave vessels (such as the aortic arch) are to be displayed, as the use of excessively thick slabs might cause the surrounding contrast-enhanced structures or background noise to "fill" the concavity, resulting in impaired depiction of the vessel borders (Prokop et al. 1997).

Such limitations can usually be overcome, at some detriment for overall panoramicity, by performing multiple MIP reconstructions of the vessel anatomy under investigation: a practical method consists of producing several thin-slab MIP images of a vessel and its branches (or just some of them or even a single branching vessel) with different orientations. In addition, the representation of vessel branching patterns can be facilitated by doing MIP over CPR reconstructions of the main vessels.

Alternatively, it is possible to edit the dataset prior to MIP in order to remove undesired structures, either through intensity threshold- and/or region growing-based segmentation algorithms or by manual cutting on native and/or MPR images (Glockner 2001; Prokop et al. 1997). For instance, the representation of hepatic arterial vessels can be significantly improved by cropping away the retroperitoneal tissues, including the abdominal aorta and renal arteries (which would obscure the origin and course of the celiac trunk and the superior mesenteric artery), allowing to get an angiography-like MIP depiction of such vessels (Prokop et al. 1997). However, the usage of editing techniques may be time-consuming and require high technical skills, especially if advanced editing is needed. Despite this, the usage of MIP and MPR/CPR algorithms, eventually associated with image editing, is usually sufficient to achieve adequate results in vascular MRI, both in diagnostic and iconographic terms (Davis et al. 1997; Hany et al. 1998).

5.4 Shaded Surface Display (SSD)

With this algorithm. a threshold is fixed by the user defining the minimum intensity of the voxels to be imaged, while voxels with lower intensity will be discarded. The reconstructed image will be composed by surfaces corresponding to the interface between voxels with intensity above the threshold and void voxels (i.e., with intensity below the threshold), obtained by simulating the illumination of surfaces from an external light source in a given user-defined direction. The result of this process is a three-dimensional (3D) representation of the voxels with above-threshold intensity, in contrast with MIP and MPR techniques, which yield two-dimensional (2D) images because of their planar nature (Fig. 5.3) (Davis et al. 1997; Prokop et al. 1997).

SSD has the advantage of relatively modest computational requirements, as only a part of the dataset voxels is used and their representation on the output image is binary. However, its main weakness is represented by the difficulty to choose a threshold that allows to effectively separate parietal from extraparietal voxels, which is key to generating nondistorted 3D images. In fact, setting too low a threshold leads to including voxels external to the vessel wall to the reconstructed image, reducing the apparent size of vascular ostia or causing "floating objects" to appear inside the vascular cavities under examination. Choosing an excessively low threshold may also lead to superimposition of background tissues or noise, resulting in a more or less "speckled" appearance of the surface contours with potential loss of details. Conversely, too high a threshold leads to surface erosion because of exclusion of parietal voxels, resulting in enlargement of vessel ostia or hole-shaped areas of signal void. Setting a good threshold may be challenging, especially on images with low SNR or inhomogenous intravascular signal, as often happens with aneurysms or cardiac cavities. For these reasons, the SSD technique is inadequate for taking vascular

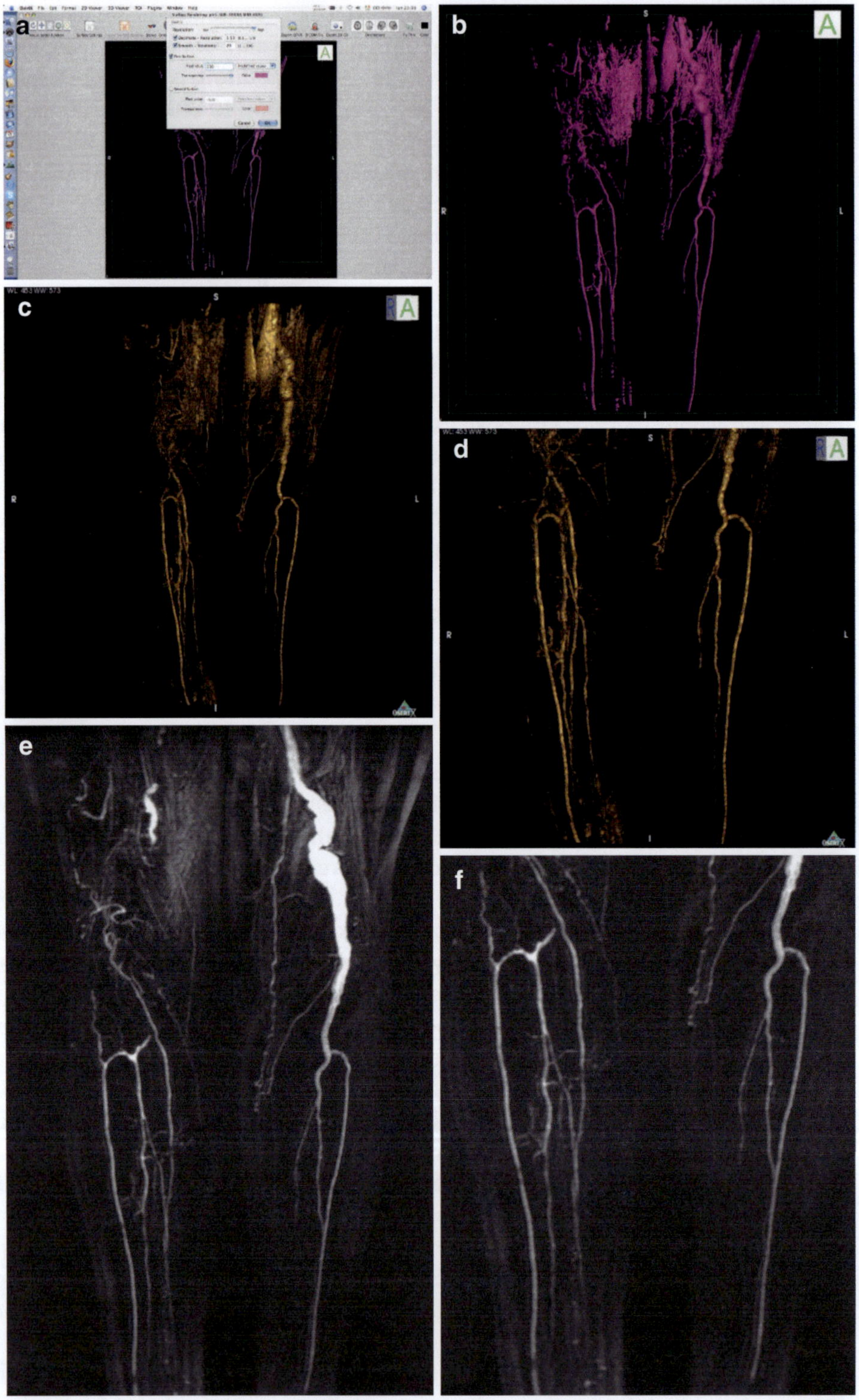

Fig. 5.3. Patient with bilateral femoro-popliteal aneurysm (right thrombosed). (**a, b**) SSD reconstruction; (**c, d**) VR reconstruction; (**e, f**) MIP reconstruction. This latter allows a more accurate assessment of infrapopliteal vessels, mainly due to elimination of nonmaximal intensity voxels associated with stationary tissues

measurements (which are usually performed on MPR reconstructions), as well as for depicting thin vessels, while it can be useful to get a 3D perspective representation of vascular anatomy. SSD is seldom performed today, as it has been superseded by the more robust and powerful volume rendering (VR) algorithms.

5.5 Volume Rendering (VR)

While MIP is based on the extraction of maximum intensity voxels among those crossed by an array of parallel lines, VR techniques rely on the ray casting principle by using the whole spatial and intensity information of the dataset and representing on the reconstructed image a weighted average of the intensity of all voxels distributed on each slab section. Each voxel is assigned a value of opacity, transparency, and color depending on its intensity, position, and the prospective direction in which the image volume is observed according to a predetermined transfer function that allows to generate the final image by weighing the intensity of the various voxels encountered by each ray. The user contributes to the definition of the transfer function by choosing a color and a transparency value and setting at least two voxel intensity thresholds, corresponding to the minimum and maximum brightness of visualizable points on the reconstructed image. It is possible to define linear transfer functions (implying that the ratio between the intensity of two voxels on native images and those of the corresponding voxels on the VR image is constant), as well as nonlinear (e.g., logarithmic) ones. A user can also define his own transfer functions, such as complex ones with multiple thresholds, in order to enhance or attenuate the contribution to the VR image of voxels with intensities within a given range (Calhoun et al. 1999; Fishman et al. 2006; Lell et al. 2006, 2007).

In vascular imaging, the choice of suitable transfer functions allows to emphasize even modest contrast differences among different vessels (for example, between arteries and veins close to each other). It is also evident, however, that the need for the user to define the proper transfer function makes VR reconstructions more operator-dependent than MIP techniques, which do not have such requirement. In addition, VR algorithms are quite computationally intensive (in terms of both processor and graphics hardware load), as they provide a 3D visualization of the whole dataset, although the continuous improvement of image-processing hardware makes this issue less important than in the past.

In comparison with MIP techniques, VR algorithms offer the advantage of greater panoramicity and anatomical fidelity, owing to the fact that the whole dataset information is maintained. In particular, VR reconstructions tend to be superior to thick-slab MIP images for the evaluation of vascular stenoses, which are depicted without overestimation due to flow dispersion phenomena (Fig. 5.4) (Baskaran et al. 2002; Mallouhi et al. 2002, 2003). Moreover, unlike MIP images, VR reconstructions yield the geometric perception of depth, thus allowing to effectively separate contiguous or tortuous vessels that would appear superimposed on MIP images and making it easier to achieve a panoramic representation of aneurysms or cardiac cavities (Fishman et al. 2006; Lell et al. 2006). Another advantage of VR over MIP is the possibility to simultaneously display vessels and stationary tissues within the same image volume, which can be useful, for example, for the depiction of hypervascular neoplasms together with their feeding vessels.

On the other hand, the MIP technique (especially thin-slab MIP) should be preferred for the evaluation of small vessels (such as, for example, infrapopliteal arteries in patients with lower limb arteriopathy), because the selection of the sole maximum intensity voxels usually permits a more detailed and continuous vision of such structures, even on images with a low SNR and/or CNR (Fishman et al. 2006). Thin-slab MIP also tends to be more widely used than VR for the evaluation of vascular stenoses and aneurysms because of its greater simplicity, speed, and reproducibility (Fink et al. 2003; Persson et al. 2004; Runck et al. 2008). In practice, both VR and MIP techniques can be employed in an integrated manner in order to optimize the visualization of the various aspects of vascular pathology (Fishman et al. 2006).

5.6 Virtual Endoscopy (VE)

By applying SSD or VR algorithms in order to visualize vessels from an intraluminal, rather than extraluminal light source, it is possible to achieve endoscopy-like views of the vascular lumen. SSD-based VE techniques require the definition of an intensity threshold and generate surfaces formed by the interfaces between lower intensity voxels (which identify the non-contrast-enhanced extraluminal compartment) and higher intensity ones, corresponding to the intraluminal signal. VR-based VE techniques are more robust and require, in analogy with the tools for adjustment of visualization window settings for digital images in general, the definition of a central intensity value (window level) together with an intensity range (window width) corresponding to the spectrum of signal intensities of voxels to be displayed with different opacities on the VE view.

Regardless of whether an SSD- or VR-based VE algorithm is used, the correct choice of the above mentioned parameters is crucial to achieve a good

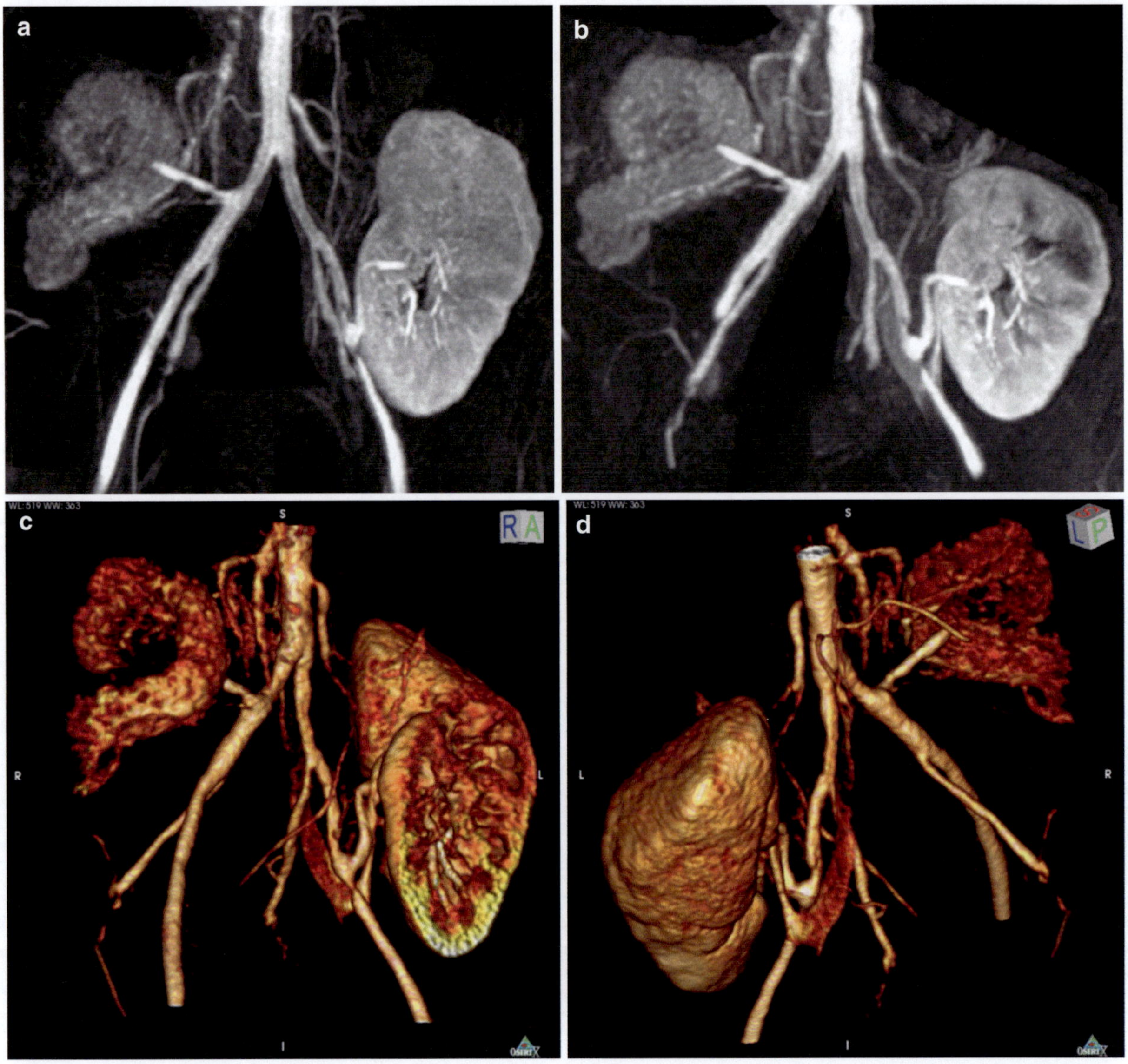

Fig. 5.4. Patient with renal transplant and ectasia of the arterial anastomosis associated with stenosis of post-anastomotic left external iliac artery. (**a**, **b**) On thick-slab MIP reconstructions the grade of the stenosis is overestimated and it is impossible to correctly evaluate patency of the venal rein due to the higher intensity of voxels of the overlying arterial graft. (**c**, **d**) VR image allows a more accurate grading of the left external iliac artery stenosis, as well as to assess patency of the renal vein, which can be separated from the renal artery owing to the usage of an appropriate transfer curve and to the possibility to generate perspective 3D views

separation between intra- and extraluminal signal, avoiding artifacts due to surface erosion or dilation (which occur when the lower threshold is too high or too low compared with the mean intensity of intravascular voxels, respectively).

VE reconstructions play an important role for the evaluation of anatomic variants of the vascular tree such as early branching vessels, for which it may be difficult to decide between the presence of a short common trunk and direct origin of vessels from adjacent ostia. Another similar vascular application of VE is the study of cardiac cavities or aneurysms, in which it is usually relevant to establish the number and position of vascular ostia (Cirillo et al. 2005; Davis et al. 1996). On VE images, it is also possible to make intraluminal geometric measurements (such as distance or section area) which cannot be performed by means of extraluminal postprocessing techniques (Fig. 5.5). Once again, however, the reliability of such measurements depends significantly on the threshold chosen for endoscopic visualization, which reflects the quality of the MRA dataset. In this sense, it is important

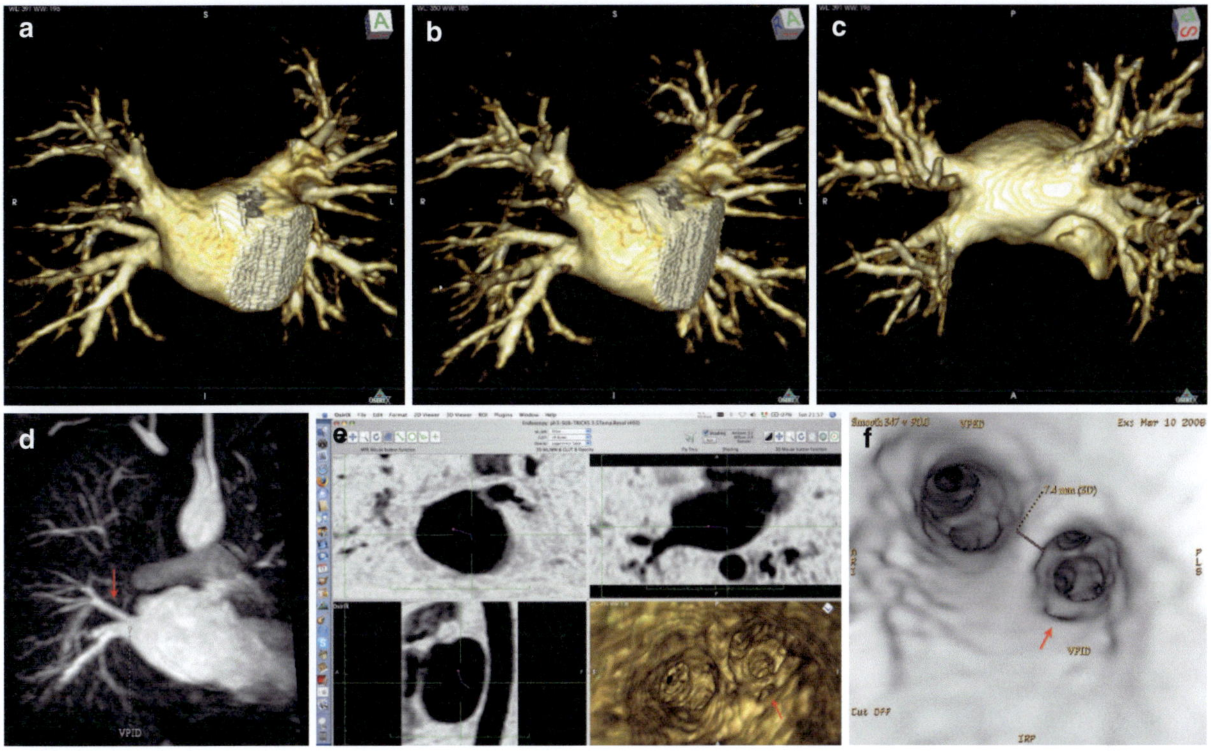

Fig. 5.5. (**a–c**) VR images of the left atrium and pulmonary veins (PV) in patient scheduled for percuteneous radiofrequency ablation of atrial fibrillation. The right inferior PV has an early branching pattern (*red arrow*), confirmed by thin-slab MIP oriented along the course of the vessel (**d**). VE views (**e, f**) allow optimal depiction of the PV ostia; in (**f**) the distance between the ostia of the superior and inferior PV is measured

that the signal of the territories under examination is homogenous and with a high contrast compared with stationary tissues, so to allow a faithful depiction of the details of the parietal structure.

References

Baskaran V, Pereles FS, Nemcek AA Jr, et al. (2002) Gadolinium-enhanced 3D MR angiography of renal artery stenosis: a pilot comparison of maximum intensity projection, multiplanar reformatting, and 3D volume-rendering postprocessing algorithms. Acad Radiol 9(1):50–59

Calhoun PS, Kuszyk BS, Heath DG, et al. (1999) Three-dimensional volume rendering of spiral CT data: theory and method. Radiographics 19(3):745–764

Cirillo S, Tosetti I, Gaita F, et al. (2005) Magnetic resonance angiography of the pulmonary veins before and after radiofrequency ablation for atrial fibrillation. Radiol Med 109(5–6):488–499

Davis CP, Hany TF, Wildermuth S, et al. (1997) Postprocessing techniques for gadolinium enhanced three-dimensional MR angiography. Radiographics 17(5):1061–1077

Davis CP, Ladd ME, Romanowski BJ, et al. (1996) Human aorta: preliminary results with virtual endoscopy based on three-dimensional MR imaging data sets. Radiology 199(1):37–40

Fink C, Hallscheidt PJ, Hosch WP, et al. (2003) Preoperative evaluation of living renal donors: value of contrast-enhanced 3D magnetic resonance angiography and comparison of three rendering algorithms. Eur Radiol 13(4):794–801

Fishman EK, Ney DR, Heath DG, et al. (2006) Volume rendering versus maximum intensity projection in CT angiography: what works best, when, and why. Radiographics 26(3): 905–922

Glockner JF (2001) Three-dimensional gadolinium-enhanced MR angiography: applications for abdominal imaging. Radiographics 21(2):357–370

Hany TF, Schmidt M, Davis CP, et al. (1998) Diagnostic impact of four postprocessing techniques in evaluating contrast-enhanced three-dimensional MR angiography. AJR 170(4):907–912

Kabul HK, Hagspiel KD (2006) Cross-sectional vascular imaging with CT and MR angiography. J Nucl Cardiol 13(3):385–401

Lell MM, Anders K, Uder M, et al. (2006) New techniques in CT angiography. Radiographics 26 (Suppl 1):S45–S62

Lell M, Fellner C, Baum U, et al. (2007) Evaluation of carotid artery stenosis with multisection CT and MR imaging: influence of imaging modality and postprocessing. AJNR 28(1):104–110

Mallouhi A, Felber S, Chemelli A, et al. (2003) Detection and characterization of intracranial aneurysms with MR angiography: comparison of volume-rendering and maximum-intensity-projection algorithms. AJR 180(1):55–64

Mallouhi A, Schocke M, Judmaier W, et al. (2002) 3D MR angiography of renal arteries: comparison of volume

rendering and maximum intensity projection algorithms. Radiology 223(2):509–516
Persson A, Dahlström N, Engellau L, et al. (2004) Volume rendering compared with maximum intensity projection for magnetic resonance angiography measurements of the abdominal aorta. Acta Radiol 45(4):453–459
Prince MR, Grist TM, Debatin JF (2003) 3D Contrast MR angiography. Springer, Berlin
Prokop M, Shin HO, Schanz A, et al. (1997) Use of maximum intensity projections in CT angiography: a basic review. Radiographics 17(2):433–451
Runck F, Steiner RP, Bautz WA, et al. (2008) MR imaging: influence of imaging technique and postprocessing on measurement of internal carotid artery stenosis. AJNR 29(9): 1736–1742
Sun Y, Parker DL (1999) Performance analysis of maximum intensity projection algorithm for display of MRA images. IEEE Trans Med Imaging 18(12):1154–1169
Westenberg JJ, van der Geest RJ, Wasser MN, et al. (2000) Vessel diameter measurements in gadolinium contrast-enhanced three-dimensional MRA of peripheral arteries. Magn Reson Imaging 18(1):13–22

Clinical Applications

Radiologic Vascular Anatomy

6

Claudio Vignali, Valentina Vallini, and Francesca Bianchi

CONTENTS

6.1 The Arteries *47*
6.1.1 The Arteries of the Head, Neck, and Thorax *48*
6.1.1.1 The Thoracic Aorta *48*
6.1.1.2 The Pulmonary Arteries *52*
6.1.2 The Abdominal Arteries *53*
6.1.2.1 Visceral Branches *53*
6.1.2.2 Parietal Branches *56*
6.1.2.3 Terminal Branches *56*
6.1.3 The Arteries of the Limb *57*
6.1.3.1 Upper Limb *57*
6.1.3.2 Lower Limb *59*

6.2 The Veins *62*
6.2.1 Veins of the Head, Neck, and Thorax *63*
6.2.1.1 The Superior Vena Cava *63*
6.2.1.2 The Pulmonary Veins *65*
6.2.1.3 The Veins of the Heart *65*
6.2.2 The Veins of the Abdomen and Pelvis *65*
6.2.2.1 The Portal System of Veins *67*
6.2.3 The Veins of the Limbs *69*
6.2.3.1 The Veins of the Upper Limb *69*
6.2.3.2 The Veins of the Lower Limb *70*

References *71*

Claudio Vignali, MD
Valentina Vallini, MD
Francesca Bianchi, MD
Diagnostic and Interventional Radiology, University of Pisa, Via Roma 67, 56100 Pisa, Italy

ABSTRACT

The distribution of the systematic arteries is like a highly ramified tree, the common trunk of which, formed by the aorta, commences at the left ventricle, while the smallest ramifications extend to the peripheral parts of the body and the contained organs.

The veins convey the blood from the capillaries of the different parts of the body to the heart. They consist of two distinct sets of vessels, the pulmonary and systemic. The Portal vein, an appendage to the systemic venous system, is confined to the abdominal cavity, and returns the venous blood from the spleen and the viscera of digestion to the liver.

6.1 The Arteries

The distribution of the systematic arteries is like a highly ramified tree, the common trunk of which, formed by the aorta, commences at the left ventricle, while the smallest ramifications extend to the peripheral parts of the body and the contained organs.

The arteries, in their distribution, communicate with one another, forming what are called anastomoses, and these communications are very free between the large as well as between the smaller branches. The anastomosis between trunks of equal size is found where great activity of the circulation is requisite, as in the brain and the intestinal arteries. In the limbs the anastomoses are most numerous and of largest size around the joints. The smaller branches of arteries anastomose more frequently than the larger and between the smallest twigs these anastomoses become so numerous as to constitute a close network that pervades nearly every tissue of the body.

Throughout the body generally the larger arterial branches pursue a fairly straight course, but in certain situations they are tortuous. Thus the external maxillary artery in its course over the face, and the arteries of the lips, are extremely tortuous to accommodate themselves to the movements of the parts. The uterine arteries are also tortuous, to accommodate themselves to the increase of size which the uterus undergoes during pregnancy.

6.1.1 The Arteries of the Head, Neck, and Thorax

6.1.1.1 The Thoracic Aorta

The *thoracic aorta* commences at the aortic orifice of the left ventricle, ascendes for a short distance, arches backward and to the left side, and then descends within the thorax on the left side of the vertebral column. Through the aortic hiatus in the diaphragm, passes into the abdominal cavity becoming abdominal aorta.

It is divided in several portions: the ascending aorta, the arch of the aorta, and the descending aorta.

The a*scending aorta* is about 5 cm in length and 3 cm in diameter. It commences at the upper part of the base of the left ventricle, on a level with the lower border of the third costal cartilage. At its origin it presents a dilatation called the aortic bulb for the presence of three small dilatations, the aortic sinuses, opposite the segments of the aortic valve.

The only branches of the ascending aorta are the *coronary arteries* which supply the heart. They arise near the commencement of the aorta immediately above the attached margins of the semilunar valves. The right coronary artery arises from the right aortic sinus. It passes at first between the conus arteriosus and the right auricula and then runs in the right portion of the coronary sulcus, until the diaphragmatic surface of the heart as far as the posterior longitudinal sulcus, down which it is continued to the apex of the heart as the posterior descending branch. It gives off a large marginal branch which follows the acute margin of the heart and supplies branches to both surfaces of the right ventricle. It also gives twigs to the right atrium and to the part of the left ventricle which adjoins the posterior longitudinal sulcus. The left coronary artery, larger than the right, arises from the left aortic sinus and divides into two branches: (a) the anterior descending branch descends in the anterior longitudinal sulcus and it gives branches to both ventricles; (b) the circumflex branch follows the left part of the coronary sulcus, reaching nearly as far as the posterior longitudinal sulcus and anastomosing with branches of the right coronary artery. It gives branches to the left atrium and ventricle.

The cursive script begins at the level of the upper border of the second sternocostal articulation of the right side, and runs at first upward, backward, and to the left in front of the trachea, then it passes downward on the left side of the body of the fourth thoracic vertebra, at the border of which it becomes continuous with the descending aorta. Its upper border is usually about 2.5 cm below the superior border of the manubrium sterni.

The branches given off from the arch of the aorta are three, generally: the innominate artery, the left common carotid artery and the left subclavian artery (Fig. 6.1).

The number of the primary branches may be reduced to one, or more commonly two because the left common carotid can arise with the innominate artery. More rarely the carotid and subclavian arteries of the left side arising from a left innominate artery. The number may be increased also to four, from the right carotid and subclavian arteries arising directly

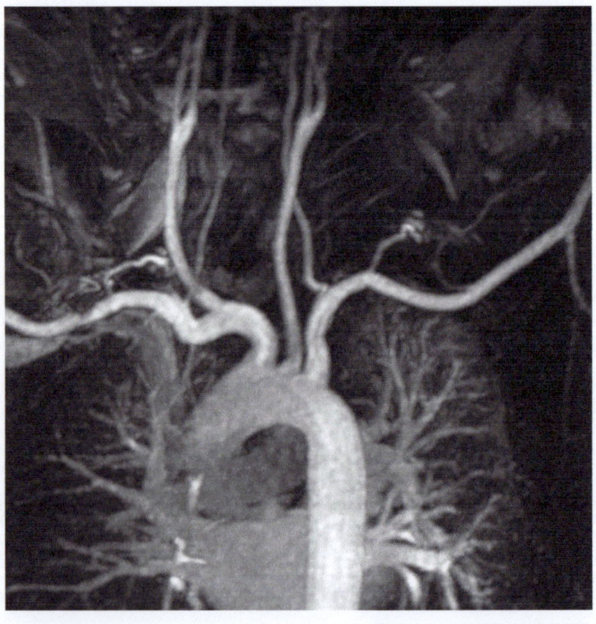

Fig. 6.1. MIP (maximum intensity projection). The branches arising from the arch of the aorta are three: the innominate artery, the left common carotid artery and the left subclavian artery

from the aorta, the innominate being absent. In most of these latter cases the right subclavian has been found to arise from the left end of the arch; in other cases it is the second or third branch given off, instead of the first. Another form in which there are four primary branches is when the left vertebral artery arises from the arch of the aorta between the left carotid and subclavian arteries. In other cases, the two carotids may be joined in a common trunk, and the subclavians arise separately from the arch, the right subclavian generally arising from the left end of the arch.

In some instances other arteries spring from the arch of the aorta. Of these the most common are the bronchial, one or both; but the internal mammary and the inferior thyroid have been seen to arise from this vessel (Gray 1918).

When the aorta arches over to the right side, the three branches have an arrangement the reverse of what is usual: the innominate artery is on the left and the right carotid and subclavian arise separately.

The *innominate* or *brachiocephalic artery* is the largest branch of the arch of the aorta, and is from 4 to 5 cm in length.

It arises, on a level with the upper border of the second right costal cartilage, on a plane anterior to the origin of the left carotid. It ascends obliquely upward, backward, and to the right to the level of the upper border of the right sternoclavicular articulation, where it divides into the right common carotid and right subclavian arteries.

The innominate artery usually gives off no branches. Occasionally a small branch, the thyreoidea ima, arises from it. Sometimes it gives off a thymic or bronchial branch.

The *common carotid arteries* differ in their mode of origin.

Normally the right begins at the bifurcation of the innominate artery behind the sternoclavicular joint while the left springs from the arch of the aorta.

The left common carotid varies in its origin more than the right. In the majority of abnormal cases it arises with the innominate artery; if that artery is absent, the two carotids arise usually by a single trunk.

At the level of the upper border of the thyroid cartilage each artery divides into two branches, the external carotid, supplying the exterior part of the head, the face, and the greater part of the neck and the internal carotid, supplying the vast majority of the tissues within the cranial and orbital cavities (Fig. 6.2).

In the majority of abnormal cases the division occurs higher than usual, the artery dividing opposite or even above the hyoid bone.

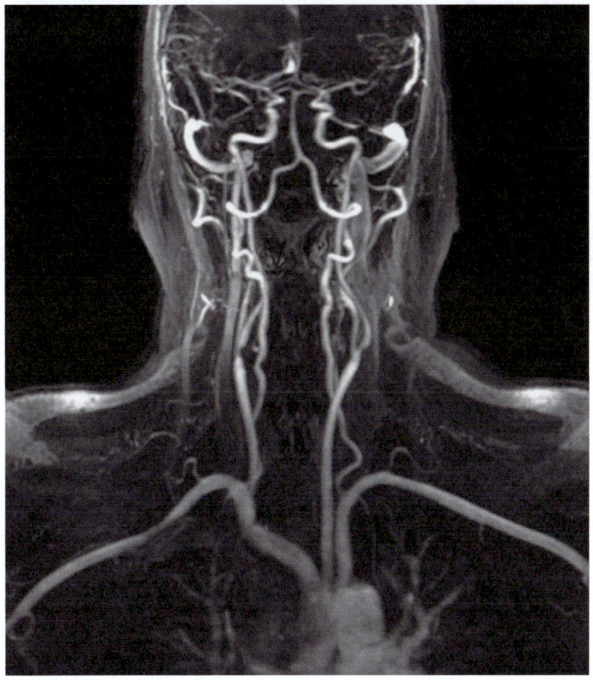

Fig. 6.2. MIP. The common carotid arteries divide into two branches, the external carotid artery and the internal carotid artery

At the bifurcation each common carotid has a dilatation which extent to the proximal tract of the internal carotid, the carotic bulb.

The common carotid usually gives off no branch previous to its bifurcation, but it occasionally gives origin to the superior thyroid or its laryngeal branch, the ascending pharyngeal, the inferior thyroid, or, more rarely, the vertebral artery.

The *external carotid artery* begins opposite the upper border of the thyroid cartilage, and, taking a slightly curved course, arrives behind the neck of the mandible, where it divides into the terminal branches, the superficial temporal and the internal maxillary arteries.

The external carotid artery is covered by the skin, superficial fascia, platysma, deep fascia, and anterior margin of the sternocleidomastoideus muscle.

The branches of the external carotid artery are: superior thyroid artery, ascending pharyngeal artery, lingual artery, occipital artery, posterior auricular artery, external maxillary artery.

The *internal carotid artery* considering its course and its relations may be divided into four portions: cervical, petrous, cavernous, and cerebral.

The cervical portion of the internal carotid begins at the bifurcation of the common carotid, opposite the upper border of the thyroid cartilage, and runs perpendicularly upward, in front of the transverse processes of the upper three cervical vertebre, to the carotid canal in the petrous portion of the temporal bone. The cervical portion of the internal carotid gives off no branches.

The petrous portion of the artery is when the internal carotid artery enters the canal in the petrous portion of the temporal bone. It first ascends a short distance, then curves forward and medialward, and again ascends as it leaves the canal to enter the cavity of the skull between the lingula and petrosal process of the sphenoid. The artery is separated from the bony wall of the carotid canal by a prolongation of dura mater.

In the cavernous portion the artery is situated between the layers of the dura mater forming the cavernous sinus, but covered by the lining membrane of the sinus. It at first ascends toward the posterior clinoid process, then passes forward by the side of the body of the sphenoid bone, and again curves upward on the medial side of the anterior clinoid process, and perforates the dura mater.

In the cerebral portion, The Artery, After having perforated the dura mater on the medial side of the anterior clinoid process passes to the anterior perforated substance at the medial extremity of the lateral cerebral fissure, where it gives off its terminal or cerebral branches.

The branches of the internal carotid artery are: caroticotympanic artery and artery of the pterygoid canal from the petrous portion, cavernous, hypophyseal, anterior meningeal, ophthalmic arteries from the cavernous portion, anterior cerebral, middle cerebral, posterior communicating and choroidal artery from the cerebral portion.

- The ophthalmic artery arises from the internal carotid, just as the vessel is emerging from the cavernous sinus, on the medial side of the anterior clinoid process, and enters the orbital cavity through the optic foramen. It then reaches the medial wall of the orbit, and divides into two terminal branches, the frontal and dorsal nasal.
- The anterior cerebral artery arises from the internal carotid, at the medial extremity of the lateral cerebral fissure. It passes forward and medialward across the anterior perforated substance; here it comes into close relationship with the opposite artery, to which it is connected by a short trunk, the anterior communicating artery. From this point the two vessels run side by side in the longitudinal fissure, curve around the genu of the corpus callosum, and turning backward continue along the upper surface of the corpus callosum to its posterior part, where they end by anastomosing with the posterior cerebral arteries.

The anterior communicating artery connects the two anterior cerebral arteries across the commencement of the longitudinal fissure. Sometimes this vessel is wanting, the two arteries joining together to form a single trunk, which afterward divides or it may be wholly, or partially, divided into two.

- The middle cerebral artery, the largest branch of the internal carotid, runs at first in the lateral cerebral or Sylvian fissure and then on the surface of the insula, where it divides into a number of branches which are distributed to the lateral surface of the cerebral hemisphere.
- The posterior communicating artery runs backward from the internal carotid, and anastomoses with the posterior cerebral artery, a branch of the basilar. It varies in size, being sometimes small, and occasionally so large that the posterior cerebral may be considered as arising from the internal carotid rather than from the basilar. It is frequently larger on one side than on the other. From its posterior half are given off a number of small branches for the medial surface of the thalami and the walls of the third ventricle.

The *subclavian artery* on the right side arises from the innominate artery behind the right sternoclavicular articulation; on the left side it springs from the arch of the aorta. The two vessels, therefore, in the first part of their course, differ in length, direction, and relation with neighboring structures. Each subclavian artery can be divided into three parts: (a) the first portion extends from the origin of the vessel to the medial border of the scalenus anterior; (b) the second lies behind this muscle; (c) the third extends from the lateral margin of the muscle to the outer border of the first rib, where it becomes the axillary artery.

The origin of the right subclavian from the innominate artery takes place, in some cases, above the sternoclavicular articulation, and less frequently, below that joint. The artery may arise as a separate trunk from the arch of the aorta, and in such cases it may be either the first, second, third, or even the last branch derived from that vessel; in the majority, however, it is the first or last, rarely the second or third. When it is the first branch, it occupies the ordinary position of the innominate artery; when the second or third, it gains its usual position by passing behind the right carotid; and when the last branch, it arises from the left extremity of the arch, and passes obliquely toward the right side, usually behind the trachea, esophagus, and right carotid, sometimes between the esophagus and trachea, to the upper border of the first rib, whence it follows its ordinary course. In very rare instances, this vessel arises from the thoracic aorta, as low down as the fourth thoracic vertebra.

The left subclavian is occasionally joined at its origin with the left carotid.

The branches of the subclavian artery are: vertebral artery, thyrocervical trunk, internal mammary artery, costocervical artery.

The *vertebral artery* is the first branch of the subclavian, and arises from the upper and back part of the first portion of the vessel and ascends through the foramina in the transverse processes of the upper six cervical vertebre. It then winds behind the superior articular process of the atlas and, entering the skull through the foramen magnum, unites, at the lower border of the pons, with the vessel of the opposite side to form the basilar artery.

The branches of the vertebral artery may be divided into two sets: those given off in the neck (spinal and muscular arteries), and those within the cranium (meningeal, posterior spinal, anterior spinal, posterior inferior cerebellar, medullary arteries).

The *basilar artery*, is a single trunk formed by the junction of the two vertebral arteries. It extends from the lower to the upper border of the pons, lying in its median groove, under cover of the arachnoid. It ends by dividing into the two posterior cerebral arteries (Fig. 6.3).

Its Branches are pontine, anterior inferior cerebellar, internal auditory, superior cerebellar and posterior cerebral arteries.

The *posterior cerebral artery* is the largest branch. Passing lateralward, parallel to the superior cerebellar artery, and receiving the posterior communicating from the internal carotid, it winds around the cerebral peduncle, and reaches the tentorial surface of the occipital lobe of the cerebrum, where it breaks up into branches for the supply of the temporal and occipital lobes.

At the base of the brain the internal carotid and the vertebral artery form a remarkable anastomosis known as the *arterial circle of Willis* (Fig. 6.4). It is formed in front by the anterior cerebral arteries, branches of the internal carotid, which are connected together by the anterior communicating, behind by the two posterior cerebral arteries, branches of the basilar, which are connected on either side with the internal carotid by the posterior communicating.

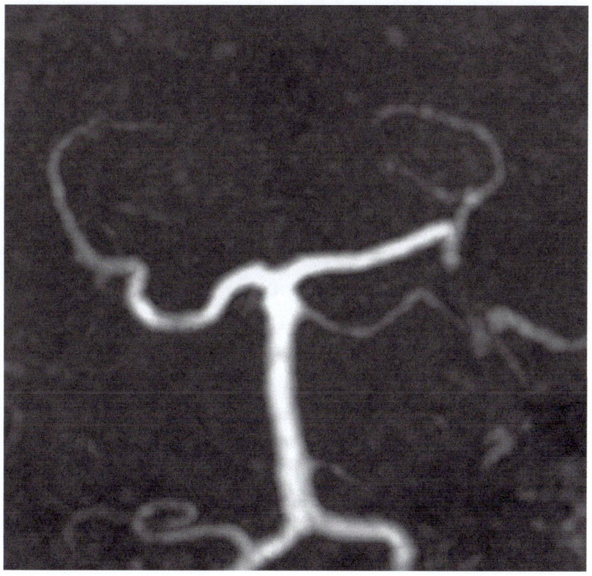

Fig. 6.3. MIP. The basilar artery ends by dividing into the two (left and right) posterior cerebral arteries

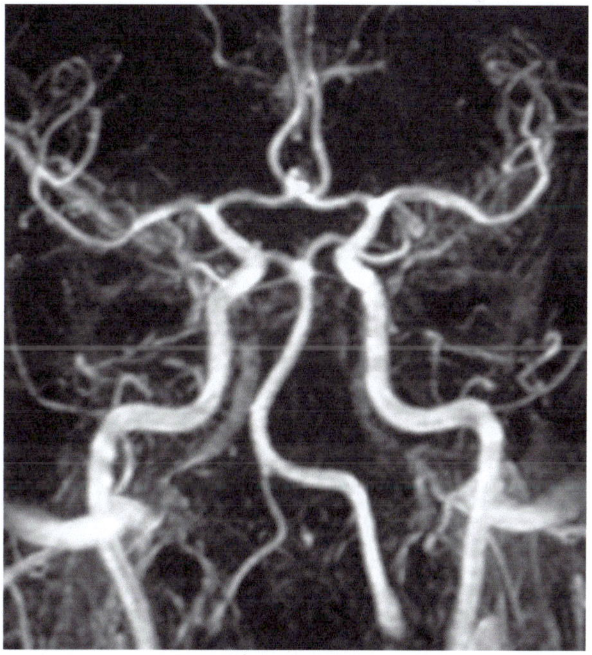

Fig. 6.4. MIP. Arterial circle of Willis

The *thyrocervical trunk* is a short thick trunk, which arises from the front of the first portion of the subclavian artery, close to the medial border of the scalenus anterior, and divides almost immediately into three branches, the inferior thyroid, transverse scapular, and transverse cervical.

The *internal mammary artery* arises from the under surface of the first portion of the subclavian, opposite the thyrocervical trunk. It descends behind the cartilages of the upper six ribs at a distance of about 1 cm from the margin of the sternum, and at the level of the sixth intercostal space divides into the musculophrenic and superior epigastric arteries.

The *thoracic descending aorta* is contained in the posterior mediastinal cavity.

It begins at the fourth thoracic vertebra where it is continuous with the aortic arch, and ends in front of the twelfth at the aortic hiatus in the diaphragm.

Branches of the thoracic aorta: (a) Visceral (pericardial arteries, bronchial arteries, esophageal arteries, mediastinal arteries); (b) Parietal (intercostal arteries, subcostal arteries, superior phrenic artery).

Visceral Branches

The *pericardial arteries* consist of a few small vessels which are distributed to the posterior surface of the pericardium.

The *bronchial arteries* vary in number, size, and origin.

Generally there is one bronchial artery on the right and two bronchial arteries on the left. The right bronchial artery can arises from the first aortic intercostal or from the upper left bronchial artery.

Each vessel runs on the back part of its bronchus, dividing and subdividing along the bronchial tubes, supplying them, the lung's parenchyma, the bronchial lymph glands, and the visceral pleura.

The *esophageal arteries* are four or five in number. They form a chain of anastomoses along the esophagus, anastomosing with the esophageal branches of the inferior thyroid arteries above, and with ascending branches from the left inferior phrenic and left gastric arteries below.

The *mediastinal branches* are numerous small vessels which supply the mediastinal lymph glands.

Parietal Branches

The *intercostal arteries* are usually nine pairs.

They arise from the back of the aorta, and run to the lower nine intercostal spaces the first two spaces supplied by the highest intercostal artery, a branch of the costocervical trunk of the subclavian.

Each artery divides into an anterior and a posterior branch.

The anterior branch crosses the corresponding intercostal space obliquely toward the angle of the upper rib, and thence is continued forward in the costal groove; it anastomoses in front with the intercostal branch of the internal mammary. The first aortic intercostal artery anastomoses with the intercostal branch of the costocervical trunk. The lower two intercostal arteries are continued anteriorly from the intercostal spaces into the abdominal wall, and anastomose with the subcostal, superior epigastric, and lumbar arteries. The anterior rami give off the following branches: collateral intercostals, lateral cutaneous, muscular, mammary.

The posterior branch runs backward through a space which is bounded above and below by the necks of the ribs, medially by the body of a vertebra, and laterally by an anterior costotransverse ligament.

It gives off a spinal branch which enters the vertebral canal through the intervertebral foramen and is distributed to the medulla spinalis and its membranes and the vertebrae. It then courses over the transverse process, supplies branches to the muscles of the back and cutaneous branches.

The *Adamkiewicz* or *great anterior radiculomedullary artery* is the largest anterior segmental medullary artery. It typically arises from a left posterior intercostal artery and supplies the lower two-thirds of the spinal cord via the anterior spinal artery. In some percentage of people, the artery of Adamkiewicz arises from a lumbar vessel.

The *subcostal arteries*, so named because they lie below the last ribs, constitute the lowest pair of branches derived from the thoracic aorta, and are in series with the intercostal arteries.

Each passes along the lower border of the twelfth rib, and it anastomoses with the superior epigastric, lower intercostal, and lumbar arteries. Each subcostal artery gives off a posterior branch which has a similar distribution to the posterior ramus of an intercostal artery.

The *superior phrenic arteries* are small and arise from the lower part of the thoracic aorta; they are distributed to the posterior part of the upper surface of the diaphragm.

6.1.1.2
The Pulmonary Arteries

The *pulmonary arteries* convey the venous blood from heart to the lungs.

The pulmonary trunk begins at the base of the right ventricle. It is short and wide ~5 cm in length and 3 cm in diameter. It then branches into two

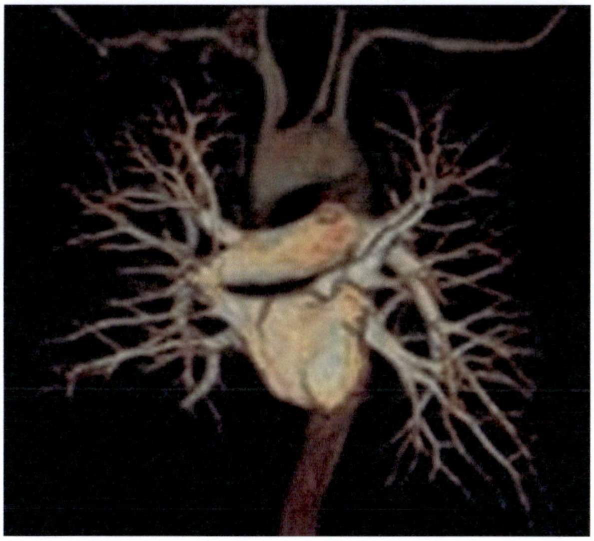

Fig. 6.5. VR (volume rendering). The pulmonary arteries with the left atrium and the pulmonary veins

pulmonary arteries, left and right (Fig. 6.5), which accompany the bronchial tubes and end in a dense capillary network in the walls of the alveoli.

6.1.2 The Abdominal Arteries

The *abdominal aorta* begins at the aortic hiatus of the diaphragm and, descending in front of the vertebral column, ends at the level of the fourth lumbar vertebra, by dividing into the two common iliac arteries.

Variations in the abdominal aorta are not common. The most common place for its bifurcation, is opposite the lower border of the body of the fourth lumbar vertebra 75% but it may divide opposite the disc between L4 and L5 or rarely opposite the body of the fifth lumbar vertebra. A higher division is less common (9%) than a lower division (11%). The following rare variations of the abdominal aorta have been described: (1) the aorta may pass through the esophageal hiatus (opening) of the diaphragm; (2) it may lie on the right side of the vena cava, the vein passing over the upper part of the aorta to gain access to the caval hiatus; (3) the aorta may have a vena cava on each side, the left vein passing across the upper part of the artery to open into the right vein just below the caval opening (Bergman et al. 1996).

The branches of the abdominal aorta may be divided into three sets: (a) Visceral (celiac trunk, superior mesenteric artery, inferior mesenteric artery, middle suprarenal artery, renal artery, internal spermatic or ovarian artery); (b) Parietal (inferior phrenic artery, middle sacral artery); (c) Terminal (common iliac artery).

6.1.2.1 Visceral Branches

The *celiac trunk* is a short thick trunk, which arise from the front of the aorta, just below the aortic hiatus of the diaphragm.

The celiac trunk conventionally divides after 1–2 cm into three large branches, the left gastric, the common hepatic, and the splenic (Fig. 6.6); this pattern is seen in 65–75% of individuals and it occasionally gives off one of the inferior phrenic arteries. It may be absent and the branches usually arising from it then arise from the aorta. The trunk may be shorter or longer than usual and in the latter case, the branches commonly arise separately from the trunk instead of radiating from a single point. The celiac trunk may lack one or more of its main branches. In such cases, the branch in question may arise from the aorta or the superior mesenteric, either independently or in conjunction with another branch. Variations have been reported to occur with the following frequencies (approximate): (1) hepatosplenogastric trunk, 90%; (2) hepatosplenic trunk (hepatic and splenic), 3.5%; (3) hepatosplenomesenteric trunk (hepatic, splenic, and superior

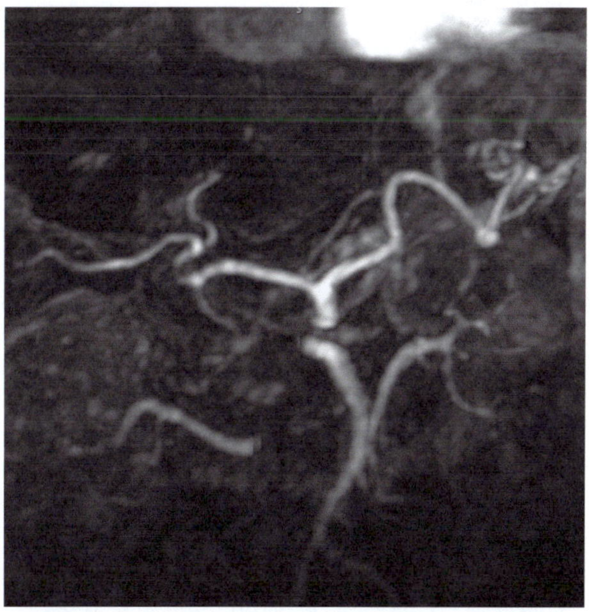

Fig. 6.6. MIP. The celiac trunk and the superior mesenteric artery arise from the abdominal aorta

mesenteric), 0.5%; (4) hepatogastric trunk (hepatic and left gastric), 1.5%; (5) splenogastric trunk (splenic and left gastric), 5.5% (BERGMAN et al. 1996).

The *left gastric artery* is the smallest of the three branches of the celiac artery. It runs from left to right, along the lesser curvature of the stomach to the pylorus, between the layers of the lesser omentum; It gives branches to both surfaces of the stomach and anastomoses with the right gastric artery. The left gastric may arise directly from the abdominal aorta, and there may be two left gastric arteries instead of one.

The *common hepatic artery* in the adult is intermediate in size between the left gastric and lienal. It crosses the portal vein anteriorly and ascends between the layers of the lesser omentum, and in front of the epiploic foramen, to the porta hepatis, where it divides into two branches, right and left, which supply the corresponding lobes of the liver. Its branches are: right gastric, gastroduodenal, and cystic. The common hepatic artery usually arises from the celiac trunk, but may also arise directly from the aorta or from the left gastric, gastroduodenal, right renal or splenic artery. In ~30% of persons all or part of the hepatic arterial supply arises from the superior mesenteric artery. The common hepatic artery may be doubled, tripled, or missing (12% of cases) and it is almost always located between the bile duct and the portal vein. The left hepatic artery arises from the common hepatic in 89% of cases; in 11% of cases the vessel is may arise from the left gastric. Accessory right or left hepatic arteries can be found approximately in 8% of subjects (BERGMAN et al. 1996).

The *right gastric artery* arises from the hepatic artery (~53%, the most frequent origin) above the pylorus, descends to the pyloric end of the stomach and anastomosing with the left gastric artery.

The *gastroduodenal artery* is a short but large branch, which descends, near the pylorus, between the superior part of the duodenum and the neck of the pancreas, and divides at the lower border of the duodenum into two branches, the right gastroepiploic and the superior pancreaticoduodenal. The gastroduodenal artery most commonly arises from the common hepatic artery almost always (75%) before the division of the hepatic artery into its right and left branches.

The *right gastroepiploic artery* runs from right to left along the greater curvature of the stomach, anastomosing with the left gastroepiploic branch of the lienal artery. This vessel gives off numerous branches, some of which ascend to supply both surfaces of the stomach, while others descend to supply the greater omentum and anastomose with branches of the middle colic.

The *superior pancreaticoduodenal artery* descends between the contiguous margins of the duodenum and pancreas. It supplies both these organs, and anastomoses with the inferior pancreaticoduodenal branch of the superior mesenteric artery, and with the pancreatic branches of the lienal artery.

The *cystic artery*, usually a branch of the right hepatic, divides into two branches, one of which ramifies on the free surface, the other on the attached surface of the gall-bladder. There is only one cystic artery in 88% of individuals. Of these individuals, the cystic artery arises from the right hepatic artery in 82%; from the proper hepatic artery in 3%; from the left hepatic artery, distal to the origin of the right hepatic artery, in 2%; and from the gastroduodenal in 1%.

The *lienal* or *splenic artery*, the largest branch of the celiac artery, is remarkable for the tortuosity of its course. It passes along the upper border of the pancreas; it divides into branches, some of which enter the hilus of that organ to be distributed to the tissues of the spleen; some are given to the pancreas, while others pass to the greater curvature of the stomach. The tortuous course of the splenic artery is considered so variable that no two arteries are alike, but the tortuosity of the artery is absent in infants and children. In congenital absence of the spleen (very rare), the splenic artery terminates in the pancreas.

The *superior mesenteric artery* arises from the front of the aorta about 1.25 cm below the celiac artery (Fig. 6.6), and is crossed at its origin by the neck of the pancreas.

It is a large vessel which supplies the whole length of the small intestine, except the superior part of the duodenum; it also supplies the cecum and the ascending part of the colon and about one-half of the transverse part of the colon. The variations in this artery are very common and numerous. The artery may be doubled, arising from the aorta as two trunks; sometimes it arises from, or in common with the celiac trunk. In some cases, it gives rise to the left colic and superior hemorrhoidal (superior rectal), thus wholly or partly replacing the inferior mesenteric artery.

Its branches are: inferior pancreaticoduodenal, ileocolic, intestinal, right colic and middle colic.

The *inferior mesenteric artery* supplies the left half of the transverse part of the colon, the whole of the descending and iliac parts of the colon, the sigmoid colon, and the greater part of the rectum. It is smaller than the superior mesenteric, and arises from the aorta, about 3 or 4 cm above its division into the common iliacs and close to the lower border of the inferior part of the duodenum. It crosses the left common iliac artery and is continued into the lesser pelvis under the name of the *superior hemorrhoidal artery*, which descends between the two layers of the sigmoid mesocolon and ends on the upper part of the rectum.

Its branches are: left colic, sigmoid, superior hemorrhoidal.

Duplication of this artery has been reported. The inferior mesenteric artery may be absent entirely, its branches being given by the superior mesenteric artery or it may be just absent from its usual location. This artery sometimes arises from the superior mesenteric by a common trunk with the middle colic or the left common iliac artery. Usually, the origin of the artery from the aorta was at the level of L3, in a few cases at the level of the intervertebral disc between L3 and L4, or at the level of L4. In a smaller number of cases, origin of the artery at the level of the intervertebral disc between L2 and L3 or L4 and L5 was observed.

The *renal arteries* are two large trunks, which arise from the side of the aorta, immediately below the superior mesenteric artery. The right is longer than the left, on account of the position of the aorta; it passes behind the inferior vena cava and the right renal vein. The left is somewhat higher than the right; it lies behind the left renal vein and is crossed by the inferior mesenteric vein.

Both renal arteries usually have two divisions which pass anterior and posterior to the renal pelvis (Fig. 6.7).

Each vessel gives off some small *inferior suprarenal branches* to the suprarenal gland, the ureter, and the surrounding cellular tissue and muscles.

In approximately two-thirds of individuals a single right and left renal artery is found. Irregularites of the renal vessels are found in about 35% of the cases. The most common is the presence of an additional vessel (~28%) and supernumerary vessels are more frequent on the left side than the right. Multiple renal arteries occur unilaterally in 32% and bilaterally in 12%. An independent superior renal pole artery observed in up to 7% and an inferior pole artery is ~5%. Superior renal polar arteries are usually single. They arise as separate branches from the aorta in 7% of cases, or as branches of the renal artery in 12% of cases. There may be two or three unilaterally or bilaterally with different aortic or renal sources. Superior renal polar arteries can also arise from the inferior phrenic or from the superior mesenteric artery. Some superior renal polar arteries provide several branches to the suprarenal gland, dividing before entering the superior pole of the kidney. Inferior renal polar arteries are usually single and arise from the aorta (95.5%), or the renal artery (1.4%). The inferior polar arteries are sometimes doubled (Fig. 6.8), with one arising from the aorta and the other from the renal, or the pair from either source (Gray 1918).

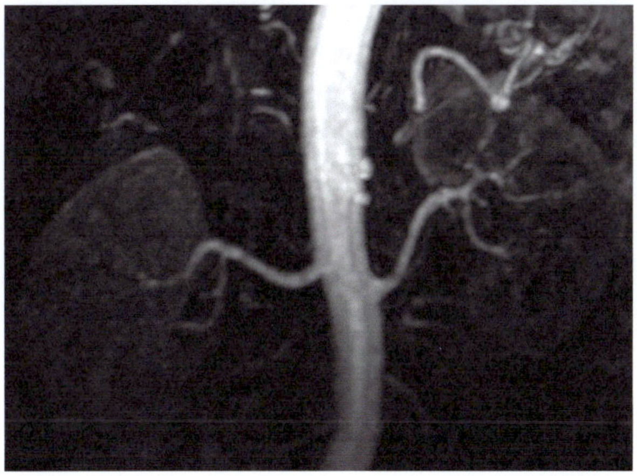

Fig. 6.7. MIP. The renal arteries with the anterior and posterior divisions

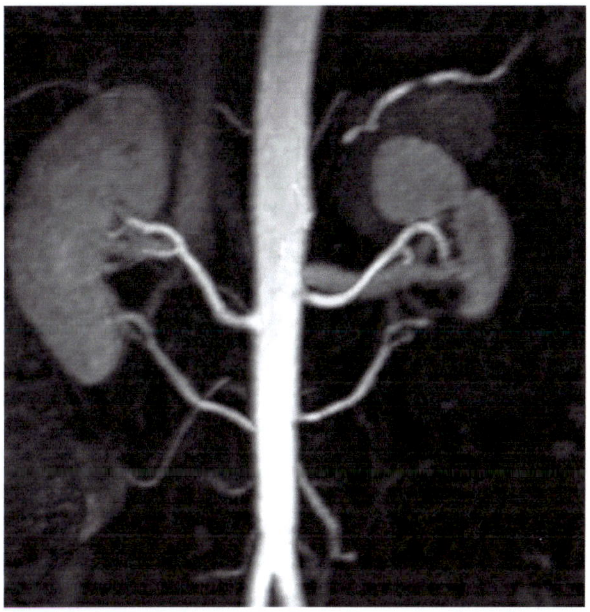

Fig. 6.8. MIP. Doubled inferior renal polar artery arising from the aorta

Aberrant renal arteries are common in fused kidneys (horseshoe- dumbbell- or disc-shaped): two to six such arteries are characteristically longer or shorter than usual. The unusual vessels may originate from the aorta, gonadal, common iliac, middle sacral, external or internal iliac, or superior or inferior mesenteric arteries. There are reports in the literature of seven, eight, and ten arteries supplying horseshoe-shaped kidneys, but such findings, are very rare.

The *internal spermatic arteries* are distributed to the testes. They are two slender vessels of considerable length, and arise from the front of the aorta a little below the renal arteries. The right spermatic may run behind, instead of in front of the inferior vena cava. Each crosses obliquely over the ureter and the lower part of the external iliac artery to reach the abdominal inguinal ring, through which it passes, along the inguinal canal to the scrotum, where it becomes tortuous, and divides into several branches. The internal spermatic artery supplies one or two small branches to the ureter, and in the inguinal canal gives one or two twigs to the cremaster. If one or both testicular arteries are missing, the testes are supplied by branches from the vesical or prostatic arteries passing under the arch of the pubis.

The *ovarian arteries* are the corresponding arteries in the female to the internal spermatic in the male. They supply the ovaries, are shorter than the internal spermatics, and do not pass out of the abdominal cavity. Small branches are given to the ureter and the uterine tube, and one passes on to the side of the uterus, and unites with the uterine artery. Other offsets are continued on the round ligament of the uterus, through the inguinal canal, to the integument of the labium majus and groin. The gonadal arteries usually arise from the anterior surface of the aorta below the renal vein, but in 15% of cases, the gonadal arteries arise from the renal artery, from one of the branches of the renal artery, or from a supernumerary or "accessory" renal artery. More rarely, a gonadal artery can originate from a suprarenal, inferior phrenic, superior mesenteric, lumbar, common iliac, or internal iliac. In some cases (17%), the gonadal arteries are doubled on one side; less commonly, they are doubled on both sides.

6.1.2.2
Parietal Branches

The *inferior phrenic arteries* are two small vessels (the left and the right phrenic), which supply the diaphragm and present much variety in their origin. They may arise, by a common trunk from the aorta or the celiac trunk in 18–30% of cases, or as independent branches from these same sources in 62%. Other sources may be the hepatic, left gastric, renal, suprarenal, or superior mesenteric arteries in about 8% of cases. Variations in the origin of the phrenic arteries are numerous and supplementary phrenic vessels are common. When independent, the right and left phrenics usually arise asymmetrically. They run obliquely upward and lateralward under surface by the diaphragti. Near the back part of the central tendon each vessel divides into a medial and a lateral branch. Each vessel gives off *superior suprarenal branches* (in ~96%) to the suprarenal gland of its own side.

The *middle suprarenal artery* is usually a single vessel (85% of cases) and it often arises from the aorta, but may also originate from the inferior phrenic, renal, superior renal polar, or from the celiac trunk. The artery may be absent, multiple, or may not even supply the suprarenal gland.

The *lumbar arteries* are in series with the intercostals. They are usually four in number on either side, and arise from the back of the aorta, opposite the bodies of the upper four lumbar vertebrae.

The first lumbar may be joined at its origin with the lowest intercostal artery; the third and fourth lumbar, or less often the second and third lumbars, may arise from the aorta as a common stem. One or both of the last pair of lumbar arteries may arise in common with the middle sacral.

The *middle sacral artery* is a small vessel, which arises from the back of the aorta, a little above its bifurcation. It descends in the middle line in front of the fourth and fifth lumbar vertebrae, the sacrum and coccyx. From it, minute branches pass to the posterior surface of the rectum.

6.1.2.3
Terminal Branches

The abdominal aorta divides, on the left side of the body of the fourth lumbar vertebra, into the two *common iliac arteries*. Each artery is about 5 cm in length. They diverge from the termination of the aorta, pass downward and lateralward, and divide, into two branches, the external iliac and hypogastric arteries, The former supplies the lower extremity the latter, the viscera and parietes of the pelvis. The right common iliac artery is somewhat longer than the left. The left common iliac artery, laterally, is in relation with the Psoas major and it is crossed at its point of bifurcation by the ureter.

The common iliac arteries give off small branches to the peritoneum, Psoas major, ureters, and the surrounding areolar tissue, and occasionally give origin to the iliolumbar, or accessory renal arteries. The point of origin varies according to the bifurcation of the aorta.

In about 80% of the cases the aorta bifurcated within 1.25 cm above or below the level of the crest of the ilium; more frequently below than above. The point of division is subject to great variety. In 68% of individuals the division occurs between the last lumbar vertebra and the upper border of the sacrum; being above that point in one case out of eight, and below it in one case out of six. The left common iliac artery divides lower down more frequently than the right.

The relative lengths, also, of the two common iliac arteries vary. The length of the arteries varied, from 3.5 to 7.5 cm; in about half of the remaining cases the artery was longer, and in the other half, shorter; the minimum length being less than 1.25 cm, the maximum, 11 cm. In rare instances, the right common iliac has been found wanting, the external iliac and hypogastric arising directly from the aorta (Gray 1918).

The *hypogastric artery* supplies the walls and viscera of the pelvis, the buttock, the generative organs, and the medial side of the thigh. It is a short, thick vessel, smaller than the external iliac, and about 4 cm in length. It arises at the bifurcation of the common iliac, opposite the lumbosacral articulation, and, passing downward to the upper margin of the greater sciatic foramen, divides into two large trunks, an anterior and a posterior. The place of division of the hypogastric varies between the upper margin of the sacrum and the upper border of the greater sciatic foramen. The right and left hypogastric arteries in a series of cases often differed in length, but neither seemed constantly to exceed the other. The branches of the hypogastric artery are: (a) from the anterior trunk (superior, middle and inferior vesical, middle hemorrhoidal, obturator, internal pudendal, inferior gluteal, uterine and vaginal (in the female); (b) from the posterior trunk (iliolumbar, lateral sacral, superior gluteal).

The *external iliac artery* is larger than the hypogastric, and passes obliquely downward and lateralward along the medial border of the Psoas major, from the bifurcation of the common iliac to a point beneath the inguinal ligament, midway between the anterior superior spine of the ilium and the symphysis pubis, where it becomes the femoral artery. Besides several small branches to the Psoas major and the neighboring lymph glands, the external iliac gives off two branches of considerable size: the inferior epigastric and the deep iliac circumflex.

The *inferior epigastric artery* arises from the external iliac, immediately above the inguinal ligament. It curves forward in the subperitoneal tissue, and ascends between the Rectus abdominis and the posterior lamella of its sheath. It finally divides into numerous branches. The origin of the inferior epigastric may take place from any part of the external iliac between the inguinal ligament and a point 6 cm above it; or it may arise below this ligament, from the femoral. It frequently springs from the external iliac, by a common trunk with the obturator. Sometimes it arises from the obturator, the latter vessel being furnished by the hypogastric, or it may be formed of two branches, one derived from the external iliac, the other from the hypogastric.

The *deep iliac circumflex artery* arises from the lateral aspect of the external iliac nearly opposite the inferior epigastric artery. It ascends between the Obliquus internus and Transversus muscles, supplying them, and anastomosing with the lumbar and inferior epigastric arteries.

6.1.3 The Arteries of the Limb

6.1.3.1 Upper Limb

The artery which supplies the upper extremity continues as a single trunk from its commencement down to the elbow; but different portions of it have received different names, according to the regions through which they pass. That part of the vessel which extends from its origin to the outer border of the first rib is termed the subclavian; beyond this point to the lower border of the axilla it is named the axillary; and from the lower margin of the axillary space to the bend of the elbow it is termed brachial; here the trunk ends by dividing into two branches the radial and ulnar.

The *axillary artery* is the continuation of the subclavian, commences at the outer border of the first rib, and ends at the lower border of the tendon of the Teres major, where it takes the name of brachial. Its direction varies with the position of the limb; thus the vessel is nearly straight when the arm is directed at right angles with the trunk, concave upward when the arm is elevated above this, and convex upward and lateralward when the arm lies by the side. At its origin the artery is very deeply situated, but near its termination is superficial, being covered only by the skin and fascia. The branches of the axillary are: highest thoracic artery, thoracoacromial artery, lateral thoracic artery, subscapular artery, posterior humeral circumflex artery, anterior humeral circumflex artery.

Sometimes the axillary artery divides into the radial and ulnar arteries, and occasionally it gives origin to the volar interosseous artery of the forearm. Occasionally the subscapular, humeral circumflex, and profunda Brach arteries arise from a common trunk, and when this occurs the branches of the brachial plexus surround this trunk instead of the main vessel.

The *brachial artery* commences at the lower margin of the tendon of the Teres major, and, passing down the arm, ends about 1 cm below the bend of the elbow, where it divides into the radial and ulnar arteries. At first the brachial artery lies medial to the humerus; but as it runs down the arm it gradually gets in front of the bone, and at the bend of the elbow it lies midway between its two epicondyles. The brachial artery may leave the medial border of the Biceps brachii, and descend toward the medial epicondyle of the humerus; in such cases it usually passes behind the supracondylar process of the humerus, from which a fibrous arch is in most cases thrown over the artery; it then runs beneath or through the substance of the Pronator teres, to the bend of the elbow. Occasionally, the artery is divided for a short distance at its upper part into two trunks, which are united below. Frequently the artery divides at a higher level than usual, and the vessels concerned in this high division are three: radial, ulnar, and interosseous. Most frequently the radial is given off high up, the other limb of the bifurcation consisting of the ulnar and interosseous; in some instances the ulnar arises above the ordinary level, and the radial and interosseous form the other limb of the division; occasionally the interosseous arises high up. Sometimes, long slender vessels, connect the brachial or the axillary artery with one of the arteries of the forearm, or branches from them. The branches of the brachial artery are, arteria profunda brachii, nutrient artery, superior ulnar collateral artery, inferior ulnar collateral artery, muscular branches.

The vessels engaged in the anastomoses may be conveniently divided into those situated in front of and those behind the medial and lateral epicondyles of the humerus. The branches anastomosing in front of the medial epicondyle are: the anterior branch of the inferior ulnar collateral, the anterior ulnar recurrent, and the anterior branch of the superior ulnar collateral. Those behind the medial epicondyle are: posterior branch of the inferior ulnar collateral, the posterior ulnar recurrent, and the posterior branch of the superior ulnar collateral. The branches anastomosing in front of the lateral epicondyle are: the radial recurrent and the terminal part of the profunda brachii. Those behind the lateral epicondyle are: the inferior ulnar collateral, the interosseous recurrent, and the radial collateral branch of the profunda brachii. There is also an arch of anastomosis above the olecranon, formed by the interosseous recurrent joining with the inferior ulnar collateral and posterior ulnar recurrent.

The *radial artery* appears, from its direction, to be the continuation of the brachial, but it is smaller in caliber than the ulnar. It commences at the bifurcation of the brachial, just below the bend of the elbow, and passes along the radial side of the forearm to the wrist. It then winds backward, around the lateral side of the carpus, beneath the tendons of the Abductor pollicis longus and Extensores pollicis longus and brevis to the upper end of the space between the metacarpal bones of the thumb and index finger. Finally it passes forward between the two heads of the first Interosseous dorsalis, into the palm of the hand, where it crosses the metacarpal bones and at the ulnar side of the hand unites with the deep volar branch of the ulnar artery to form the deep volar arch.

The origin of the radial artery is, in nearly one case in eight, higher than usual; more often it arises from the axillary or upper part of the brachial than from the lower part of the latter vessel. In the forearm it deviates less frequently from its normal position than the ulnar. It has been found lying on the deep fascia instead of beneath it.

The branches of the radial artery may be divided into three groups, corresponding with the three regions in which the vessel is situated: (a) in the forearm (radial recurrent, muscular, volar carpal, superficial volar); (b) at the wrist (dorsal carpal, first dorsal metacarpal); (c) in the hand (princeps pollicis, volaris indicis radialis, volar metacarpal, perforating, recurrent).

The *ulnar artery* the larger of the two terminal branches of the brachial, begins a little below the bend of the elbow, and, passing obliquely downward, reaches the ulnar side of the forearm at a point about midway between the elbow and the wrist. It then runs along the ulnar border to the wrist, crosses the transverse carpal ligament on the radial side of the pisiform bone, and immediately beyond this bone divides into two branches, which enter into the formation of the superficial and deep volar arches.

The ulnar artery may arise about 5–7 cm below the elbow, but more frequently higher, the brachial being more often the source of origin than the axillary. Variations in the position of this vessel are more common than in the radial. When its origin is normal, the course of the vessel is rarely changed. When it arises high up, it is almost invariably superficial to the Flexor muscles in the forearm, lying commonly beneath the fascia, more rarely between the fascia and integument. In a few cases, its position is subcutaneous in the upper part of the forearm, and subaponeurotic in the lower part.

The branches of the ulnar artery may be arranged in the following groups: (a) in the forearm (anterior recurrent, posterior recurrent, common interosseous, muscular); (b) at the wrist (volar carpal, dorsal carpal); (c) in the hand (deep volar, superficial volar arch).

6.1.3.2 Lower Limb

The artery which supplies the greater part of the lower extremity is the direct continuation of the external iliac. It runs as a single trunk from the inguinal ligament to the lower border of the Popliteus, where it divides into two branches, the anterior and posterior tibial. The upper part of the main trunk is named the femoral, the lower part the popliteal.

The Femoral Artery

The *femoral artery* begins immediately behind the inguinal ligament, midway between the anterior superior spine of the ilium and the symphysis pubis, and passes down the front and medial side of the thigh. It ends at the junction of the middle with the lower third of the thigh, where it passes through an opening in the Adductor magnus to become the popliteal artery. The vessel, at the upper part of the thigh, lies in front of the hip-joint; in the lower part of its course it lies to the medial side of the body of the femur, and between these two parts, where it crosses the angle between the head and body, the vessel is some distance from the bone. The first 4 cm of the vessel is enclosed in a fibrous sheath, the femoral sheath. In the upper third of the thigh the femoral artery is contained in the femoral triangle (Scarpa's triangle), and in the middle third of the thigh, in the adductor canal (Hunter's canal).

Several cases are recorded in which the femoral artery divided into two trunks below the origin of the profunda, and became reunited near the opening in the Adductor magnus, so as to form a single popliteal artery. A few cases have been recorded in which the femoral artery is absent, its place being supplied by the inferior gluteal artery which accompanied the sciatic nerve to the popliteal fossa. The external iliac in these cases is small, and terminate in the profunda.

The branches of the femoral artery are: superficial epigastric, deep external pudendal, superficial iliac circumflex, muscular, superficial external pudendal, profunda femoris, highest genicular.

The *superficial epigastric artery* arises from the front of the femoral artery about 1 cm below the inguinal ligament. It distributes branches to the superficial subinguinal lymph glands, the superficial fascia, and the integument; it anastomoses with branches of the inferior epigastric, and with its fellow of the opposite side.

The *superficial iliac circumflex artery*, the smallest of the cutaneous branches, arises close to the preceding, and it distributes branches to integument of the groin, the superficial fascia, and the superficial subinguinal lymph glands, anastomosing with the deep iliac circumflex, the superior gluteal and lateral femoral circumflex arteries.

The *superficial external pudendal artery* arises from the medial side of the femoral artery and it distributes to he integument on the lower part of the abdomen, the penis and scrotum in the male, and the labium majus in the female, anastomosing with branches of the internal pudendal.

The *deep external pudendal artery*, more deeply seated than the preceding, is distributed, in the male, to the integument of the scrotum and perineum, in the female to the labium majus; its branches anastomose with the scrotal (or labial) branches of the perineal artery.

Muscular branches are supplied by the femoral artery to the Sartorius, Vastus medialis, and Adductores.

The Profunda Femoris Artery

The *profunda femoris artery* is a large vessel arising from the lateral and back part of the femoral artery, from 2 to 5 cm below the inguinal ligament. At first it lies lateral to the femoral artery; it then runs behind it to the medial side of the femur, and, passing downward behind the Adductor longus, ends at the lower third of the thigh in a small branch, which pierces the Adductor magnus, and is distributed on the back of the thigh to the hamstring muscles. The terminal part of the profunda is sometimes named the "fourth perforating artery."

This vessel sometimes arises from the medial side, and, more rarely, from the back of the femoral artery. In three-fourths of a large number of cases it arises from 2.25 to 5 cm below the inguinal ligament; in a few cases the distance is less than 2.25 cm; more rarely, opposite the ligament, or above the inguinal ligament, from the external iliac. Occasionally the distance between the origin of the vessel and the inguinal ligament exceeds 5 cm (Gray 1918).

The profunda gives off the following branches: lateral femoral circumflex, perforating, medial femoral circumflex, muscular.

The *lateral femoral circumflex artery* arises from the lateral side of the profunda, passes horizontally behind the Sartorius and Rectus femoris, and divides into ascending, transverse, and descending branches.

The *medial femoral circumflex artery* arises from the medial and posterior aspect of the profunda, and winds around the medial side of the femur, passing first between the Pectineus and Psoas major, and then between the Obturator externus and the Adductor brevis.

Numerous *muscular branches* arise from the profunda; some of these end in the Adductores, others pierce the Adductor magnus, give branches to the hamstrings, and anastomose with the medial femoral circumflex artery and with the superior muscular branches of the popliteal.

The Popliteal Artery

The *popliteal artery* is the continuation of the femoral, and courses through the popliteal fossa (Fig. 6.9). It extends from the opening in the Adductor magnus, at the junction of the middle and lower thirds of the thigh, downward and lateralward to the intercondyloid fossa of the femur, and then vertically downward to the lower border of the Popliteus, where it divides into anterior and posterior tibial arteries (Fig. 6.10). Occasionally the popliteal artery divides into its terminal branches opposite the knee-joint. The anterior tibial under these circumstances usually passes in front of the Popliteus. The artery sometimes divides into the anterior tibial and peroneal, the posterior tibial being wanting, or

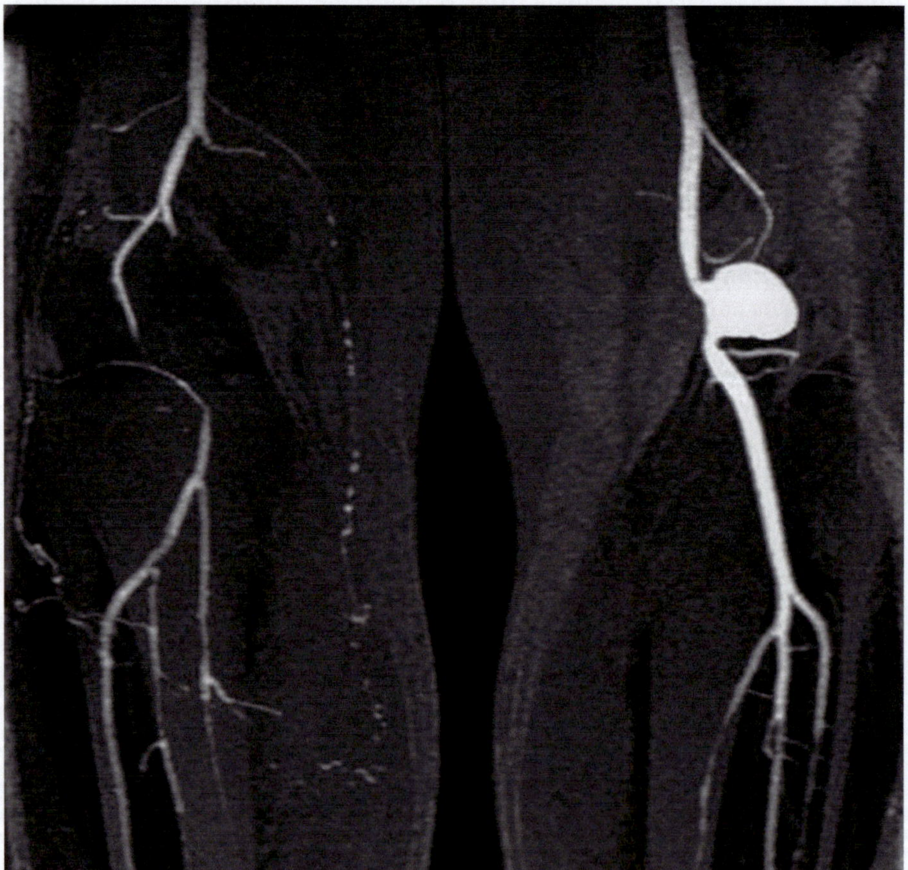

Fig. 6.9. MIP. The popliteal arteries: popliteal aneurysm (on the *right* of the picture), an arterial obstruction (on the *left* of the picture)

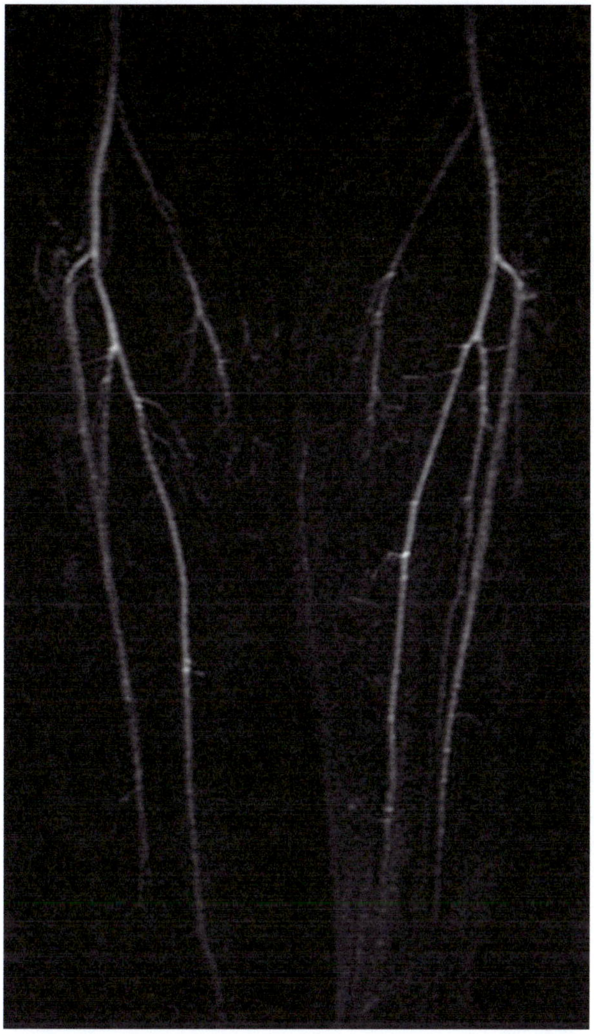

Fig. 6.10. MIP. The popliteal artery divides into anterior and posterior tibial arteries

very small. Occasionally it divides into three branches, the anterior and posterior tibial, and peroneal. Around and above the patella, and on the contiguous ends of the femur and tibia, There is a network of vessels forming a superficial and a deep plexus.

The Anterior Tibial Artery

The *anterior tibial artery* commences at the bifurcation of the popliteal, at the lower border of the Popliteus, passes forward between the two heads of the Tibialis posterior, and through the aperture above the upper border of the interosseous membrane, to the deep part of the front of the leg. It then descends on the anterior surface of the interosseous membrane, gradually approaching the tibia. At the lower part of the leg it lies on this bone, and then on the front of the ankle-joint, where it is more superficial, becomes the dorsalis pedis.

The arteries around the ankle-joint anastomose freely with one another and form networks below the corresponding malleoli.

The *arteria dorsalis pedis*, the continuation of the anterior tibial, passes forward from the ankle-joint along the tibial side of the dorsum of the foot to the proximal part of the first intermetatarsal space, where it divides into two branches, the first dorsal metatarsal and the deep plantar.

The Posterior Tibial Artery

The *posterior tibial artery* begins at the lower border of the Popliteus, opposite the interval between the tibia and fibula; it extends obliquely downward, and, as it descends, it approaches the tibial side of the leg, lying behind the tibia, and in the lower part of its course is situated midway between the medial malleolus and the medial process of the calcaneal tuberosity. Here it divides, beneath the origin of the Adductor hallucis, into the *medial plantar arteries* and the *lateral plantar arteries.*

The branches of the posterior tibial artery are: peroneal, posterior medial malleolar, nutrient, communicating, muscular, medial calcaneal.

The Peroneal Artery

The *peroneal artery* is deeply seated on the back of the fibular side of the leg. It arises from the posterior tibial, about 2.5 cm below the lower border of the popliteus, passes obliquely toward the fibula, and then descends along the medial side of that bone, contained in a fibrous canal between the Tibialis posterior and the Flexor hallucis longus, or in the substance of the latter muscle (Fig. 6.11). It then runs behind the tibiofibular syndesmosis and divides into lateral calcaneal branches which ramify on the lateral and posterior surfaces of the calcaneus. The peroneal artery may arise 7 or 8 cm below the Popliteus, or from the posterior tibial high up, or even from the popliteal. Its size is more frequently increased than diminished; and then it either reinforces the posterior tibial by its junction with it, or altogether takes the place of the posterior tibial in the lower part of the leg and foot, the

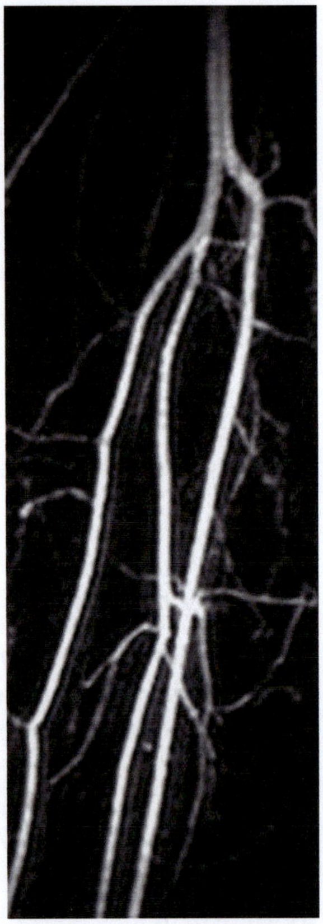

Fig. 6.11. MIP. The peroneal artery arises from the posterior tibial artery

latter vessel only existing as a short muscular branch. The branches of the peroneal are: muscular, perforating, nutrient, communicating, lateral calcaneal.

6.2 The Veins

The veins convey the blood from the capillaries of the different parts of the body to the heart. They consist of two distinct sets of vessels, the pulmonary and systemic.

The *pulmonary veins,* unlike other veins, contain arterial blood, which they return from the lungs to the left atrium of the heart.

The *systemic veins* return the venous blood from the body generally to the right atrium of the heart.

The *portal vein,* an appendage to the systemic venous system, is confined to the abdominal cavity, and returns the venous blood from the spleen and the viscera of digestion to the liver. This vessel ramifies in the substance of the liver and there breaks up into a minute network of capillary-like vessels, from which the blood is conveyed by the hepatic veins to the inferior vena cava.

The veins commence by minute plexuses which receive the blood from the capillaries. The branches arising from these plexuses unite together into trunks, and these, in their passage toward the heart, constantly increase in size as they receive tributaries, or join other veins. The veins are larger and altogether more numerous than the arteries; hence, the entire capacity of the venous system is much greater than that of the arterial. The capacity of the pulmonary veins, however, only slightly exceeds that of the pulmonary arteries. The veins are cylindrical like the arteries; their walls, however, are thin and they collapse when the vessels are empty, and the uniformity of their surfaces is interrupted at intervals by slight constrictions, which indicate the existence of valves in their interior. They communicate very freely with one another, especially in certain regions of the body and these communications exist between the larger trunks as well as between the smaller branches. Thus, between the venous sinuses of the cranium, and between the veins of the neck, where obstruction would be attended with imminent danger to the cerebral venous system, large and frequent anastomoses are found. The same free communication exists between the veins throughout the whole extent of the vertebral canal, and between the veins composing the various venous plexuses in the abdomen and pelvis, e.g., the spermatic, uterine, vesical, and pudendal.

The systemic venous channels are subdivided into three sets, superficial and deep veins, and venous sinuses.

The *superficial veins* are found between the layers of the superficial fascia immediately beneath the skin; they return the blood from these structures, and communicate with the deep veins by perforating the deep fascia.

The *deep veins* accompany the arteries, and are usually enclosed in the same sheaths with those vessels. With the smaller arteries – as the radial, ulnar, brachial, tibial, peroneal – they exist generally in pairs, one lying on each side of the vessel, and are called *venae comitantes*. The larger arteries – such as the axillary, subclavian, popliteal, and femoral – have usually only one accompanying vein. In certain organs of the body, however, the deep veins do not accompany the arteries; for instance, the veins in the skull and vertebral canal, the hepatic veins in the liver, and the larger veins returning blood from the bones.

Venous sinuses are found only in the interior of the skull, and consist of canals formed by a separation of the two layers of the dura mater. Their outer coat consists of fibrous tissue, their inner of an endothelial layer continuous with the lining membrane of the veins.

6.2.1 Veins of the Head, Neck, and Thorax

6.2.1.1 The Superior Vena Cava

The *superior vena cava* forms in the thorax by the confluence of innominate veins or venous brachiocefalic trunks right and left; these are formed from the union of the internal jugular vein and the subclavian vein of each side (Fig. 6.12).

The only tributary of the superior vena cava is the azygos vein.

The *azygos vein* begins opposite the first or second lumbar vertebra, by a branch called the ascending lumbar vein. It enters the thorax through the aortic hiatus in the diaphragm, and passes along the right side of the vertebral column to the fourth thoracic vertebra, where it arches forward and ends in the superior vena cava. Its tributaries: Are the right subcostal and intercostal veins, the upper three or four of these latter opening by a common stem, the highest superior intercostal vein. It receives the hemiazygos veins, several esophageal, mediastinal, and pericardial veins, and, near its termination, the right bronchial vein.

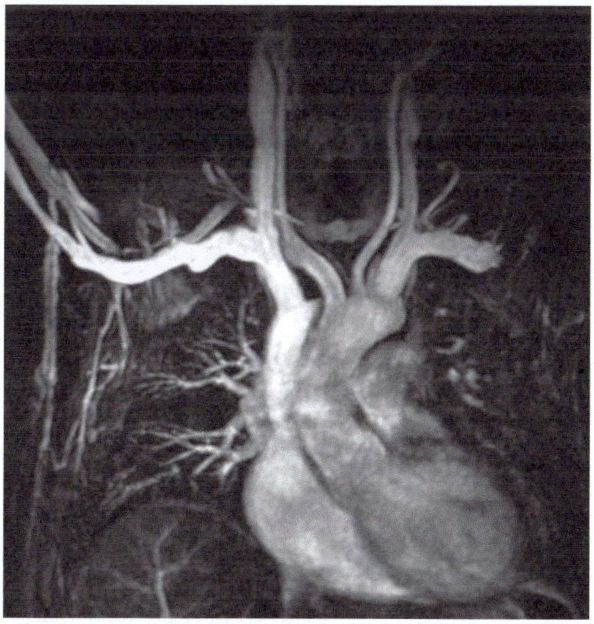

Fig. 6.12. MIP. The superior vena cava and the confluence of the right brachiocefalic trunk

The *hemiazygos vein* begins in the left ascending lumbar or renal vein. It enters the thorax, through the left crus of the diaphragm, and, ascending on the left side of the vertebral column, as high as the ninth thoracic vertebra, passes across the column, behind the aorta, esophagus, and thoracic duct, to end in the azygos vein. It receives the lower four or five intercostal veins and the subcostal vein of the left side, and some esophageal and mediastinal veins.

The *accessory hemiazygos vein* descends on the left side of the vertebral column, and varies inversely in size with the highest left intercostal vein. It receives veins from the three or four intercostal spaces between the highest left intercostal vein and highest tributary of the hemiazygos; the left bronchial vein sometimes opens into it. It either crosses the body of the eighth thoracic vertebra to join the azygos vein or ends in the hemiazygos.

In the obstruction of the superior vena cava, the azygos and hemiazygos veins are one of the principal means by which the venous circulation is carried on, connecting the superior and inferior vena cava, and communicating with the common iliac veins by the ascending lumbar veins and with many of the tributaries of the inferior vena cava.

The *venous brachiocefalic trunk* has the following tributaries: inferior thyroid veins, internal mammary vein, superior phrenic veins, vertebral vein, posterior jugular vein, timic, pericardic, anterior mediastinal veins.

The *inferior thyroid veins* two or frequently three or four in number, arise in the venous plexus on the thyroid gland, communicating with the middle and superior thyroid veins. They form a plexus in front of the trachea, behind the sternothyreoidei muscles. From this plexus, a left vein descends and joins the left innominate trunk, and a right vein passes obliquely downward to open into the right innominate vein, just at its junction with the superior vena cava; sometimes the right and left veins open by a common trunk in the latter situation. These veins receive esophageal, tracheal and inferior laryngeal veins, and are provided with valves at their terminations in the innominate veins.

The *internal mammary vein* is a trunk which runs up on the medial side of the artery and ends in the corresponding innominate vein.

The *vertebral vein* is formed in the suboccipital triangle, from numerous small tributaries which spring from the internal vertebral venous plexuses and issue from the vertebral canal above the posterior arch of the atlas. They unite with small veins from the deep muscles at the upper part of the back of the neck, and form a vessel which enters the foramen in the transverse process of the atlas, and descends, forming a dense plexus around the vertebral artery, in the canal formed by the foramina transversaria of the cervical vertebrae. This plexus ends in a single trunk, which emerges from the foramen transversarium of the sixth cervical vertebra, and opens at the root of the neck into the back part of the innominate vein near its origin.

The *internal jugular vein* collects the blood from the brain, from the superficial parts of the face, and from the neck. It is directly continuous with the transverse sinus, and begins in the posterior compartment of the jugular foramen, at the base of the skull. It runs down the side of the neck in a vertical direction, lying at first lateral to the internal carotid artery, and then lateral to the common carotid, and at the root of the neck unites with the subclavian vein to form the innominate vein. The left vein is generally smaller than the right, and each contains a pair of valves, which are placed about 2.5 cm above the termination of the vessel.

This vein receives in its course the inferior petrosal sinus, the common facial, lingual, pharyngeal, superior and middle thyroid veins, and sometimes the occipital.

The *subclavian vein* is the continuation of the axillary vein. It extends from the outer border of the first rib to the sternal end of the clavicle, where it unites with the internal jugular to form the innominate vein. It is in relation, in front, with the clavicle and the subclavius muscle; behind and above, with the subclavian artery, from which it is separated medially by the scalenus anterior muscle and the phrenic nerve; below, it rests in a depression on the first rib and upon the pleura. Its tributary is the external jugular vein.

The *external jugular vein* receives the greater part of the blood from the exterior of the cranium and the deep parts of the face, being formed by the junction of the posterior division of the posterior facial with the posterior auricular vein. It commences in the substance of the parotid gland, on a level with the angle of the mandible, and runs perpendicularly down the neck at the posterior border of the sternocleidomastoideus. It ends in the subclavian vein, lateral to or in front of the scalenus anterior muscle. The external jugular vein varies in size, bearing an inverse proportion to the other veins of the neck. This vein receives the occipital vein and, near its termination, the transverse cervical, transverse scapular and anterior jugular veins. In the substance of the parotid, a large branch of communication from the internal jugular joins it.

The *anterior jugular vein* begins near the hyoid bone by the confluence of several superficial veins from the submaxillary region. It descends between the median line and the anterior border of the sternocleidomastoideus, and, at the lower part of the neck, open into the termination of the external jugular, or, in some instances, into the subclavian vein. Most frequently there are two anterior jugulars, a right and a left; but sometimes only one. Its tributaries are some laryngeal veins, and occasionally a small thyroid vein. Just above the sternum the two anterior jugular veins communicate by a transverse trunk, the venous jugular arch, which receive tributaries from the inferior thyroid veins.

The veins of the brain possess no valves, and their walls, owing to the absence of muscular tissue, are extremely thin. They pierce the arachnoid membrane and the inner or meningeal layer of the dura mater, and open into the cranial venous sinuses. They may be divided into two sets, cerebral and cerebellar.

The *cerebral veins* are divisible into external and internal groups according as they drain the outer surfaces or the inner parts of the hemispheres. The external veins are the superior, inferior, and middle cerebral. The internal cerebral veins drain the deep parts of the hemisphere and are two in number; each is formed near the interventricular foramen by the union of the terminal and choroid veins. They run backward parallel with one another, between the layers of the tela chorioidea of the third ventricle, and beneath the splenium of the corpus callosum, where they unite to form a short trunk, the great cerebral vein; just before their union each receives the corresponding basal vein.

The *cerebellar veins* are placed on the surface of the cerebellum, and are disposed in two sets, superior and inferior.

The *sinuses of the dura mater* are venous channels which drain the blood from the brain. They are devoid of valves, and are situated between the two layers of the dura mater and lined by endothelium continuous with that which lines the veins.

They may be divided into two groups: a postero-superior group, at the upper and back part of the skull, and an antero-inferior group, at the base of the skull. The postero-superior group comprises: the superior sagittal, the straight, the inferior sagittal, the two transverse, the occipital. The antero-inferior group of sinuses comprises: the two cavernous, the two superior petrosal, the two intercavernous, the two inferior petrosal, the basilar plexus.

6.2.1.2
The Pulmonary Veins

The *pulmonary veins* return the arterialized blood from the lungs to the left atrium of the heart. They are four in number, two from each lung (Fig. 6.13). They commence in a capillary network upon the walls of the air sacs, where they are continuous with the capillary ramifications of the pulmonary artery. These vessels uniting successively form a single trunk for each lobe, three for the right, and two for the left lung. The vein from the middle lobe of the right lung generally unites with that from the upper lobe, so that ultimately two trunks from each lung are formed; they open separately into the upper and back part of the left atrium. Occasionally the three veins on the right side remain separate. Not infrequently the two left pulmonary veins end by a common opening. At the root of the lung, the superior pulmonary vein lies in front of and a little below the pulmonary artery; the inferior is situated at the lowest part of the hilus of the lung and on a plane posterior to the upper vein.

6.2.1.3
The Veins of the Heart

Most of the veins of the heart open into the *coronary sinus*. This is a wide venous channel situated in the posterior part of the coronary sulcus, and covered by muscular fibers from the left atrium. It ends in the right atrium between the opening of the inferior vena cava and the atrioventricular orifice, where there is a semilunar valve, the valve of the coronary sinus (valve of Thebesius).

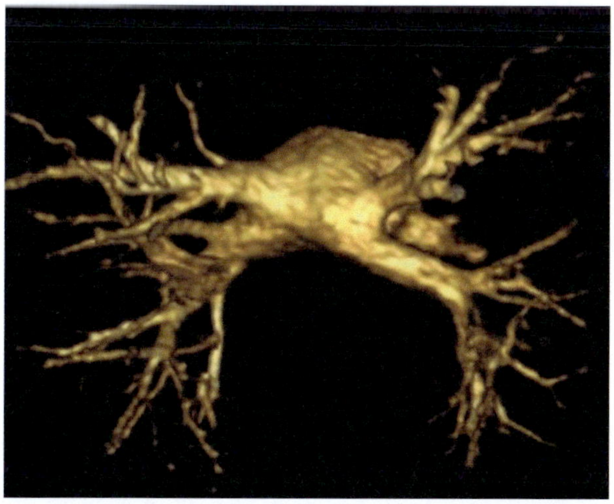

Fig. 6.13. VR. The pulmonary veins: four in number, two from each lung

Its tributaries are the great, small, and middle cardiac veins, the posterior vein of the left ventricle, and the oblique vein of the left atrium.

The *great cardiac* or *left coronary vein* begins at the apex of the heart and ascends along the anterior longitudinal sulcus to the base of the ventricles. It then curves to the left in the coronary sulcus, and reaching the back of the heart, opens into the left extremity of the coronary sinus. It receives tributaries from the left atrium and from both ventricles: one, the left marginal vein, is of considerable size, and ascends along the left margin of the heart.

The *small cardiac* or *right coronary vein* runs in the coronary sulcus between the right atrium and ventricle, and opens into the right extremity of the coronary sinus. It receives blood from the back of the right atrium and ventricle. The right marginal vein ascends along the right margin of the heart and joins it in the coronary sulcus, or opens directly into the right atrium.

The *middle cardiac vein* commences at the apex of the heart, ascends in the posterior longitudinal sulcus, and ends in the coronary sinus near its right extremity.

The following cardiac veins do not end in the coronary sinus:

- The anterior cardiac veins, comprising three or four small vessels which collect blood from the front of the right ventricle and open into the right atrium.
- The right marginal vein frequently opens into the right atrium, and is therefore sometimes regarded as belonging to this group.
- The veins of Thebesius, consisting of a number of minute veins which arise in the muscular wall of the heart; the majority open into the atria, but a few end in the ventricles (Gray 1918).

6.2.2
The Veins of the Abdomen and Pelvis

The *external iliac vein*, the upward continuation of the femoral vein, begins behind the inguinal ligament, and, passing upward along the brim of the lesser pelvis, ends opposite the sacroiliac articulation, by uniting with the hypogastric vein to form the common iliac vein. It frequently contains one, sometimes two, valves and it may be doubled or lie lateral to the external iliac artery.

The external iliac vein receives the inferior epigastric, deep iliac circumflex, and pubic veins.

The *hypogastric vein* begins near the upper part of the greater sciatic foramen, passes upward behind and slightly medial to the hypogastric artery and, at the brim of the pelvis, joins with the external iliac to form the common iliac vein. The tributaries of the hypogastric vein correspond with the branches of the hypogastric artery. It receives: (a) the gluteal, internal pudendal, and obturator veins, which have their origins outside the pelvis; (b) the lateral sacral veins, which lie in front of the sacrum; and (c) the middle hemorrhoidal, vesical, uterine, and vaginal veins, which originate in venous plexuses connected with the pelvic viscera. The internal iliac vein may receive an unusual vessel representing a confluence of the internal pudendal, obturator, gluteal, and sciatic veins and occasionally both internal iliac veins unite and open as a common trunk at the confluence of the right and left external iliac veins to form the vena cava.

The *common iliac veins* are formed by the union of the external iliac and hypogastric veins, in front of the sacroiliac articulation; passing obliquely upward toward the right side, they end upon the fifth lumbar vertebra, by uniting with each other at an acute angle to form the inferior vena cava. The *right common iliac* is shorter than the left, nearly vertical in its direction, and ascends behind and then lateral to its corresponding artery. The *left common iliac,* longer than the right and more oblique in its course, is at first situated on the medial side of the corresponding artery, and then behind the right common iliac. Each common iliac receives the iliolumbar, and sometimes the lateral sacral veins. The left receives, in addition, the middle sacral vein. No valves are found in these veins. One of the veins may be absent, in which case the external and internal veins join the opposite common iliac to form the vena cava and a connection between the portal and right common iliac has been reported. The left common iliac vein, instead of joining with the right in its usual position, occasionally ascends on the left side of the aorta as high as the kidney, where, after receiving the left renal vein, it crosses over the aorta, and then joins with the right vein to form the vena cava. In these cases, the two common iliacs are connected by a small communicating branch at the spot where they are usually united.

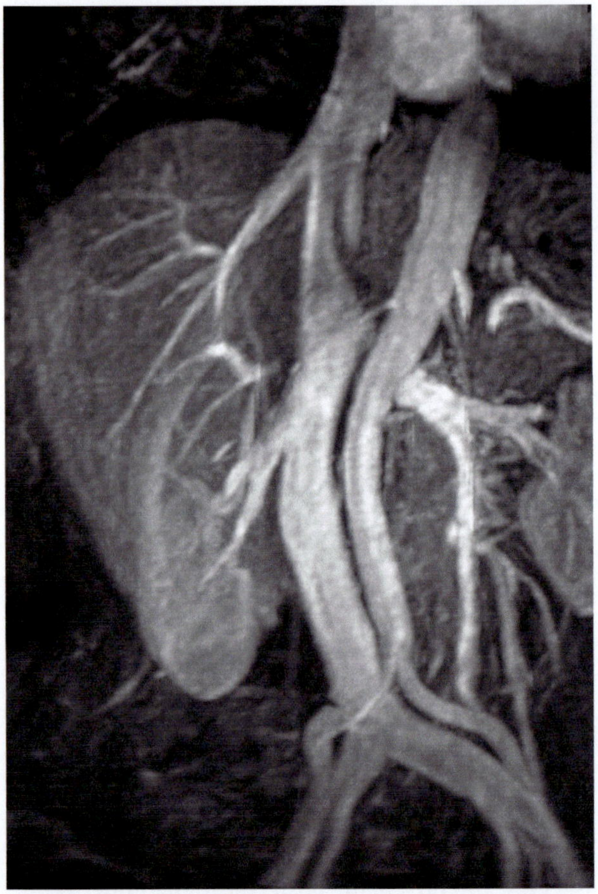

Fig. 6.14. MIP. The inferior vena cava

The *inferior vena cava* returns to the heart the blood from the parts below the diaphragm. It is formed by the junction of the two common iliac veins (Fig. 6.14), on the right side of the fifth lumbar vertebra. It ascends along the front of the vertebral column, on the right side of the aorta and, having reached the liver, is continued in a groove on its posterior surface. It then perforates the diaphragm between the median and right portions of its central tendon and subsequently inclines forward and medialward for about 2.5 cm, and, piercing the fibrous pericardium, passes behind the serous pericardium to open into the lower and back part of the right atrium. In front of its atrial orifice is a semilunar valve, termed the "valve of the inferior vena cava": this is rudimentary in the adult, but is of large size and exercises an important function in the fetus.

This vessel is sometimes placed on the left side of the aorta, as high as the left renal vein, and, after receiving this vein, crosses over to its usual position on the right side. It may be placed altogether on the left side of the aorta, and in such a case the abdominal and thoracic viscera, together with the great vessels, are all transposed. The inferior vena cava may be doubled (2.2–3.0% of cases) or it may be absent. If the inferior vena cava is absent, the blood from the lower limbs may pass through the diaphram into the superior vena cava by way of a large vein in the location of the ascending lumbar and azygos veins. As a result, the hepatic veins drain directly into the right atrium through the normal caval opening in the diaphragm. Occasionally the inferior vena cava joins the azygos vein, which is then of large size (Gray 1918).

The inferior vena cava receives the following veins: Lumbar, Right Spermatic or Ovarian, Renal, Suprarenal, Inferior Phrenic and Hepatic.

The *lumbar veins* four in number on each side, collect the blood by dorsal tributaries from the muscles and integument of the loins, and by abdominal tributaries from the walls of the abdomen, where they communicate with the epigastric veins. At the vertebral column, they receive veins from the vertebral plexuses, and then pass forward, around the sides of the bodies of the vertebra, beneath the Psoas major, and end in the back part of the inferior venacava. The left lumbar veins are longer than the right, and pass behind the aorta. The lumbar veins are connected together by a longitudinal vein which passes in front of the transverse processes of the lumbar vertebræ, and is called the; it forms the most frequent origin of the corresponding vein ascending lumbar azygos or hemiazygos vein, and serves to connect the common iliac, iliolumbar, and azygos or hemiazygos veins of its own side of the body.

The *spermatic veins* emerge from the back of the testis, and receive tributaries from the epididymis; they unite and form a convoluted plexus, called the "pampiniform plexus," which constitutes the greater mass of the spermatic cord; the vessels composing this plexus are very numerous, and ascend along the cord, in front of the ductus deferens. Below the subcutaneous inguinal ring they unite to form three or four veins, which pass along the inguinal canal, and, entering the abdomen through the abdominal inguinal ring, coalesce to form two veins, which ascend on the Psoas major, behind the peritoneum, lying one on either side of the internal spermatic artery. These unite to form a single vein, which opens on the right side into the inferior vena cava, at an acute angle; on the left side into the left renal vein, at a right angle. The spermatic veins are provided with valves. The left spermatic vein passes behind the iliac colon, and is thus exposed to pressure from the contents of that part of the bowel.

The *ovarian veins* correspond with the spermatic in the male. They form a plexus in the broad ligament near the ovary and uterine tube, and communicate with the uterine plexus. They end in the same way as the spermatic veins in the male. Valves are occasionally found in these veins. Like the uterine veins, they become much enlarged during pregnancy.

The *renal veins* are of large size, and placed in front of the renal arteries. The left is longer than the right, and passes in front of the aorta, just below the origin of the superior mesenteric artery. It receives the left spermatic and left inferior phrenic veins, and, generally, the left suprarenal vein. It opens into the inferior vena cava at a slightly higher level than the right. The renal veins may be multiple in about 28% of cases but they show less variation than do the renal arteries. The left renal vein may be retroaortic (3%) or circumaortic (17%). The right renal vein may be doubled, even though the left renal vein is usually single(Bergman et. al. 1996).

The *suprarenal veins* are two in number: the right ends in the inferior vena cava; the left, in the left renal or left inferior phrenic vein.

The *inferior phrenic veins* follow the course of the inferior phrenic arteries; the right ends in the inferior vena cava; the left is often represented by two branches, one of which ends in the left renal or suprarenal vein, while the other passes in front of the esophageal hiatus in the diaphragm and opens into the inferior vena cava.

The *hepatic veins* commence in the substance of the liver, in the terminations of the portal vein and hepatic artery, and are arranged in two groups, upper and lower. The upper group usually consists of three large veins, which converge toward the posterior surface of the liver, and open into the inferior vena cava, while that vessel is situated in the groove on the back part of the liver. The veins of the lower group vary in number, and are of small size; they come from the right and caudate lobes. They are destitute of valves. A common variant of the hepatic veins is the presence of a varying number of accessory hepatic veins draining blood from the lateral and posterior segments of right hepatic lobe directly into the inferior vena cava. This occurs in ~6% of subjects.

6.2.2.1 The Portal System of Veins

The *portal system* includes all the veins which drain the blood from the abdominal part of the digestive tube (with the exception of the lower part of the rectum) and from the spleen, pancreas, and gall-bladder. From these viscera the blood is conveyed to the liver by the portal vein. In the liver this vein ramifies like an artery and ends in capillary-like vessels termed sinusoids, from which the blood is conveyed to the inferior vena cava by the hepatic veins. In the adult the portal vein and its tributaries are destitute of valves.

The *portal vein* is about 8 cm in length, and is formed at the level of the second lumbar vertebra by the junction of the superior mesenteric and lienal veins (Fig. 6.15), the union of these veins taking place in front of the inferior vena cava and behind the neck of the pancreas.

At the porta hepatis, the portal vein divides into a right and a left branch, which accompany the corresponding branches of the hepatic artery into the substance of the liver. The right branch of the portal vein enters the right lobe of the liver, but before doing so generally receives the cystic vein. The left branch, longer but of smaller caliber than the right, gives branches to the caudate lobe, and then enters the left lobe of the liver. The portal vein is rarely variable; it is among the most constant blood vessels (including arteries and veins). It has also been reported to be doubled. Branches forming the portal vein are also variable. Accessory splenic veins have been reported by several investigators. The splenic vein may anastomose with the renal or with the superior mesenteric vein. The superior mesenteric vein has been reported opening into the renal vein.

The tributaries of the portal vein are: Lienal, Superior Mesenteric, Coronary, Pyloric, Cystic, and Paraumbilical veins.

The *lienal vein* commences by five or six large branches which return the blood from the spleen. These unite to form a single vessel, which passes from left to right, grooving the upper and back part of the pancreas, below the lineal artery, and ends behind the neck of the pancreas by uniting at a right angle with the superior mesenteric to form the portal vein. The lienal vein is of large size, but is not tortuous like the artery.

The lineal vein receives the short gastric veins, the left gastroepiploic vein, the pancreatic veins, and the inferior mesenteric veins.

The *inferior mesenteric vein* returns blood from the rectum and the sigmoid, and descending parts of the colon. It begins in the rectum as the superior hemorrhoidal vein, which has its origin in the hemorrhoidal plexus, and through this plexus communicates with the middle and inferior hemorrhoidal veins. The inferior mesenteric vein receives the sigmoid veins from the sigmoid colon and iliac colon, and the left colic vein from the descending colon and left colic flexure.

The *superior mesenteric vein* returns the blood from the small intestine, from the cecum, and from

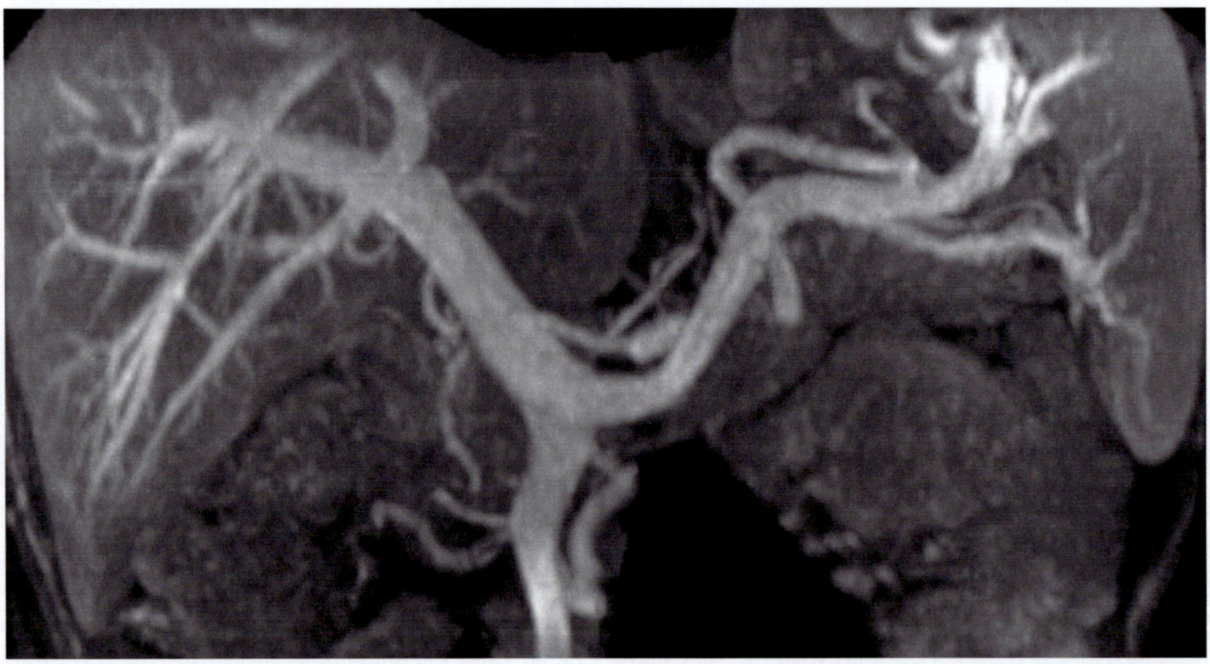

Fig. 6.15. MIP. The confluence of the lienal and the superior mesenteric veins into the portal vein

the ascending and transverse portions of the colon. It begins in the right iliac fossa and ascends between the two layers of the mesentery on the right side of the superior mesenteric artery. Behind the neck of the pancreas it unites with the lienal vein to form the portal vein. Besides the tributaries which correspond with the branches of the superior mesenteric artery, the intestinal, ileocolic, right colic, and middle colic veins, the superior mesenteric vein is joined by the right gastroepiploic and pancreaticoduodenal veins.

Collateral venous circulation to relieve portal obstruction in the liver may be effected by communications between: (a) the gastric veins and the esophageal veins which often project as a varicose bunch into the stomach, emptying themselves into the hemiazygos vein; (b) the veins of the colon and duodenum and the left renal vein; (c) the accessory portal system of Sappey, branches of which pass in the round and falciform ligaments (particularly the latter) to unite with the epigastric and internal mammary veins, and through the diaphragmatic veins with the azygos; a single large vein, shown to be a parumbilical vein, may pass from the hilus of the liver by the round ligament to the umbilicus, producing there a bunch of prominent varicose veins known as the caput medusae; (d) the veins of Retzius, which connect the intestinal veins with the inferior vena cava and its retroperitoneal branches; (e) the inferior mesenteric veins, and the hemorrhoidal veins that open into the hypogastrics; (f) very rarely the ductus venosus remains patent, affording a direct connection between the portal vein and the inferior vena cava (Gray 1918).

6.2.3 The Veins of the Limbs

The veins of the extremities are divided into two sets, superficial and deep with diffuse anastomoses. The superficial veins are placed immediately beneath the integument between the two layers of superficial fascia. The deep veins accompany the arteries. Both sets are provided with valves, which are more numerous in the deep than in the superficial veins.

6.2.3.1 The Veins of the Upper Limb

The *superficial veins* of the upper extremity are the digital, metacarpal, cephalic, basilic, median.

The *dorsal digital veins* pass along the sides of the fingers and are joined to one another by oblique communicating branches. Those from the adjacent sides of the fingers unite to form three *dorsal metacarpal veins*, which end in a dorsal venous network opposite the middle of the metacarpus. The radial part of the network is joined by the dorsal digital vein from the radial side of the index finger and by the dorsal digital veins of the thumb, and is prolonged upward as the cephalic vein. The ulnar part of the network receives the dorsal digital vein of the ulnar side of the little finger and is continued upward as the basilic vein. A communicating branch frequently connects the dorsal venous network with the cephalic vein about the middle of the forearm.

The volar digital veins on each finger are connected to the dorsal digital veins by oblique intercapitular veins. They drain into a venous plexus which is situated over the thenar and hypothenar eminences and across the front of the wrist.

The *cephalic vein* begins in the radial part of the dorsal venous network and winds upward around the radial border of the forearm, receiving tributaries from both surfaces. Below the front of the elbow it gives off the vena mediana cubiti, which receives a communicating branch from the deep veins of the forearm and passes across to join the basilic vein. The cephalic vein then ascends in front of the elbow in the groove between the brachioradialis and the biceps brachii. It crosses superficial to the musculocutaneous nerve and ascends in the groove along the lateral border of the biceps brachii. In the upper third of the arm it passes between the pectoralis major and deltoideus, where it is accompanied by the deltoid branch of the thoracoacromial artery. It pierces the coracoclavicular fascia and, crossing the axillary artery, ends in the axillary vein just below the clavicle. Sometimes it communicates with the external jugular vein by a branch which ascends in front of the clavicle.

The accessory cephalic vein arises either from a small tributory plexus on the back of the forearm or from the ulnar side of the dorsal venous network; it joins the cephalic below the elbow. In some cases the accessory cephalic springs from the cephalic above the wrist and joins it again higher up. A large oblique branch frequently connects the basilic and cephalic veins on the back of the forearm.

The *basilic vein* begins in the ulnar part of the dorsal venous network. It runs up the posterior surface of the ulnar side of the forearm and inclines forward to the anterior surface below the elbow, where it is joined by the vena mediana cubiti. It ascends obliquely in the groove between the biceps brachii and pronator teres and crosses the brachial artery, from which it is separated by the lacertus fibrosus. It then runs upward along the medial border of the biceps brachii, perforates the deep fascia and ends in the medial brachial vein.

The *median antibrachial vein* drains the venous plexus on the volar surface of the hand. It ascends on the ulnar side of the front of the forearm and ends in the basilic vein or in the vena mediana cubiti; in a small proportion of cases it divides into two branches, one of which joins the basilic, the other the cephalic, below the elbow.

The *deep veins* follow the course of the arteries. They are generally arranged in pairs, and are situated one on either side of the corresponding artery, and connected at intervals by short transverse branches.

The superficial and deep volar arterial arches are each accompanied by a pair of venæ comitantes which constitute respectively the superficial and deep volar venous arches, and receive the veins corresponding to the branches of the arterial arches; thus the common volar digital veins, formed by the union of the proper volar digital veins, open into the superficial, and the volar metacarpal veins into the deep volar venous arches. The dorsal metacarpal veins receive perforating branches from the volar metacarpal veins and end in the radial veins and in the superficial veins on the dorsum of the wrist.

The *deep veins of the forearm* are the venæ comitantes of the radial and ulnar veins and constitute respectively the upward continuations of the deep and superficial volar venous arches; they unite in front of the elbow to form the brachial veins. The radial veins are smaller than the ulnar and receive the dorsal metacarpal veins. The ulnar veins receive tributaries from the deep volar venous arches and communicate with the superficial veins at the wrist; near the elbow they receive the volar and dorsal interosseous veins and send a large communicating branch (profunda vein) to the vena mediana cubiti.

The *brachial veins* are placed one on either side of the brachial artery, receiving tributaries corresponding with the branches given off from that vessel. Near the lower margin of the subscapularis, they join the axillary vein; the medial one frequently joins the basilic vein.

These deep veins have numerous anastomoses, not only with each other, but also with the superficial veins.

The *axillary vein* begins at the lower border of the teres major, by the confluence of the two brachial veins and ends at the outer border of the first rib as the subclavian vein. Close to its termination, It receives the cephalic vein its other tributaries correspond with the branches of the axillary artery. It lies on the medial side of the artery, which it partly overlaps; between the two vessels are the medial cord of the brachial plexus, the median, the ulnar, and the medial anterior thoracic nerves. It is provided with a pair of valves opposite the lower border of the subscapularis; valves are also found at the ends of the cephalic and subscapular veins.

6.2.3.2 The Veins of the Lower Limb

The *superficial veins* of the lower extremity are the great and small saphenous veins and their tributaries.

On the dorsum of the foot the dorsal digital veins receive, in the clefts between the toes, the intercapitular veins from the plantar cutaneous venous arch and join to form short common digital veins which unite across the distal ends of the metatarsal bones in a dorsal venous arch. Proximal to this arch is an irregular venous network which receives tributaries from the deep veins and is joined at the sides of the foot by a medial and a lateral marginal vein, formed mainly by the union of branches from the superficial parts of the sole of the foot. On the sole of the foot the superficial veins form a plantar cutaneous venous arch which extends across the roots of the toes and opens at the sides of the foot into the medial and lateral marginal veins. Proximal to this arch is a plantar cutaneous venous network which is especially dense in the fat beneath the heel; this network communicates with the cutaneous venous arch and with the deep veins, but is chiefly drained into the medial and lateral marginal veins.

The *great saphenous vein* the longest vein in the body, begins in the medial marginal vein of the dorsum of the foot and ends in the femoral vein about 3 cm below the inguinal ligament. It ascends in front of the tibial malleolus and along the medial side of the leg in relation with the saphenous nerve. It runs upward behind the medial condyles of the tibia and femur and along the medial side of the thigh and, passing through the fossa ovalis, ends in the femoral vein. At the ankle it receives branches from the sole of the foot through the medial marginal vein; in the leg it anastomoses freely with the small saphenous vein, communicates with the anterior and posterior tibial veins and receives many cutaneous veins; in the thigh it communicates with the femoral vein and receives numerous tributaries; those from the medial and posterior parts of the thigh frequently unite to form a large accessory saphenous vein which joins the main vein at a variable level. Near the fossa ovalis it is joined by the superficial epigastric, superficial iliac circumflex, and superficial external pudendal veins. A vein, named the thoracoepigastric, runs along the lateral aspect of the trunk between the superficial epigastric vein below and the lateral thoracic vein above and establishes an important communication between the femoral and axillary veins.

The valves in the great saphenous vein vary from ten to twenty in number; they are more numerous in the leg than in the thigh.

The *small saphenous vein* begins behind the lateral malleolus as a continuation of the lateral marginal vein; it first ascends along the lateral margin of the tendocalcaneus, and then crosses it to reach the middle of the back of the leg. Running directly upward, it perforates the deep fascia in the lower part of the popliteal fossa, and ends in the popliteal vein, between the heads of the gastrocnemius. It communicates with the deep veins on the dorsum of the foot, and receives numerous large tributaries from the back of the leg. Before it pierces the deep fascia, it gives off a branch which runs upward and forward to join the great saphenous vein. The small saphenous vein possesses from nine to twelve valves, one of which is always found near its termination in the popliteal vein. In the lower third of the leg the small saphenous vein is in close relation with the sural nerve, in the upper two-thirds with the medial sural cutaneous nerve.

The *deep veins* of the lower extremity accompany the arteries and their branches. The *plantar digital veins* arise from plexuses on the plantarsurfaces of the digits, and, after sending intercapitular veins to join the dorsal digital veins, unite to form four *metatarsal veins;* these run backward in the metatarsal spaces, communicate, by means of perforating veins, with the veins on the dorsum of the foot, and unite to form the *deep plantar venous arch* which lies alongside the plantar arterial arch. From the deep plantar venous arch the medial and lateral plantar veins run backward close to the corresponding arteries and, after communicating with the great and small saphenous veins, unite behind the medial malleolus to form the posterior tibial veins.

The *posterior tibial veins* accompany the posterior tibial artery, and are joined by the *peroneal veins.*

The *anterior tibial veins* are the upward continuation of the venæ comitantes of the dorsalis pedis artery. They leave the front of the leg by passing between the tibia and fibula, over the interosseous membrane, and unite with the posterior tibial, to form the popliteal vein.

The *popliteal vein* is formed by the junction of the anterior and posterior tibial veins at the lower border of the Popliteus; it ascends through the popliteal fossa to the aperture in the Adductor magnus, where it becomes the femoral vein. It receives tributaries corresponding to the branches of the popliteal artery, and it also receives the small saphenous vein.

The *femoral vein* accompanies the femoral artery through the upper two-thirds of the thigh. In the lower part of its course it lies lateral to the artery, higher up, it is behind it, and at the inguinal ligament, it lies on its medial side. It receives numerous muscular tributaries, and about 4 cm below the inguinal ligament is joined by the deep femoral vein. Near its termination it is joined by the great saphenous vein.

The *deep femoral vein* receives tributaries corresponding to the perforating branches of the profunda artery, and through these establishes communications with the popliteal vein below and the inferior gluteal vein above. It also receives the medial and lateral femoral circumflex veins.

References

Bergman RA, Afifi AK, Miyauchi R (1996). Illustrated encyclopedia of human anatomic variation: opus II: cardiovascular system. Available at www.anatomyatlases.org

Gray H (1918) Anatomy of the human body. Philadelphia, Lea & Febiger. Available at Bartleby.com, 2000. www.bartleby.com/107/

Neck Vessels

7

Mirco Cosottini and Ilaria Pesaresi

CONTENTS

7.1 Technical Considerations *74*

7.2 Subclavian Arteries *77*

7.3 Carotid Arteries *78*
7.3.1 Atherosclerotic Disease *78*
7.3.2 Dissection *82*

7.4 Vertebral Arteries *83*
7.4.1 Atherosclerosis *83*
7.4.2 Dissection *83*

References *84*

ABSTRACT

The technology advancement including high performance gradients, parallel imaging, more efficient methods in sampling k-space, intravascular contrast agents, and higher magnetic field strengths, allows to improve time and spatial resolution of contrast-enhanced MRA. Moreover, technical refinements contribute to a renaissance of nonenhanced MRA techniques.

Clinical applications of MRA in studying neck vessels benefit of both faster acquisitions and sharpened images. Arterial steno-occlusive disease of supra-aortic vessels may lead to cerebral infarction that is one of the main worldwide problem of public health. Therefore, the principal MRA application concerns the atherosclerosis of carotid bifurcation. In this field, MRA is a competitor of CTA as a confirmatory noninvasive diagnostic test, aiming to replace DSA in the diagnostic workup of patients with carotid stenosis. In this chapter, we discuss the role and limitations of MRA in grading carotid stenosis before revascularization. Indication to surgical or endovascular revascularization could be set with MRA with a low misclassification rate in case of high-grade carotid stenosis while inconclusive data are available for moderate stenosis.

Because of its short acquisition time, neck vessels MRA can be easily implemented in an MRI protocol of patients with cerebrovascular pathology also in the acute phase of the disease. Typical examples are carotid or vertebral dissections as the main cause of juvenile stroke. In these cases, MRA in conjunction with conventional MR images allows a prompt diagnosis avoiding invasive procedures both at the onset and in the follow-up for monitoring recanalization or eventual pseudoaneurismatic dilation.

Mirco Cosottini, MD
Unit of Neuroradiology, Department of Neurosciences, University of Pisa, Via Roma 67, 56100 Pisa, Italy
Ilaria Pesaresi, MD
Diagnostic and Interventional Radiology, Department of Oncology, Transplants, and Advanced Technologies in Medicine, University of Pisa, Via Roma 67, 56100 Pisa, Italy

Since the introduction of dedicated coils covering the upper thorax, MRA constitutes a robust tool for the panoramic evaluation of the origin of epiaortic vessel, a vascular district traditionally difficult to be explored noninvasively. Stenosis of vertebral origin, subclavian steal, or thoracic outlet syndrome can now be detected with an objective noninvasive technique.

7.1 Technical Considerations

In this paragraph we report some technical considerations to focus on some proper aspects of the neck vessels examination that are extensively reported in Chap. 3.

The introduction of CE-MRA technique in the evaluation of neck vessels provided diagnostic gain in several pathologies. Although some investigators insist that TOF MR angiography remains the most accurate technique in evaluating carotid stenosis (Townsend et al. 2003), the main advantage of CEMRA was the limitation of artifacts with respect to flow-based MRA techniques. The basic principle of CE-MRA is the T1 shortening effect of a paramagnetic contrast media passing as a bolus through the vasculature (Prince et al. 1993). The contrast between stationary tissues and vessels depend on shortening the T1 time of blood. As a consequence CEMRA is less sensitive to intravoxel dephasing from turbulent flow and does not suffer from loss of signal due to saturation effects. Moreover the short acquisition time limits the motion artifacts that causes degradation of images.

The crucial point in obtaining an adequate CEMRA examination is the timing of acquisition with respect to the contrast bolus injection. The main goal of CEMRA acquisition is the sampling of the center of k-space during the peak arterial concentration of contrast media. CEMRA of neck vessels is actually acquired with elliptical centric encoding that helps to fill the centre of k-space in a short period of time at the beginning of the scan. Elliptical centric encoding permits sampling the low spatial frequencies (image contrast) quickly during the arterial phase of bolus passage and reducing the venous contamination of the image. By using a centric or elliptical encoding that fills the k-space at the beginning of the acquisition, the scan delay corresponds to the circulation time. The calculation of the acquisition time may be obtained with a bolus test. Nevertheless for acquisitions with an elliptical centric encoding an automatic or semiautomatic method to detect the bolus arrival may be preferable; indeed a fast k-space center filling determines a narrow temporal window to collect the arterial passage of contrast bolus.

The bolus geometry is the other important factor in executing CEMRA besides the acquisition timing (Wetzel and Bongartz 2000). The bolus geometry in the vasculature depends not only on physiological parameters of the patient but also on the contrast amount and injection rate. The main parameters characterizing the bolus geometry (Fig. 7.1) are the maximum peak (as the maximal contrast media concentration in the vessel of interest), the bolus arrival time (as the time from injection to reach the 80% of the maximum peak of concentration), and the bolus length (as the time during which the intravascular concentration of contrast media remains over 80% of maximum). Generally a fixed dose of contrast media is administered for CEMRA acquisition of the neck vessels. A standard injection dose of 20 mL of gadolinium chelate facilitates the exam conduction. Adopting an injection rate of 2.5 mL/s we can obtain bolus geometry with a constant bolus length in the arterial phase of 5–7 s. With the actually permitted acquisition time, a bolus length of this order achieves at least the filling of the central part of the k-space, guaranteeing a sufficient contrast resolution of the angiogram. Lower amount of contrast media or higher injection rates determine a shorter bolus length, while higher contrast doses or lower injection rates cause a longer bolus length. An important consequence of this consideration is that bolus length can be optimized on the bases of the scan duration. A faster acquisition time (shorter duration of k-space center filling) needs a lower contrast dose to fill the entire center of k-space; on the other hand a longer acquisition time needs a higher amount of contrast media. Summarizing faster is cheaper.

An inadequate timing in CEMRA acquisition determines artifacts that deteriorate the image quality (Fig. 7.1). A precocious acquisition, before the bolus reach the maximum in the vessel of interest, causes the sampling of the central k-space before the arterial phase of contrast distribution, while the periphery of k-space (high spatial frequencies) is sampled during the arterial phase. The obtained angiogram will be affected by ringing artifact (Maki's effect) (Maki et al. 1996) and will have poor contrast but high resolution (Fig. 7.1a). A delayed acquisition, after the bolus reach the maximum in the vessel of

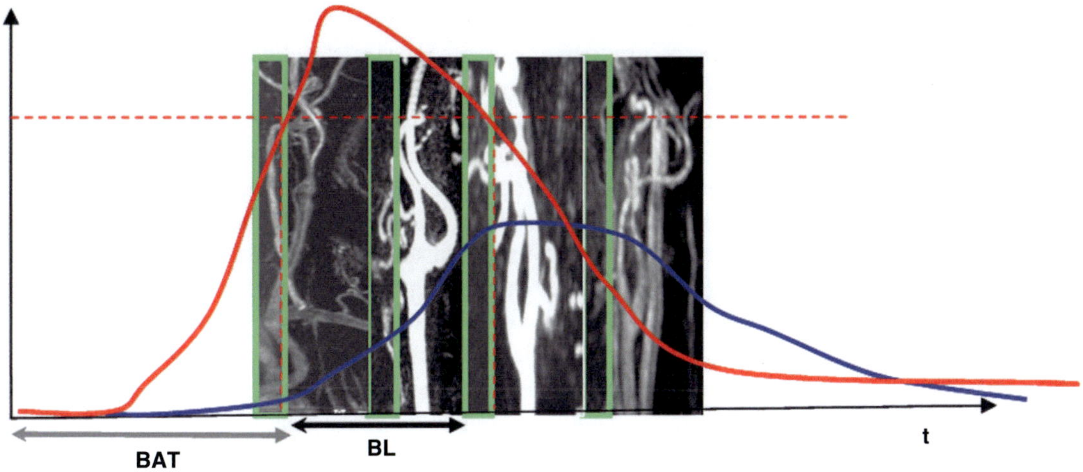

Fig. 7.1. Bolus geometry in arteries (*red line*) and veins (*blue line*). Bolus arrival time (BAT) is the time from injection to the time 80% of the maximum peak of concentration is reached. Bolus length (BL) is the time during which the intravascular concentration of contrast media remains over 80% of maximum. Examples of mistimed acquisition of CEMRA. Each image represent the acquisition time of CEMRA. The *green box* is the time during which the center of k-space is collected while the remain portion of the image is the time during which the periphery of k-space is collected. (*a*, ringing artefact): the central k-space is collected before reaching BL while the periphery of k-space is sampled during the arterial phase. (*c*, blurring artefact): the center of k-space is correctly sampled during BL while the periphery of k-space is sampled when the contrast concentration falls down within the arteries. (*d*, venous superimposition): the entire k-space is collected during the maximum of venous enhancement. (*b*, optimal acquisition time): a good quality angiogram is obtained if the centre and periphery of k-space are collected during bolus length

interest, causes the collection of the central k-space after the arterial phase and during the venous phase of contrast distribution. The obtained angiogram will be a good quality venogram with superimposition of arterial and venous structures (Fig. 7.1d). A third error occurs if the center of k-space is correctly sampled when the bolus reach the maximum in the vessel of interest but the bolus length is too short and solely low spatial frequencies are sampled during the arterial phase. Since the periphery of k-space is sampled when the contrast concentration falls down within the arteries, the obtained angiogram will be affected by a blurring artifact and will have good contrast but a low spatial resolution (Fig. 7.1c).

Automatic or semiautomatic methods for bolus detection, facilitates the exam conduction reducing the artifacts linked to the incorrect acquisition timing. A further technique that frees CEMRA from circulation time calculation and avoids mistimed examinations is the time resolved CEMRA. This technique has the distinct advantage of improving the temporal resolution of CEMRA, which is one of its main limitations. In circles with high circulation rate such as cerebral or renal arteries, the temporal resolution may be a limitation in clinical application of CEMRA. Time resolved MRA was introduced to often improve the temporal resolution in spite of spatial resolution. Smaller matrixes and volumes of acquisition, higher bandwidth, rectangular field of view, and partial Fourier and zero filling techniques were implemented. The technology advancement implementing gradients with faster slew rates permits lower TE (less than 2 ms) and TR (less that 4 ms) improving time resolution. Moreover time resolution of CEMRA is improved by more efficient methods in sampling k-space. Three dimension acquisitions reconstructed with a novel oversampling of the center of k-space are available for clinical use. An example is the TRICKS technique (time-resolved intravascular contrast kinetics) that samples the center of k-space every 2 s providing well enhanced serial images of which at least one captured the arterial phase of contrast distribution.

To reduce acquisition time, parallel imaging techniques partially substitute gradient-encoded steps with coil-encoding. An array of multiple receiving coils with specific spatial sensitivity is employed to detect the electromagnetic signal. In order to shorten the acquisition time, some portion of the gradient-encoding steps are omitted and the lacking spatial information is obtained by combining the signals from each receiving coils on the basis of their specific spatial sensitivity functions (coil-encoded steps). Depending on the parallel reconstruction algorithm, the coil-encoded spatial information can be utilized in the k-space domain (SMASH, GRAPPA algorithms) or in the image domain (SENSE algorithm). The acceleration factor defines the acquisition time gain due to parallel acquisition, calculated by the ratio between the numbers of gradient steps in the unaccelerated acquisition with respect to the accelerated one. The drawback of parallel acquisition is a degradation in SNR (signal-to-noise ratio), mainly based on the degree of k-space undersampling and the coil array geometry known as the g factor. In CEMRA acquisition of high number of receiver channels, optimized coil geometry, and an adequate acceleration factor (four) provide better sensitivity profiles resulting in g factors closer to one.

Moreover the introduction in the clinical practice of high field magnet (3 T) permits improved SNRs which is advantageous when highly accelerated parallel imaging is applied, obtaining high resolution images with a lower in plane resolution to CTA or DSA with a comparable diagnostic value (NAEL et al. 2007).

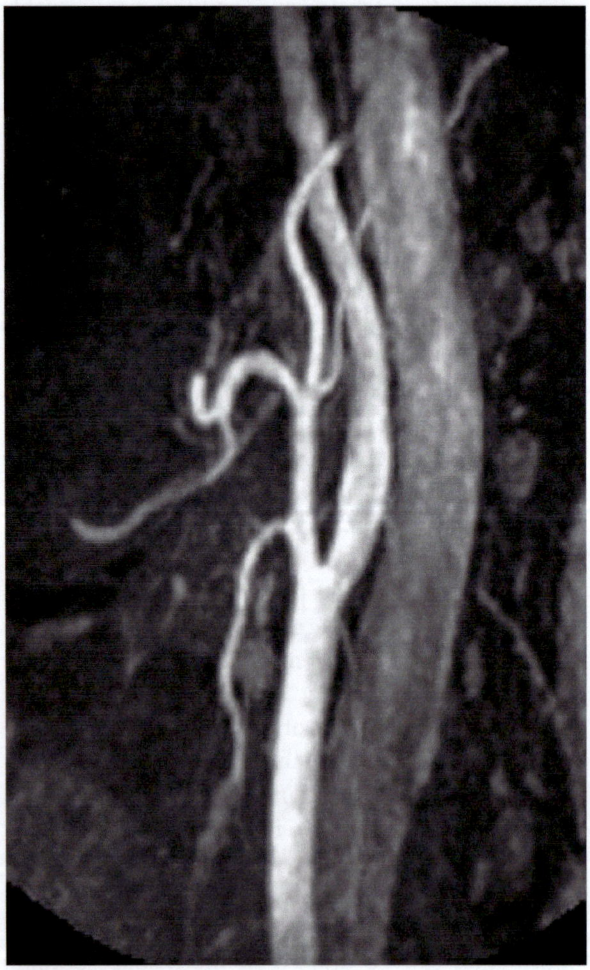

Fig. 7.2. Nonenhanced MRA obtained with SSFP sequence. Courtesy Ghedin GE Medical systems

Among the technical improvements in CEMRA of carotid arteries, development in the field of contrast media deserve mention. Standard extracellular contrast media are currently used for almost all CE-MRA applications; however the rapid extravasation of the contrast medium limits acquisition time and therefore spatial resolution as well as contrast-to-noise ratio. To improve spatial resolution of CEMRA, it is necessary to prolong imaging time and extend the arterial phase of contrast distribution. Intravascular contrast media allows both a first-pass contrast-enhanced MRA and high-resolution T1-weighted imaging approach in the equilibrium phase extending acquisition windows.

Several preclinical studies tested paramagnetic or superparamagnetic compounds synthesized to obtain a blood pool agent with a selective intravascular distribution (BREMERICH et al. 2007). To date only the Gadolinium chelates with a protein binding are clinically tested and approved for humans. Gd-BOPTA had weak and transient interaction with serum proteins – particularly albumin; its greater intravascular T1 relaxivity has been shown to produce a higher vascular peak enhancement of longer duration than that achieved with other Gadolinium chelates at the same dose (KNOPP et al. 2002). Higher signal to noise ratio between vessel and signal background may improve the evaluation of carotid arteries (ANZALONE et al. 2006). Gadofosveset trisodium is an intravascular contrast agent approved for clinical application in MRA. Due to strong hydrophilic profile which allows a reversible noncovalent binding with albumin, this contrast agent increases its intravascular half life permitting ultra-high resolution in the equilibrium phase with promising results on the evaluation of neck vessels (BLUEMKE et al. 2001).

In evaluating carotid arteries, CEMRA replaced nonenhanced MRA. Actually solely PC sequences can be implemented in a clinical MR protocol to obtain information about flow direction or quantification. Besides considerations regarding the level of diagnostic accuracy, long acquisition time and image artifacts limited the use of classical flow-based techniques in favor of CEMRA. Nevertheless, recent improvement in MR hardware and software permitting faster acquisition and concerns about gadolinium based contrast agents contribute to the renaissance of nonenhanced MRA techniques (Fig. 7.2). Two new families of nonenhanced MRA techniques are becoming commercially available: EEG gated partial-Fourier fast spin echo technique and balanced steady-state free precession imaging with or without arterial spin labeling. While EEG gated partial-Fourier fast spin echo method and balanced steady-state free precession imaging without arterial spin labeling are implemented in studying different vascular districts, nonenhanced MRA with arterial spin labeling may also be used for carotid imaging. The ALS-based MRA can be obtained by subtracting tag-on and tag-off images (Nishimura et al. 1987) by placing an inversion pulse to the upstream slab. The tagged blood flows to the vessels of interest after a time delay (depending on blood velocity and distance between tagged and imaged area) and can be imaged with partial-Fourier fast spin echo or balanced steady-state free precession sequence. A bright blood angiogram can also be obtained with a single acquisition by using spatially nonselective and selective tagged pulses. If images are acquired after a proper delay time after nonselective pulse, stationary tissue can be selectively suppressed at their null point while selectively tagged protons of blood reached the imaged area (Miyazaki and Lee 2008).

7.2 Subclavian Arteries

Atherosclerotic stenosis is the most frequent pathology in the subclavian arteries. With the advent of endovascular treatment, the correct grading of the stenosis has assumed an important clinical impact. In order to determine the success of endovascular treatment it is fundamental to know the site, the length, and the morphology of the lesion.

CEMRA revealed an optimal agreement with DSA findings in different types of subclavian diseases constituting a robust tool in the noninvasive evaluation of epiaortic vessels. The method identifies correctly the presence of stenoses, with a slight tendency to overestimation. For the steno-occlusive disease of the sublavian arteries, agreement with conventional angiography, is excellent ($k = 0.85$), with a sensitivity of 90%, specificity of 95%, and diagnostic accuracy of 93% (Cosottini et al. 2000).

Subclavian artery stenosis may cause arm claudication, paresthesias, weakness, and pain and also a subclavian steal. Subclavian steal is due to reversal flow in a vertebral artery associated with a prevertebral tract obstruction or stenosis of the subclavian artery. Subclavian steal may result in vertebrobasilar insufficiency. In diagnosing subclavian steal, CE-MRA is often associated with flow-based MRA techniques such as phase contrast or time of flight. In the first case, a phased display axial 2D PC sequence sensitized to flow in slice direction is helpful to depict the reverse flow direction in the vertebral artery homolateral to the obstructed/stenosed subclavian artery. In the second case, a 2D TOF sequence implemented with spatial band saturation can be repeated, alternatively placing the saturation band above and below the slices to obtain selective spin saturation that flow downward or upward.

Reperfusion of the subclavian artery in the postvertebral tract by the inverted vertebral flow may be due to subclavian obstruction or severe stenosis. Because endovascular treatment yields poor results with occlusions especially for occluded tracts longer than 5 cm, a preoperative diagnosis of subclavian occlusion is important. Differential diagnosis between high grade stenosis and occlusion may be problematic with CEMRA. Due to the tendency of overestimation of CEMRA, it is possible to misinterpret the occluded arteries that are reperfused by inverted flow in the vertebral artery as an overgraded severe stenosis (Fig. 7.3).

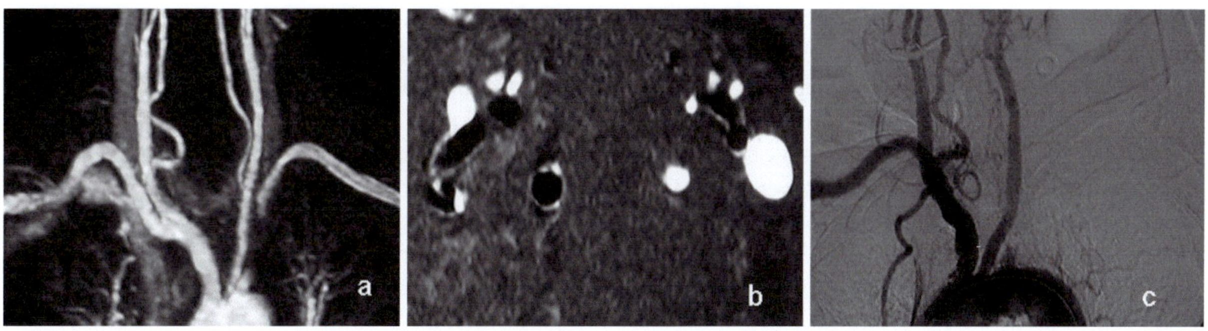

Fig. 7.3. CEMRA (**a**) reveals a proximal obstruction of the left subclavian artery with postvertebral enhancement. Distal reperfusion by reversal blood flow in left vertebral artery is demonstrated with 2D PC sequence with phase display (**b**). Subclavian obstruction can be erroneously interpreted as an overestimated severe stenosis. DSA confirms the left subclavian obstruction (**c**)

Diagnostic gain in solving the dilemma between occlusion and overgraded severe stenosis comes from technical development in the fields of dedicated multichannel coils and parallel imaging, which improves the spatial resolution. A further possibility may source from increasing temporal resolution. The main limit of CEMRA compared to DSA, is the scarce temporal resolution that does not permit appreciation of the progressive flowing of the contrast medium into the arterial tree enough to see the reperfusion of the occluded subclavian artery directly through the vertebral artery. Time resolved MRA, especially with implementation of more efficient methods in sampling k-space, allows sampling of every few seconds the contrast media passes through vertebral and subclavian vessels. It permits adoption of the same radiological semiology in CEMRA examinations as that in DSA.

Thoracic outlet syndrome (TOS) refers to clinical manifestations due to impingement of the brachial plexus and subclavian vessels as they pass from the thoracic cavity to the axilla. Approximately 98% of patients with TOS have compression of the brachial plexus, while a minority referred symptoms due to arterial or venous impingement.

The vascular or neurovascular compression is provoked by anatomical abnormalities and/or eliciting maneuver and is characterized by a functional subclavian stenosis causing transient paresthesias, weakness, and pain of the upper limb. Repetitive arterial trauma results in focal stenosis, subsequent poststenotic dilatation, and ultimately microembolization with tissue loss.

There is no generally accepted diagnostic protocol for the investigation of TOS although DSA is considered the mainstain technique. CT or MRI can be used to determine the functional anatomy of the thoracic outlet and can be implemented with angiographic acquisition to demonstrate the effect of vascular compression. MRI eventually acquired during provocative maneuver into the scanner showed the cause of impingement by reformatting the 3D data set and other conventional sequences (Charon et al. 2004).

Presently, CE MRA offers extensive vessel coverage, is less prone to artifact, and frequently demonstrates the underlying cause of compression as compared to flow-based MRA sequences. CT angiography can be preferred to MRA for its ability to depict bony thorax, scalene muscles insertion expecially in case of aberrant cervical ribs.

7.3 Carotid Arteries

7.3.1 Atherosclerotic Disease

Stroke and transient ischemic attack exceed coronary heart disease (Rothwell et al. 2005) constituting one of the main problems of public health and social assistance for survivals. Arterial steno-occlusive disease of supra-aortic vessels is frequent in the general population with a potential risk of thrombo-embolic and hemodynamic cerebral infarctions.

Based upon data of multicenter trials, carotid endarterectomy (CEA) is a proven treatment in carotid artery stenosis and is considered the most effective method of preventing stroke occurrence in patients with symptomatic and asymptomatic high-grade carotid artery disease (NASCET 1991; Barnett et al. 1998). CEA may prevent stroke in symptomatic patients with >70% stenosis, or even with stances of 50–69%, and in asymptomatic patients with at least 60% stenosis (ACAS 1995). For high risk surgical patients, percutaneous carotid transluminal angioplasty and stenting (CAS) are emerging with encouraging results as alternative method to CEA (Roubin et al. 2001).

For effective surgical treatment, correct classification of atherothrombotic lesions is essential and an accurate carotid imaging is important for secondary stroke prevention.

Currently, digital subtraction angiography (DSA) is considered the standard of reference in detecting and grading carotid stenosis due to the excellent spatial (matrix size 1,024 × 1,024) and temporal resolution (3–6 frames/s). Nevertheless the procedural risk of stroke for this method is not negligible and reported ranging between 0.7 and 1% (Kuntz et al. 1995).

A recent review of 19,826 consecutive patients who underwent cerebral catheter angiography (Kaufmann et al. 2007) showed an overall neurologic complication rate of 2.63%.

In order to reduce the risks and discomfort of invasive procedures, new and noninvasive diagnostic tools such as spiral CT, enhanced magnetic resonance angiography (MRA), and color Doppler ultrasound were developed.

Noninvasive techniques are changing the diagnostic algorithm in patients with arterial occlusive disease. There is a distinct trend towards surgery without DSA, adopting ultrasound as a screening tool in combination with a confirmatory test (CTA or MRA). DSA is increasingly reserved for endovascular intervention or for cases in which noninvasive test results are equivocal.

CEMRA was evaluated as a highly safe, feasible tool for assessing patients with suspected carotid atherosclerosis, showing high accuracy and agreement with DSA in detecting and grading ICA stenosis. CEMRA has the advantages of being noninvasive, more tolerated (U-King Im et al. 2004c), less expensive(U-King Im et al. 2004a), less time-consuming as compared with DSA. With respect to conventional flow-based MRA techniques, CEMRA provides better image quality, higher level of diagnostic confidence, and more interobserver agreement (Mitra et al. 2006).

For clinical application, the main problem in replacing an invasive tool with a noninvasive one is the diagnostic accuracy of the new method. Several works stated the high level of diagnostic accuracy of CEMRA in grading carotid stenosis (Le Clerc et al. 1999; Remonda et al. 1998; Nederkoorn et al. 2002). A meta-analysis study (Wardlaw et al. 2006a) revised 2,541 patients with carotid stenosis evaluated with CEMRA and digital subtraction angiography reporting high sensitivity (0.94) and specificity (0.93) solely for high grade stenosis (70–99%), while inconclusive data are available for moderate stenosis (50–69%). They cautiously suggest a possible replacing of DSA by using CEMRA for high grade stenosis. A more recent meta-analysis (Debrey et al. 2008) confirms the previous high levels of diagnostic accuracy for severe stenosis (sensitivity 94.6% and specificity of 91.9%) and also for carotid occlusion (sensitivity 99.4% and specificity 99.6%) but assesses a fair sensitivity for moderate stenosis (sensitivity 65.9% and specificity 93.5%).

Results on the diagnostic accuracy that source from several dedicated studies may be biased by selection of patient population. In the majority of cases CEMRA is validated with respect to DSA after these patients are screened with CD-US for carotid stenosis (verification bias). The number of true positive as well as sensitivity is thus invariably increased. This bias has a limited role if CEMRA is considered as a screening tool. Nevertheless CEMRA is proposed to scientific community as a confirmatory test after a screening with CD-US. In this case a counteract effect on the diagnostic accuracy may source from the spectrum bias. In calculating diagnostic accuracy, the inclusion of normal or nearly normal vessels (true-negative cases) can be expected to falsely elevate the specificity of CE-MRA as a confirmatory test because calculation of specificity is heavily dependent on the number of true-negative cases. In the clinical setting CEMRA should be a tool for identification of true negative patients erroneously screened as stenotic. In the literature the specificity of MRA has likely been overstated for the inclusion of true negative patients in the statistical analysis (Layton et al. 2007).

A further problem is due to operators who publish diagnostic accuracy results. In several works, specialized radiologists dedicated to the carotid artery stenosis, evaluate the stenosis grade, while in clinical practice, the wide spread of noninvasive imaging modalities is available for nonspecialized radiologists. In this regard studies showing excellent concordance between MRA and DSA are most often performed in academic radiology departments while lower levels of CEMRA diagnostic accuracy are reported if the grading of stenosis is retrospectively evaluated by the radiological report in daily activity of general hospitals (Johnston et al. 2002).

CEMRA has gained increasing importance as a noninvasive test in the preoperative work-up of carotid artery stenosis (Borisch et al. 2003) The acceptance of this MRA technique by surgeons is due to the reported high diagnostic accuracy, panoramic potential of the evaluation, and to the better representation of the morphology of the stenosis as compared to the angiographic flow-based methods.

Nevetheless many clinicians require (Westwood et al. 2002) further clinical evidence to establish the effectiveness of MRA in diagnosing ICA stenosis and in making decisions about revascularization.

For the preoperative evaluation of atherosclerotic pathology of carotid arteries a low level (3.1%) of misclassification rate has been reported when adopting CEMRA (Cosottini et al. 2005b). Less encouraging results show that 24% of patients would have been misclassified for CEA on the basis of CEMRA results alone (sensitivity, 92%; specificity, 62%) When both duplex ultrasonography and CEMRA were performed and results were concordant, the misclassification rate decreased to 17%. More data is also needed on the accuracy of less invasive tests in combination. No studies reported on the accuracy of less

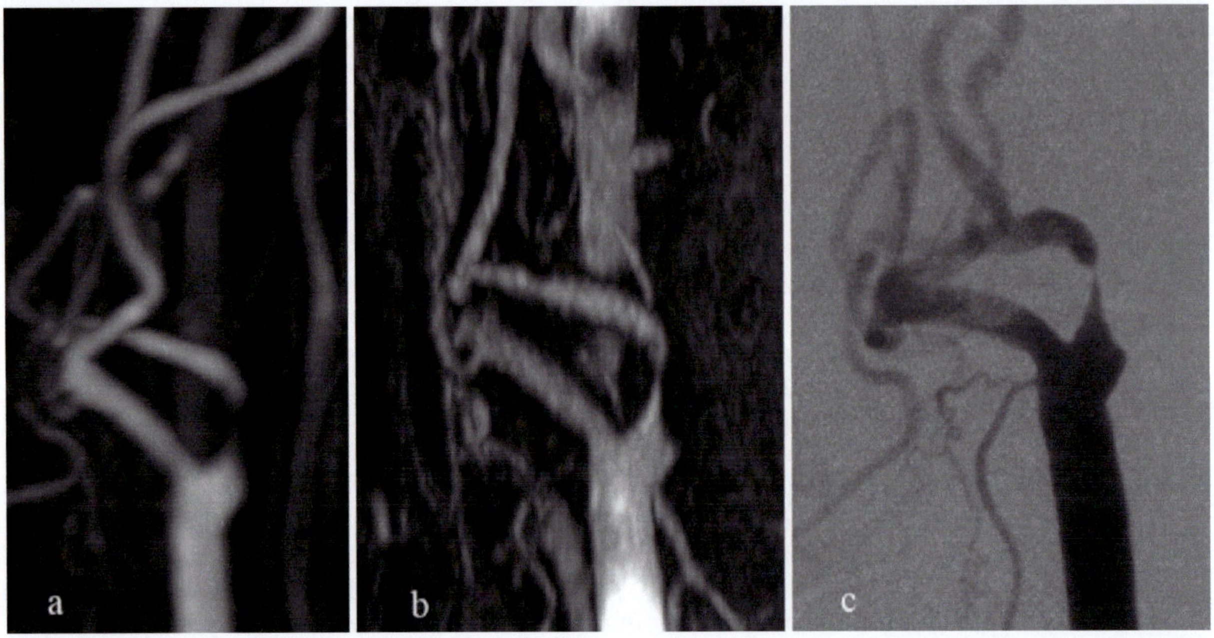

Fig. 7.4. First pass CE-MRA obtained with blood pool agent Gadofosveset trisodium (**a**). The intravascular contrast media allows the acquisition of high resolution CE-MRA at steady state (**b**) that improves the sharpness of carotid bifurcation. CE-MRA at steady state shows a more accurate grading of the stenosis reducing the tendency of first pass CE-MRA to overestimation with respect to DSA (**c**)

invasive tests in the same patients or gave an account with any reliability on how the accuracy of two less invasive tests used in combination compares with each individual test used alone (Wardlaw et al. 2006b).

The classification error of CEMRA in planning endarterectomy avoids a necessary surgical intervention and deprives patients of the surgical benefit of revascularization; it mainly exposes patients with nonsignificant stenosis to unnecessary surgical risk. Indeed the greater errors in stenosis grading source from the tendency of CEMRA to overestimate the stenosis.

The phenomenon of stenosis overestimation is well known (Leclerc et al. 1998) and CE-MRA tended to overestimate stenosis compared with DSA by a mean bias of 2.4–3.8% (U-King-Im et al. 2004b). Several factors may contribute to overestimation. The first one is the low spatial resolution of CEMRA with respect to the digital angiography as a standard of reference. Larger voxels causes a blurring of the vessels profiles especially in narrowed tracts (Cosottini et al. 2003). It has been estimated (Hoogeveen et al. 1998) that at least three pixels need to be included into the patent residual lumen at level of stenosis to obtain a correct grading. For severe stenosis with extremely narrowed residual patent lumen, small voxel size is necessary to maintain image sharpness. In clinical practice the voxel size suggested for CEMRA examination of carotid arteries is about 1 mm^3. Based upon the published data about first-pass CEMRA, it is reasonable to exploit parallel imaging capabilities in order to increase spatial resolution instead of temporal resolution.

To improve spatial resolution of CEMRA, it is necessary to prolong imaging time. Due to the prolongation of intravascular distribution, blood pools contrast agents allow extending the imaging windows, opening up the possibility of high resolution imaging in the equilibrium phase. A novel strategy in the conduct of CEMRA examination includes the possibility of obtaining angiograms in the first pass acquisition and at the equilibrium phase using high resolution sequence. In our experience, first pass CEMRA is useful for a panoramic evaluation of cerebropethal vessels while high resolution CEMRA at steady state reduces the number of over graded stenosis visualizing a more sharpened stenosis (Fig. 7.4). Overall accuracy for carotid high resolved CE-MRA performed during steady state, approximately 5 min after blood pool agent injection has been reported to be high (88–100%). Interestingly, the image quality seems to be best at relatively larger voxel size (0.8 mm) with faster acquisition (acquisition time = 40 s) than with respect to smaller voxel size (0.5 mm) with longer acquisition time (acquisition time = 1 min 40 s) because images are degraded by pulsation and motion artifacts.

Further causes of overestimation may be related to the mistimed examinations during arterial bolus passage or to spin intravoxel dephasing in CEMRA (Scheffler et al. 1998).

The postprocessing technique (MPR tended to show lower degrees of stenoses than MIP) adopted to display the carotid bifurcation has been reported (Lell et al. 2007) as a cause of stenosis overestimation as well as the method adopted from the radiologist to measure the stenosis. With respect to an objective measurement with a calliper, the subjective visual impression determines a misclassification in 11.3–15.1% of the stenosis ranging between 70 and 99% (U-King-Im et al. 2007). An automated approach to the measurement of the stenosis seems to allow accurate assessment of stenosed vessel diameter according to the North American Symptomatic Carotid Endarterectomy Trial criteria (Habibi et al. 2009).

Finally the stenosis overestimation adopting CEMRA may be related to the DSA as the standard of reference. The limited number of projections obtained with DSA may be a limitation in the evaluation of high-grade stenosis leading to underestimation of severity (Elgersma et al. 1999).

A comparison between bidimensional techniques used as gold standard with respect to a 3D method may generate a false overestimation due to an underestimation of stenosis by conventional DSA (Fellner et al. 2001). A better correlation was observed between rotational DSA and CE MRA than between rotational DSA and conventional DSA suggesting that the multiplanar capability of MRA makes it superior to conventional DSA in two or three planes in grading carotid stenosis (Anzalone et al. 2005). Although 3D imaging modalities could achieve a more accurate measurement of the stenosis, without repeating the endarterectomy trials, conventional DSA must remain the reference standard as the link between the percentage of stenosis and risk of stroke. As a consequence, noninvasive imaging modalities have to mimic and not overcome conventional DSA in grading stenosis to correctly locate the single patient in a precise risk category.

In assessing carotid artery stenosis, CT angiography is an alternative to CEMRA as a confirmatory test. The wider diffusion of CT angiography in the evaluation of the carotid arteries results from its simplicity and high spatial resolution. The axial images obtained have the advantage of being utilized in the measurement of the crosssection area reduction of the vessel or in the measurement of the plaque densities to establish fatty, calcified, or fibrous components. The main drawbacks of CT angiography are radiation exposure, nephrotoxicity, and sensitivity to artifacts from dental amalgam and vascular calcifications.

With the introduction of multisection technology, CT angiography increases its performance and can be proposed as a substitute for CEMRA due to excellent correlation with conventional angiography and the best spatial resolution among all noninvasive techniques. Accuracy of CTA has been shown to have a sensitivity in the range of 85–95% and specificity of 93–98% for detecting carotid stenoses >70% (Koelemay et al. 2004) comparable to that of CEMRA (Alvarez-Linera et al. 2003) (Fig. 7.5).

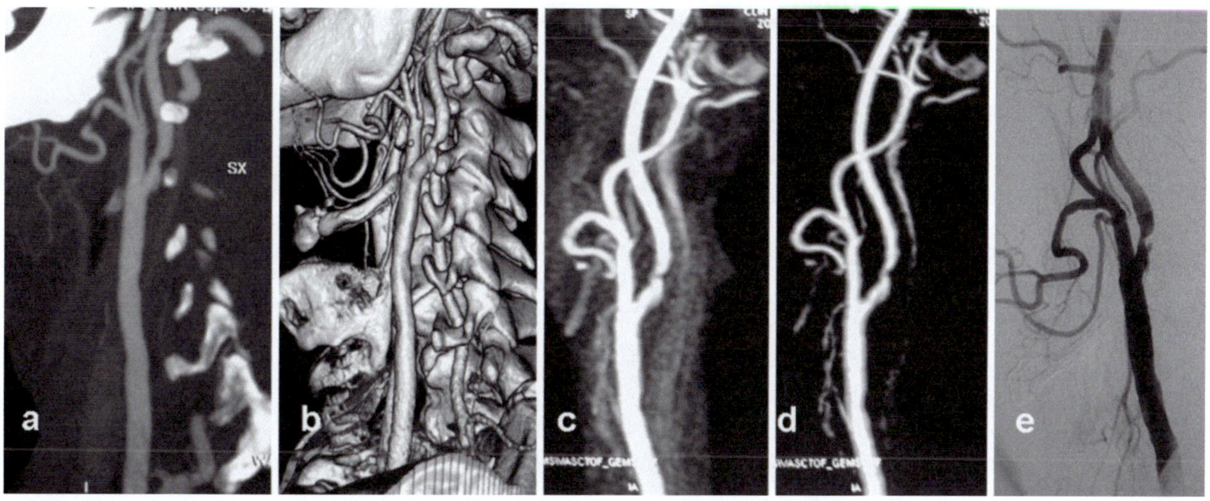

Fig. 7.5. Multiplanar and volume reconstructions of a left internal carotid stenosis obtained with CTA (**a, b**) and CE-MRA (**c, d**) with respect to the gold standard DSA (**e**). Diagnostic accuracy of CTA and MRA are comparable in detecting and grading carotid stenosis

7.3.2 Dissection

Cervical artery dissection (CAD) is a major cause of cerebral infarction in young people, and accounts for 10% of strokes under the age of 20 and 20% under the age of 30. The major cause of internal carotid dissection is trauma; the remaining cases are spontaneous. The pathogenesis of CAD remains unclear and an underlying arteriopathy responsible for the weakness of the arterial wall is postulated. Predisposing factors for spontaneous dissection include fibromuscular dysplasia (up to 15%), Marfan's syndrome, Ehlers Danlos type IV syndrome, osteogenesis imperfecta, pseudoxantoma elasticum, cystic medial necrosis, and other collagen vascular diseases (Cosottini et al. 2005a).

Ultra structural connective tissue aberrations in skin biopsies of patients without manifestations of a connective tissue disease are shown to correlate with CAD suggesting that it represents a manifestation of a genetic predisposition with a vascular phenotype (Brandt et al. 2001).

In conventional MRI, hematoma in carotid dissection appears as a crescent-shaped area around the eccentric signal void of the patent lumen. The signal intensity of the hematoma varies according to the timing of the bleeding. On T1-weighted images it is classically hypointense in the acute phase for the presence of deoxiemoglobin and in the chronic stage for the presence of methemoglobin. On the other hand hematoma is hyperintense due to the methemoglobin deposit between 7 days and 2 months. Subintimal dissection is responsible for stenotic pattern of dissection characterized by progressive luminal narrowing detectable with angiographic techniques including CEMRA. The intramural hematoma in conventional MRI can often be seen to spiral along the length of the dissected segment.

Instead the subadventitial dissection causes an aneurismal dilation when the arterial lumen communicates with the dissection cavity (Anson and Crowell 1991). This type of angiographic pattern suggests a more severe underlying arteriopathy with respect to dissection without aneurysm and is reported

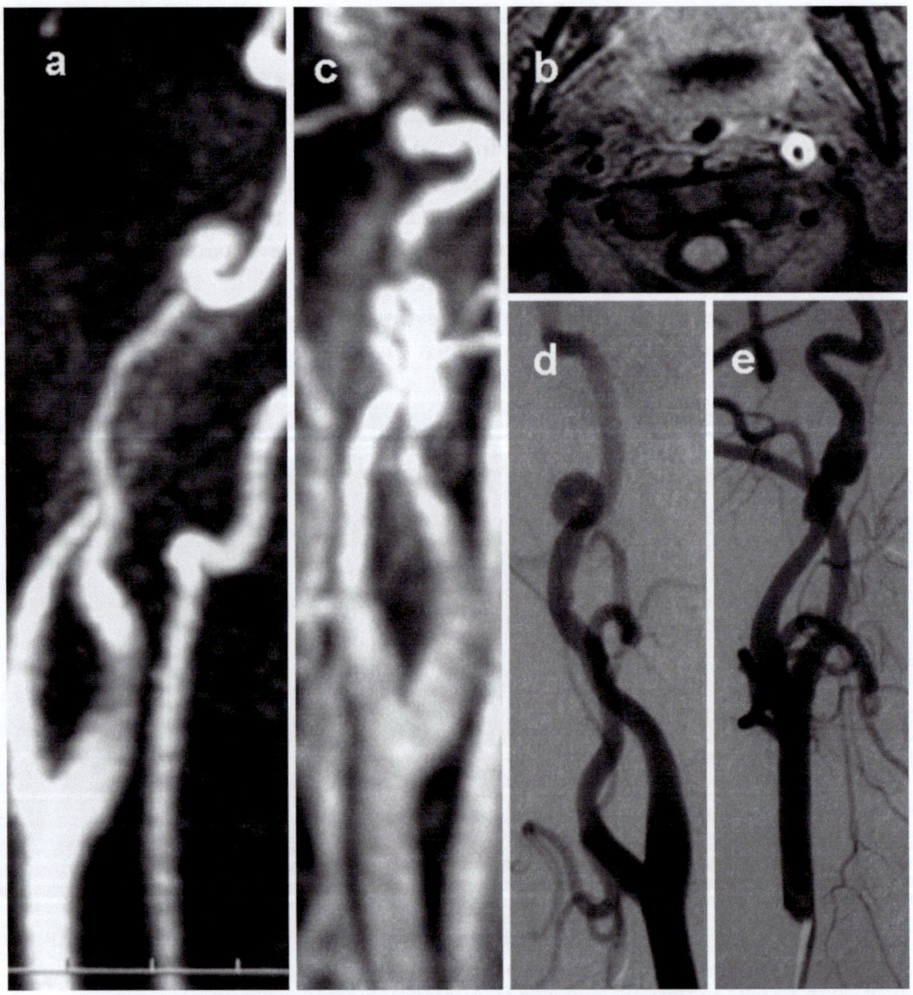

Fig. 7.6. Spontaneous left internal carotid artery dissection. CE-MRA reveals a progressive luminal narrowing in the middle portion of cervical tract (**a**) while conventional T1-weighted images with fat saturation pulse obtained in the axial plane at level of dissection shows a hyperintense ring of metahemoglobin (**b**) referred to mural haematoma. The angiogram of the right internal carotid artery (**c**) depicts a narrowed middle cervical tract and an irregular dilation referred to an aneurismal pattern of a previous asymptomatic carotid dissection. One month later DSA statesd recanalization of left internal carotid artery (**d**) and confirmed a stable pseudo aneurismal pattern of a previous dissection (**e**)

in 49.3% of spontaneous CADs (Touzè et al. 2001). Although classical flow-based MRA techniques are reported as an useful noninvasive method in the detection, follow-up, and response to treatment of CAD, their main limit remain the evaluation of vessel wall morphology such as in the case of fibromuscular dysplasia and aneurismal pattern of carotid dissection (Zuber et al. 1994). CEMRA as a luminography has a better resolution in evaluating vessel wall irregularities and is reported as a robust method to also evaluate dissections with pseudoaneurysm (Phan et al. 2001). To evaluate patients with recurrent asymptomatic dissection, DSA follow-up is generally accepted. A noninvasive method of detection that can be carried out repeatedly over many years like CEMRA may constitute a reliable method in investigating the recurrent dissection (Fig. 7.6).

7.4 Vertebral Arteries

7.4.1 Atherosclerosis

Pathologies involving the vertebral circulation have devastating consequences compared to that of carotid arteries (Fig. 7.7). Vertebral arteries atherosclerotic stenosis is associated with carotid stenosis but is generally less severe. The most frequent site of vertebral artery involvement is at the origin of the subclavian artery. CE-MRA allows imaging of the vertebral arteries from their origins to the basilar artery and has taken the place of TOF techniques, which are notoriously unreliable in the study of the vertebral ostium because of breath-motion artifacts, and in the study of the superior cervical tract, where the tortuous coursing of the vertebral artery is burdened by on-plane saturation artefacts. The sensitivity and specificity of CE-MRA for the evaluation of vertebral ostial stenosis was 100 and 85%, respectively (Randoux et al. 2003). Suboptimal results in this vascular segment are due to a more pronounced tendency to overestimate the degree of stenosis when the vessels have small dimensions and the voxel size is excessive. Overestimation may also be explained by the possible reduced concentration of the contrast medium in the poststenotic arterial lumen as a result of altered flow by the stenosis. Now after CD-US screening, a preferable second level of diagnosis is made with CEMRA. Indeed, notwithstanding advanced algorithms for bone removal, CTA application was limited by the close relationship of the vertebral arteries to the bone in the transverse foramina and at the skull base.

7.4.2 Dissection

Vertebral dissection is the penetration of blood into the vessel wall splitting the tunica media. It may occur spontaneously or following severe or blunt trauma. Complications of vertebral dissection include embolic infarct of the posterior circulation or subarachnoid hemorrhage. Conventional angiography is still considered the gold standard in the diagnosis of vertebral dissection but is gradually being replaced by MRI and MRA. As well as occur in carotid arteries, MRI can show intramural hematoma. One of the main pitfalls of conventional MRI to diagnose vertebral dissection is that the flow-related enhancement of the vertebral venous plexus can mimic the high signal of intramural hematoma (Miaux et al. 1996). Moreover conventional MRI also cannot differentiate between intramural hematoma and intraluminal

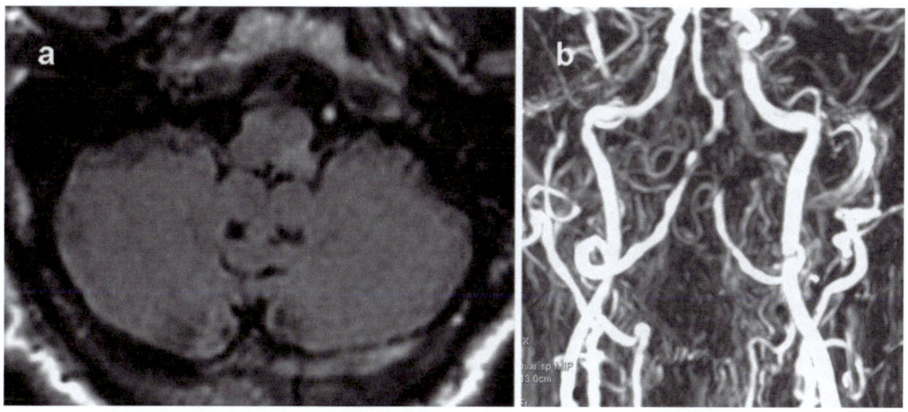

Fig. 7.7. Claude Bernard Horner syndrome with acute onset due to a posterior-lateral infarction of the medulla oblongata (**a**). Note the high signal intensity of the intracranial tract of the left vertebral artery. CE-MRA angiogram obtained in the same session (**b**) revealed an obstruction of the V4 tract and atherosclerotic parietal irregularities of the controlateral artery

thrombus. With CE-MRA it is possible to detect the level of stenosis/obstruction. CE-MRA also has the advantage of differentiating residual flow from mural hematoma. Encouraging results of combined CEMRA and MRI for the follow-up of vertebral dissection are reported especially in dissecting aneurysms type.

References

Alvarez-Linera J, Benito-Leon J, Escribano J, et al. (2003) Prospective evaluation of carotid artery stenosis: Elliptic centric contrastenhanced MR angiography and spiral CT angiography compared with digital subtraction angiography. AJNR Am J Neuroradiol 24:1012–1019

Anson J, Crowell RM (1991) Cervicocranial arterial dissection. Neurosurgery 29(1):89–96

Anzalone N, Scomazzoni F, Castellano R, et al. (2005) Carotid artery stenosis: Intraindividual correlations of 3D time-of-flight MR angiography, contrast-enhanced MR angiography, conventional DSA, and rotational angiography for detection and grading. Radiology 236(1):204–213

Anzalone N, Scotti R, Vezzulli P (2006) High relaxivity contrast agents in MR angiography of the carotid arteries. Eur Radiol 16 (Suppl 7):M27–M34

Barnett HJM, Taylor DW, Eliasziw M, et al.; for the North American Symptomatic Carotid Endarterectomy Trial Collaborators (1998) Benefit of carotid endarterectomy in patients with symptomatic moderate or severe stenosis. N Engl J Med 339:1415–1425

Bluemke DA, Stillman AE, Bis KG, et al. (2001) Carotid MR angiography: Phase II study of safety and efficacy for MS-325. Radiology 219(1):114–122

Borisch I, Horn M, Butz B, et al. (2003) Preoperative evaluation of carotid artery stenosis: Comparison of contrast-enhanced MR angiography and duplex sonography with digital subtraction angiography. AJNR Am J Neuroradiol 24:1117–1122

Brandt T, Orberk E, Weber R, et al. (2001) Pathogenesis of cervical artery dissections: Association with connective tissue abnormalities. Neurology 57(1):24–30

Bremerich J, Bilecen D, Reimer P (2007) MR angiography with blood pool contrast agents. Eur Radiol 17(12):3017–3024

Charon JP, Milne W, Sheppard DG, et al. (2004) Evaluation of MR angiographic technique in the assessment of thoracic outlet syndrome. Clin Radiol 59(7):588–595

Cosottini M, Calabrese R, Puglioli M, et al. (2003) Contrast enhanced three-dimensional (3D) MR-angiography (CEMRA) of neck vessels: Does dephasing effect alter diagnostic accuracy? Eur Radiol 13:571–581

Cosottini M, Michelassi MC, Lazzarotti G (2005a) Vessels of the neck. In: Lombardi M, Bartolozzi C (eds) MRI of the heart and vessels. Springer, Milan, pp 121–143

Cosottini M, Pingitore A, Puglioli M, et al. (2005b) Contrast enhanced three dimensional magnetic resonance angiography of atherosclerotic internal carotid stenosis as the non invasive imaging modality in revascularization decision making. Stroke 34:660–664

Cosottini M, Zampa V, Petruzzi P, et al. (2000) Contrast enhanced three dimensional MR angiography in the assessment of subclavian artery diseases. Eur Radiol 10:1737–1744

Debrey SM, Yu H, Lynch JK, et al. (2008) Artery disease: A systematic review and meta-analysis diagnostic accuracy of magnetic resonance angiography for internal carotid. Stroke 39:2237–2248

Elgersma OE, Buijs PC, Wust AF, et al. (1999) Maximum internal carotid arterial stenosis: Assessment with rotational angiography versus conventional intraarterial digital subtraction angiography. Radiology 213:777–783

Executive Committee for the Asymptomatic Carotid Atherosclerotic Study (1995) Endarterectomy for asymptomatic carotid artery stenosis. JAMA 273:1421–1428

Fellner FA, Wutke R, Lang W (2001) Imaging of Internal carotid arterial stenosis: Is the new standard non invasive? Radiology 219:8588–8589

Habibi R, Lell MM, Steiner R, et al. (2009) High-resolution 3T MR angiography of the carotid arteries: Comparison of manual and semiautomated quantification of stenosis. AJNR Am J Neuroradiol 30(1):46–52

Hoogeveen RM, Bakker CJ, Viergever MA (1998) Limits to the accuracy of vessel diameter measurement in MR angiography. J Magn Reson Imaging 8(6):1228–1235

Johnston DC, Eastwood JD, Nguyen T, et al. (2002) Contrast-enhanced magnetic resonance angiography of carotid arteries: Utility in routine clinical practice. Stroke 33:2834–2838

Kaufmann TJ, Huston J III, Mandrekar JN, et al. (2007) Complications of diagnostic cerebral angiography: Evaluation of 19,826 consecutive patients. Radiology 243(3):812–819

Knopp M, Schoenberg S, Rehm C, et al. (2002) Assessment of gadobenate dimeglumine (Gd-BOPTA) for MR angiography: Phase I studies. Invest Radiol 37:706–715

Koelemay MJ, Nederkoorn PJ, Reitsma JB, et al. (2004) Systematic review of computed tomographic angiography for assessment of carotid artery disease. Stroke 35:2306–2312

Kuntz KM, Skillman JJ, Whittemore AD, et al. (1995) Carotid endarterectomy in asymptomatic patients: Is contrast angiography necessary? A morbidity analysis. J Vasc Surg 22:706–716

Layton KF, Huston J 3rd, Cloft HJ, et al. (2007) Specificity of MR angiography as a confirmatory test for carotid artery stenosis: Is it valid? AJR Am J Roentgenol 188:1114–1116

Le Clerc X, Gauvrit JY, Nicol L, et al. (1999) Contrast enhanced MR angiography of the craniocervical vessels: A review. Neuroradiology 41:867–874

Leclerc X, Martinat P, Godefroy O, et al. (1998) Contrast-enhanced three-dimensional fast imaging with steady-state precession (FISP) MR angiography of supraaortic vessels: Preliminary results. AJNR Am J Neuroradiol 19:1405–1413

Lell M, Fellner C, Baum U, et al. (2007) Evaluation of carotid artery stenosis with multisection CT and MR imaging: Influence of imaging modality and postprocessing. AJNR Am J Neuroradiol 28(1):104–110

Maki JH, Prince MR, Londy FJ, et al. (1996) The effects of time varying intravascular signal intensity and k-space acquisition order on three-dimensional MR angiography image quality. J Magn Reson Imaging 6(4):642–651

Miaux Y, Cognard C, Martin-Duverneuil N, et al. (1996) Flow-related enhancement in the vertebral plexus mimicking an intramural hematoma. AJNR Am J Neuroradiol 17:191–192
Mitra D, Connolly D, Jenkins S, et al. (2006) Comparison of image quality, diagnostic confidence and interobserver variability in contrast enhanced MR angiography and 2D time of flight angiography in evaluation of carotid stenosis. Br J Radiol 79(939):201–207
Miyazaki M, Lee VS (2008) Nonenhanced MR angiography. Radiology 248(1):20–43
Nael K, Villablanca JP, Pope WB, et al. (2007) Supraaortic arteries: Contrast-enhanced MR angiography at 3.0 T–highly accelerated parallel acquisition for improved spatial resolution over an extended field of view. Radiology 242(2):600–609
Nederkoorn PJ, Mali WP, Eikelboom BC, et al. (2002) Preoperative diagnosis of carotid artery stenosis: Accuracy of noninvasive testing. Stroke 33:2003–208
Nishimura DG, Macovski A, Pauly JM, et al. (1987) MR angiography by selective inversion recovery. Magn Reson Med 4(2):193–202
North American Symptomatic Carotid Endarterectomy Trial Collaborators (1991) Beneficial effect of carotid endarterectomy in symptomatic patients with high grade stenosis. N Engl J Med 325:445–453
Phan T, Huston J 3rd, Bernstein MA, et al. (2001) Contrast enhanced magnetic resonance angiography of the cervical vessels: Experience with 422 patients. Stroke 32:2282–2286
Prince MR, Yucel EK, Kaufman JA, et al. (1993) Dynamic gadolinium-enhanced three-dimensional abdominal MR arteriography. J Magn Reson Imaging 3:877–881
Randoux B, Marro B, Koskas F, et al. (2003) Proximal great vessels of aortic arch: Comparison of three-dimensional gadolinium-enhanced MR angiography and digital subtraction angiography. Radiology 229:697–702
Remonda L, Heid O, Schroth G (1998) Carotid artery stenosis, occlusion, and pseudo-occlusion: First-pass, gadolinium-enhanced, threedimensional MR angiography. Preliminary study. Radiology 209:95–102
Rothwell PM, Coull AJ, Silver LE, et al.; Oxford Vascular Study (2005) Population-based study of event-rate, incidence, case fatality, and mortality for all acute vascular events in all arterial territories (Oxford Vascular Study). Lancet 366(9499):1773–1783
Roubin SG, New G, Iyer SS, et al. (2001) Immediate and late clinical outcomes of carotid artery stenting in patients with symptomatic and asymptomatic carotid artery stenosis. A 5-years prospective analysis. Circulation 103:532–537
Scheffler K, Boos M, Steinbrich W, et al. (1998) Flow sensitivity of contrast enhanced MRA. MAGMA 6(suppl 1):178
Touzé E, Randoux B, Méary E, et al. (2001) Aneurysmal forms of cervical artery dissection: Associated factors and outcome. Stroke 32(2):418–423
Townsend TC, Saloner D, Pan XM et al. (2003), Contrast material-enhanced MRA overestimates severity of carotid stenosis, compared with 3D time-of-flight MRA. J Vasc Surg 38:36–40
U-King-Im JM, Graves MJ, Cross JJ, et al. (2007) Internal carotid artery stenosis: Accuracy of subjective visual impression for evaluation with digital subtraction angiography and contrast-enhanced MR angiography. Radiology 244(1): 213–222
U-King-Im JM, Hollingworth W, Trivedi RA, et al. (2004a) Contrast-enhanced MR angiography vs intra-arterial digital subtraction angiography for carotid imaging: Activity-based cost analysis. Eur Radiol 14(4):730–735
U-King-Im JM, Trivedi RA, Graves MJ, et al. (2004b) Contrast-enhanced MR angiography for carotid disease: Diagnostic and potential clinical impact. Neurology 62:1282–1290
U-King-Im JM, Trivedi R, Cross J, et al. (2004c) Conventional digital subtraction x-ray angiography versus magnetic resonance angiography in the evaluation of carotid disease: Patient satisfaction and preferences. Clin Radiol 59(4):358–363
Wardlaw JM, Chappell FM, Best JJ, et al.; NHS Research and Development Health Technology Assessment Carotid Stenosis Imaging Group (2006a) Non-invasive imaging compared with intra-arterial angiography in the diagnosis of symptomatic carotid stenosis: A meta-analysis. Lancet 367(9521):1503–1512
Wardlaw JM, Chappell FM, Stevenson M, et al. (2006b) Best Accurate, practical and cost-effective assessment of carotid stenosis in the UK. Health Technol Assess 10:1–145
Westwood ME, Kelly S, Bramford MJ, et al. (2002) Use of magnetic resonance angiography to select candidates with recently symptomatic carotid stenosis for surgery: Systematic review. BMJ 324:198–201
Wetzel S, Bongartz GM (2000) Carotid and vertebral arteries in magnetic resonance angiography. Springer, Berlin, pp 217–234
Zuber M, Meary E, Meder JF, et al. (1994) Magnetic resonance imaging and dynamic CT scan in cervical artery dissections. Stroke 25(3):576–581

Thoracic Aorta and Pulmonary Vessels

8

Marco Di Terlizzi, Ambrosini Roberta, Rita Fossaceca, and Alessandro Carriero

CONTENTS

8.1 Introduction *87*

8.2 Thoracic Aorta *88*
8.2.1 Spectrum and Prevalence of Disease *88*
8.2.2 Normal Aortic Anatomy *88*

8.3 MR Technique *89*
8.3.1 Contrast-Enhanced Magnetic Resonance Angiography (CE-MRA) *89*
8.3.2 Contrast Injection Synchronization *90*
8.3.3 Contrast Dose and Administration *90*
8.3.4 Image Processing *91*
8.3.5 Black Blood Techniques *91*
8.3.6 Bright Blood Techniques *92*
8.3.7 Fundamental Sequences *92*

8.4 Normal MR Aortic Anatomy *93*

8.5 Aortic Diseases *93*
8.5.1 Aortic Aneurysm *93*
8.5.1.1 Clinical Findings *93*
8.5.1.2 Histopathology *93*
8.5.1.3 MR Features *94*
8.5.1.4 Differential Diagnosis *95*
8.5.2 Aortic Dissection and Intramural Hematoma *96*
8.5.2.1 Clinical Findings *96*
8.5.2.2 Histopathology *97*
8.5.2.3 MR Features *97*

8.6 Trauma *98*

8.7 Vascular Thoracic Outlet Syndrome *98*

8.8 Post EVAR Evaluation *98*

8.9 Pulmonary Vessels *100*
8.9.1 Normal MR Pulmonary Vessels Anatomy *100*
8.9.2 Bronchial Artery Anatomy *101*
8.9.3 Magnetic Resonance Imaging and Magnetic Resonance Angiography *101*
8.9.3.1 Perfusion Imaging and Ventilation *102*
8.9.3.2 Time-Resolved MRA *102*
8.9.3.3 BFFE Imaging *102*
8.9.3.4 Thrombus Imaging *103*
8.9.3.5 Blood-Pool Imaging *103*

References *103*

ABSTRACT

Magnetic Resonance technology's fast development has permitted a wider application of MR angiography for an accurate and noninvasive imaging of thoracic vessel disease. Diseases such as aortic dissection, aneurysm or pulmonary embolism can now be reliably studied with a very high degree of accuracy and reproducibility and much less feasibility limitation.

8.1 Introduction

MRI is characterized by intrinsic sensitivity to detect and highlight the phenomenon of flow, meaning proton movements. Since its debut, the initial applications of MR were aimed at the quantitative analysis of flow and they were essentially on the basis of two principles: (1) interaction between the moving protons and radiofrequencies, the so called "amplitude" or "Time of flight" (TOF) effects; (2) detection and

Marco Di Terlizzi, MD
Ambrosini Roberta, MD
Rita Fossaceca, MD
Diagnostic and Interventional Radiology Department, Corso Mazzini 18, 28100 Novara, Italy
Alessandro Carriero, MD
Diagnostic and Interventional Radiology Department, University del Piemonte Orientale "A. AVOGADRO," Corso Mazzini 18, 28100 Novara, Italy

measurement of flow phenomenon in the presence of a magnetic gradient, the so called "Phase" effects. In 1950, Suryan was the first to observe the "time of flight" effect; afterwards Bowman e Kudracev, Morse, and Singer improved the quantification of flow velocity using TOF technique by introducing the use of surface phased arrays, coils, and saturation technique.

The use of techniques based on measurement of flow phenomenon in the presence of a magnetic gradient was introduced after the TOF technique. MR images are obtained using two fundamental and relatively independent phenomena: proton excitement and saturation by radiofrequency pulse; signal sampling and proton localization by using magnetic field gradients. While stationary blood signal is dependant on T1 and T2 relaxation times, which are relatively prolonged as for other fluids, the flowing blood signal, on the opposite, is dependant on two phenomena: the time of flight effect, determined by spin wash-in/wash-out because of the emission of radiopulses and the phase shift of spin moving along a magnetic field gradient, which is dependant on the type of flow and on the gradient along the flow direction. These phenomena may cause, in different sequences, the appearance of lack of signal (flow void) or an increased signal (flow related enhancement) and are considered as the fundamentals of both MR angiography (MRA) and flow quantification technique.

8.2 Thoracic Aorta

Thoraco-abdominal aortic diseases represent the main indication for an MRA study. In fact, MRA allows representation of the vessel in various planes and may be performed even without the administration of contrast material and without using ionizing radiations. Moreover, it allows precise and high resolution depiction of arterial wall details, with an accuracy equal to that of trans-oesophageal echo and MDCT angiography. The wide field of view, the non-invasiveness and the possibility to gather information about the dynamics of aortic blood flow makes MRA the method of first choice for the elective study of aortic disease in cooperative patients. The main difference is the known limitation with respect to CTA, regarding the inability, inherent to the MR physics, to evaluate vessel wall calcifications.

8.2.1 Spectrum and Prevalence of Disease

Thoracic aortic diseases may often result in life-threatening conditions that require immediate diagnosis and treatment (AHA 2002 Heart and stroke statistical update). Such conditions remain major causes of morbidity and mortality throughout the world (Olsson et al. 2006), despite continuous progress in identification of clinical risk factors, prevention, and early diagnosis of these conditions, including aneurysms, acute aortic syndromes (i.e., aortic dissection, intramural hematoma [IMH], and penetrating ulcer), trauma, congenital abnormalities, vasculitis (Scheel et al. 2004), and steno-occlusive diseases. Once relying uniquely on digital subtraction angiography (DSA), the imaging of the aorta now benefits from the multi-planar information available with magnetic resonance angiography (MRA) (Prince et al. 1996). Volumetric and multiplanar capabilities of magnetic resonance imaging (MRI) have overcome the issue of tortuous vessels, enabling the accurate evaluation of the real transverse aortic plane (i.e., for aortic aneurysms), the assessment of intramural and perivascular abnormalities, and the evaluation of relations between the aorta and its branches.

8.2.2 Normal Aortic Anatomy

The aorta begins at the base of the left ventricle of the heart, where it is about 3 cm in diameter. After ascending for a short distance, it arches backward and to the left side, over the left pulmonary artery. It then descends on the left side of the vertebral column, passes into the abdominal cavity through the aortic hiatus in the diaphragm and ends, considerably diminished in size (1.5–1.75 cm in diameter), at the level of the lower border of the fourth lumbar vertebra; here it branches into the right and left common iliac arteries. Thoracic aorta is commonly referred to as three distinct and arbitrary sections: the ascending aorta, the arch of the aorta, and the descending aorta. The descending aorta then continues into the abdominal aorta. The length of ascending aorta is about 5 cm. At its origin three small dilatations called the aortic sinuses are present: the right, the left, and the posterior (non coronary). Part of the ascending aorta is contained within the pericardium, similar to the pulmonary artery. For this reason a rupture of the wall of the ascending aorta (i.e., an aortic dissection)

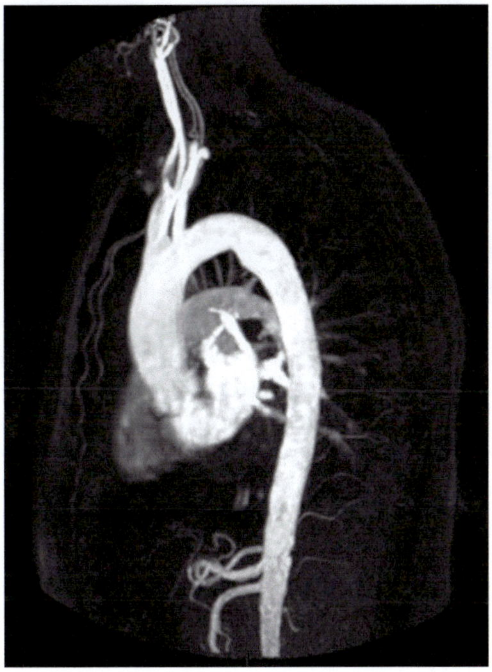

Fig. 8.1. CE-MRA MIP reconstruction in a para-sagittal plane clearly shows the aortic anatomy, the origin of supra-aortic vessels, and the aortic spindle

may lead to a life threatening hemopericardium with cardiac tamponade. The only branches of the ascending aorta are the two coronary arteries, which supply oxygen to the heart; they arise at the respective aortic sinuses. The arch of the aorta begins at the level of the upper border of the second sterno-costal joint on the right side. It runs upward, backward, and to the left in front of the trachea. It then turns backwards on the left side of the trachea and finally goes downward on the left side of the body of the fourth thoracic vertebra, at the lower border of which it becomes continuous with the descending aorta. The upper border of the aortic arch is usually about 2.5 cm below the superior border to the manubrium sterni. Three vessels branch off from the arch of the aorta: the innominate, the left common carotid, and the left subclavian arteries. The attachment site of the ductus arteriosus, immediately downward the origin of the left subclavian artery, corresponds to the aortic isthmus. Immediately beyond the ductus arteriosus the aorta may also present a fusiform dilation termed the aortic spindle. This is marked in the arch concavity by an indentation or angle. These formations may persist, to some extent, in the adult age and the average diameter of the spindle exceeds that of the isthmus by roughly 3 mm. The thoracic aorta starts after the arch of the aorta and runs down to the diaphragm whereupon the abdominal aorta begins (Fig. 8.1). Numerous vessels branch off from the thoracic aorta to supply oxygenated blood to the chest cage and the organs within the chest. Like other sections of the aorta (the ascending aorta, aortic arch, and abdominal aorta), the thoracic aorta is considered as an arbitrary anatomic entity.

8.3 MR Technique

The Patient is positioned supine on the moving table, with the specific vascular thoracic or thoracic-abdominal phased array coil wrapped around the torso (in case of lack of the specific vascular phased array coil, a generic abdominal/torso phased array coil may instead be used).

It is necessary to link the patient to the respiratory trigger system because some of the sequences used would need that synchronization. It is possible, when needed, to use some ECG-gated sequences typically used for "cardiac" imaging, for example the cine Fast Gradient Echo, which are useful for the evaluation of the aortic blood flow in case of dissection or coarctation. In this case the acquisition will need to be ECG-gated.

Up to date many techniques have been developed for imaging structures and blood flow within great vessels and pulmonary arteries. Non contrast-enhanced methods are generally referred to as "black blood" and "bright blood" techniques, depending on the signal intensity of the blood within the vessels.

Both un-enhanced and contrast-enhanced techniques are available, and in many examinations, depending on the clinical query, it is common to use a combination of the two techniques to obtain a comprehensive evaluation of the vasculature. Three-dimensional contrast-enhanced MR angiography (CE-MRA) is a fast sequence that generates a 3D data set with high signal to noise ratio images of the intravascular space during a rapid bolus of a gadolinium chelate.

8.3.1 Contrast-Enhanced Magnetic Resonance Angiography (CE-MRA)

CE-MRA is a fast, accurate, and flexible method for noninvasive imaging of the arterial and venous system (Earls et al. 1998). Since its development 10 years ago, the method has been tremendously refined (Prince 1994). CE-MRA is performed by acquiring a 3D gradient echo sequence during a rapid intravenous bolus of gadolinium. It is both reliable and relatively easy to perform, given the automation and speed of current MR systems. Gadolinium chelates can be used with relative safety because of the very low rate of allergic reactions; recently more attention has been aimed to their use in patients with renal failure because of the possibility to induce the NSF (nephrogenic systemic fibrosis) (Sadowski et al. 2007). CE-MRA depicts the arteries and, when desired, venous structures in a 10–30-s acquisition, depending on the field strength and other technical factors of the MR system used for the study. Acquisition can be performed during a breath hold if the patient is non sedated and able to comply with breath-holding instructions. For most thoracic cardiovascular examinations, a 3D spoiled gradient echo (SPGR) volume that includes the heart, pulmonary arteries, thoracic aorta, and proximal great vessels is used. The parameters for the 3D SPGR sequence are optimized to attain the highest-quality images. In general, faster is better for data acquisition with 3D CE-MRA and low repetition time (TR) and echo time (TE) values are usually selected. Excessive spin dephasing can cause signal loss at stenoses, although this can be reduced or eliminated by selecting a TE less than 3 ms. Faster data acquisitions allow the gadolinium contrast material to be injected with a higher injection rate, producing a higher arterial gadolinium concentration and optimized enhancement of the relevant anatomy. The high arterial signal to noise (S/N) ratio may then compensate for reduced T1 relaxation and signal averaging. Fast data acquisition minimizes motion artefacts and makes it easier for patients to successfully suspend breathing for the entire data acquisition window. In addition to minimizing TR and TE, one needs to select the smallest number of sections, sufficient to cover the arterial anatomy and to keep the acquisition time to a minimum. Widening the bandwidth also makes the acquisition faster, but it may also reduce S/N, especially if a very small field-of-view is used, as required for small children and infants. The signal of background tissue is also reduced or eliminated by obtaining a 3D data set "mask" before gadolinium administration to use for digital subtraction (Lee et al. 1996). The imaging volume is prescribed to depict all of the relevant anatomy while minimizing the actual acquisition time. In almost all cases, a coronal plane volume is used, the exception being adult patients in whom the thoracic aorta is the only region of interest. In this scenario, a para-sagittal imaging plane is used.

8.3.2 Contrast Injection Synchronization

In larger patients, correct timing of the contrast medium injection with peak vascular enhancement during acquisition of central k-space data is essential for good quality studies. There are several ways to synchronize contrast delivery. Fluoroscopic triggering (i.e., Bolus-Track, Philips), although not yet widely available, is reliable and fast for achieving optimal enhancement in almost every case (Riederer et al. 1999). In infants and small children, it is difficult to achieve a well-timed examination that results in a pure arterial- or venous-phase acquisition. Because the circulation time is so rapid, both arteries and veins enhance within a short period of time following contrast infusion. Reduction of the acquisition time to a point, where a timed acquisition can be effective, may sacrifice image resolution and reduce SNR. Usually simultaneous initiations of contrast infusion and data acquisition result in diagnostic CE-MRA studies that have adequate resolution and SNRs, although both venous and arterial structures are enhanced (Prince et al. 1997).

8.3.3 Contrast Dose and Administration

The dose of gadolinium is an important determinant of image quality, and widely ranging doses have been advocated. Three-dimensional CE-MRA studies have used doses of gadolinium ranging from 0.5 mmol/kg ("half dose") to 0.3 mmol/kg ("triple dose"). However, for most vascular studies a dose of 0.2 mmol/kg is routinely used (Weiger et al. 2000). In children and infants, careful attention must be given to the volume of infused gadolinium chelate in order to not exceed the clinically accepted upper limit of 0.3 mmol/kg. After the gadolinium has been injected, a saline flush of approximately twice the gadolinium volume is infused to clear the line and to ensure that the patient receives the entire gadolinium dose. Contrast may be administered either by hand or by a MR compatible power injector. For larger children or adults, a 20- or 22-gauge angio-catheter is routinely inserted in an antecubital vein prior to the start of the study. In smaller children or infants, a smaller angio-catheter can be used if needed and this can be positioned elsewhere as long as it can handle a rapid bolus of 1–2 mL/s. A long extension tube is used to connect to a power injector located inside the MR scanning room. The power injector ensures a reliable rate of contrast delivery with an infusion rate of 1–2 mL/s immediately followed by a 15–30-mL saline flush at the same infusion rate.

8.3.4 Image Processing

Interpretation of CE-MRA requires interactive examination of the 3D data sets. The reconstructed images greatly enhance diagnostic confidence. In the past, the most widely used postprocessing technique for CE-MRA was maximum-intensity projection. The diagnostic accuracy of contrast-enhanced MR angiography using the MIP technique is well described and accepted clinically (Mallouhi et al. 2002). In the thorax, sub-volume MIPs, made from only a selected portion of the original data set, are useful to exclude nonrelevant anatomy. Multiplanar reconstructions (MPR) are useful in the evaluation of thoracic vascular anomalies. Because both MIPs and volume-rendered images are projectional images, adjacent structures may overlap and obscure the relevant anatomy. MPRs are generated using any angle through the original data set. Orientations that depict the blood vessel(s) of interest in an optimal manner are selected interactively. This is a very practical method to accurately determine the size of a vessel or its angulation and relationship with respect to other structures. Volume rendering (VR) is another useful method for depicting thoracic vessels. Mallouhi et al. found that VR performed slightly better than MIP for quantification of renal stenoses greater than 50% and significantly better for severe stenoses. VR also had a substantial improvement in positive predictive value and renal vascular delineation on VR images was significantly better. In another series, MIPs were statistically less reliable for determining renal artery stenoses compared with digital subtraction angiography (Baskaran et al. 2002). The MIP algorithm selects only the voxel with the highest attenuation along a ray projected through the data set; volume-averaged voxels may be erroneously excluded from the final image, resulting in overestimation of stenoses. VR is on the basis of the percentage classification technique, which is used to estimate the probability of a material being homogeneously present in a voxel. This provides an accurate determination of the amounts of materials when the voxel consists of two or more materials, which are volume averaged. With VR, the volume-averaged voxels are included in the final image because VR calculates a weighted sum of data from all voxels along a ray projected through the data set.

8.3.5 Black Blood Techniques

Spin echo (SE) was the first sequence used for evaluating cardiac and thoracic vascular morphology. The development and introduction of ECG-gating made SE technique especially useful by substantially reducing motion artifacts (Pettigrew et al. 1999). Spin echo sequences generally provide good contrast between the vessel wall and blood. These are called black blood images because of the signal void created by flowing blood. Blood signal may appear brighter in slowly flowing areas, such as areas immediately adjacent to the vessel wall. Presaturation with radio frequency and reduction of the echo time minimize blood signal and increase contrast on gated SE images. Although widely available, SE imaging has limited temporal resolution and is degraded by respiratory and other motion-related artifacts. Shorter acquisition times are achieved with fast (or turbo) spin echo (FSE) pulse sequences, also known as rapid-acquisition relaxation enhancements (RARE) (Haddad et al. 1995). Several adjustments to the basic FSE sequence have been made, including the use of one or more inversion pulses, increased echo train length, half-Fourier reconstruction, and echo planar techniques. Single-shot FSE (SSFSE) sequences use a very long echo train in tandem with half-Fourier reconstruction (Semelka et al. 1996). The center of k-space is acquired in a short time, minimizing motion blurring. The rapid acquisition of multiple phase lines for a single TR allows for coverage of the entire heart and thorax in the time frame of one or two breath holds. SSFSE technique has the advantage of being faster than the FSE technique in the evaluation of thoracic aortic disease. The SSFSE sequence can be modified for better cardiac results by reducing the echo train length, lowering the effective TE, and using a blood-suppressed preparation method (Vignaux et al. 2001). T2-weighted inversion recovery imaging is now used as the front-line sequence for depiction of cardiac and thoracic vascular morphology. This technique uses a selective and a nonselective 180° inversion pulse followed by a long inversion time to null blood magnetization (Simonetti et al. 1996). A second selective 180° inversion pulse can also be applied to null fat. This is referred to as double or triple inversion recovery (DIR or TIR). The sequence is acquired either with breath-hold or a non-breath-hold technique and provides for excellent delineation of vessel wall or myocardial blood interfaces. It effectively nulls blood and depicts blood-vessel interfaces in an optimal way.

8.3.6 Bright Blood Techniques

Bright blood imaging depicts both morphologic and functional data. In this sequence, blood has a bright signal intensity. Multiple consecutive images are acquired and can be viewed dynamically to depict cardiac motion. Sequences include gradient echo (GRE), fast GRE, segmented k-space fast GRE, and steady state free precession (SSFP) techniques. GRE imaging is well suited for cardiac and vascular imaging because of its short echo and repetition times. Blood appears bright compared with adjacent myocardium because of time-of-flight effects, as well as the relatively long T2. Markedly turbulent blood induces a signal because of to intravoxel dephasing, which in some instances may be a "helpful" artifact for assessing areas of stenoses (Jara et al. 1999). A segmented k-space approach provides high-resolution dynamic images and can be performed much more rapidly than other techniques (Pruessmann et al. 1999). Using short echo times (2 ms) and short TRs (<10 ms), multiple lines (segments) of k-space are acquired during each cardiac cycle. In GRE techniques, only a single line of k-space is acquired per cycle. Segmented k-space fast GRE imaging remains a mainstay for dynamic vascular imaging and has been improved and adapted for fastest possible acquisition times (Reeder et al. 1999). Nevertheless, the technique is limited by the need to maintain adequate enhancement of inflowing blood. At lower TRs, now available with high-performance gradient systems, inflow enhancement of the cardiac blood pool diminishes and saturation occurs, reducing vessel wall–blood contrast. The inability to further reduce TR effectively limits achievable spatial and temporal resolution. Steady state free precession is a state-of-the-art approach to the improvement of cine imaging. Image contrast in SSFP depends on the T1/T2 ratio of tissue and, unlike GRE techniques, is less dependent on flow. SSFP is susceptible to magnetic field inhomogeneities and requires very short TRs, limiting its use until recently. With technical improvements in magnetic field homogeneity and the development of higher performance gradient systems, diagnostic SSFP images can now be obtained with limited artifacts (Plein et al. 2001). This technique is also known as BFFE (balanced fast field echo), FIESTA (fast imaging employing steady state acquisition), FISP (fast imaging with steady precession), and true FISP depending on the vendor.

8.3.7 Fundamental Sequences

- Triple plane (axial, sagittal, coronal) low resolution survey.
- Phased array coil "reference" sequence.
- Spin echo (SE) T1-weighted sequences in the near-sagittal or axial plane, going through the ascending and descending aorta, and on the axial plane; these "black blood" sequences (Fig. 8.2), allow to obtain a sharp contrast between the hypointense (dark) flowing blood and stationary tissues (aortic wall). These sequences need to be ECG-gated (Stemerman et al. 1999).
- As an alternative, inversion recovery fast spin echo (IR FSE) sequences could be used; these sequences allow to shorten the acquisition time to around 10–20 s by using additional radiofrequency pulses with multiple echoes sampling for every phase encode. So it becomes possible to acquire the entire data set in a single breath-hold, with a definite improvement in image quality.

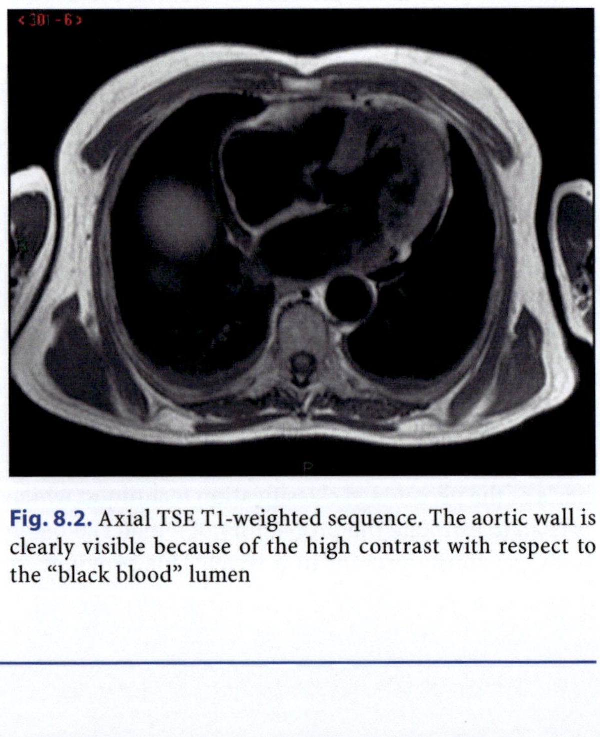

Fig. 8.2. Axial TSE T1-weighted sequence. The aortic wall is clearly visible because of the high contrast with respect to the "black blood" lumen

- T2-weighted ad IR sequences both in the near-sagittal plane, parallel to the vessel major axis and on the axial plane, perpendicular to the vessel major axis. These sequences allow to characterize the vessel wall signal for the evaluation of endovascular thrombotic apposition, intra-plaque hemorrhage, intramural hemorrhage/hematoma or vessel wall oedema in case of acute/subacute vasculitis.
- Additional sequences like GRE or fast spoiled gradient echo (FSGR) (with TR and TE <10 ms), even called BFFE (Philips), true FISP (Siemens), or FIESTA (General Electric), are extremely fast and do not need contrast medium; if acquired on near-sagittal or coronal plane these sequences allow the dynamic qualitative study of the blood flow and the modifications induced by arterial wall disease (intimal flap, coarctation, etc) (Kersting-Sommerhoff et al. 1987), eventually showing vascular stenoses jet or aortic valvular regurgitation.
- Three dimensional contrast enhanced sequences to obtain an MRA (3D Spoiled GE like FFE or FISP): these are modified 3D TOF and allow a breath-hold high contrast and spatial resolution volumetric acquisition (Persson et al. 2004); these are acquired on a near sagittal plane for the study of the thoracic aorta and on the coronal plane, when an evaluation of the thoracic-abdominal aorta and the iliac-femoral vessels is needed. To speed up the acquisition in poorly cooperative patients it is possible to widen the reception bandwidth (i.e., ±128 kHz) or using parallel imaging. Parallel Imaging technique like SENSE (Philips) (Sodickson and Manning 1997), ASSET (General Electric), or GRAPPA (Siemens) (Griswold et al. 2002) allow the simultaneous acquisition of multiple data sets by different phased array coil receiving channels, reducing acquisition time to 5–20 s, depending on the acceleration factor selected.
- Phase contrast (PC) sequences may be used to quantitatively evaluate the blood flow in a specific location, making it possible to calculate flow velocity and acceleration for the estimation of stenoses.

8.4 Normal MR Aortic Anatomy

On SE or FSE T1-weighted images the aorta shows an even diameter, with a maximum of less than 3.5 cm in the thoracic district and less than 2.5 cm in the abdominal district.

Aortic wall has to appear smooth and with an even thickness, usually less than 3 mm. In T2-weighted images, the aortic wall has to appear hypointense, in order to exclude the presence of oedema or intramural hemorrhage.

8.5 Aortic Diseases

Clinically relevant aortic pathologies may be grouped into four entities:

- Aneurism
- Dissection
- Intramural hematoma
- Trauma

8.5.1 Aortic Aneurysm

8.5.1.1 Clinical Findings

- The pathogenesis of the development of aneurysms is not fully understood.
- They are associated to atherosclerosis, without this being a direct causing agent, and to arterial hypertension.
- Usually asymptomatic, they are frequently discovered by accident during examinations performed according to other requests (Grathwohl et al. 1999; Stefens et al. 1994; Van Dyke and White 1994).
- Sometimes they are associated with genetically transmitted pathologies such as Marfan or Ehlers–Danlos syndrome, Cystic Medial necrosis.
- They are sometimes associated with infective processes (Syphilis).

8.5.1.2 Histopathology

- The clinical definition of an aneurysm is "a vessel which has a diameter at least 1.5 times lager than attended."
- It could assume different geometrical shapes; fusiform as frequently seen in atherosclerosis, or saccular as frequently seen in post-traumatic or post infective cases (so called mycotic-aneurysm).
- They could present with parietal homogeneous or irregular thrombotic stratification which can be either concentrically or eccentrically positioned.

8.5.1.3
MR Features

- FSE, BTFE, GRE and 3D FFE T1-weighted sequences allow to evaluate the maximal external vessel diameter, which is to diagnose an aneurysm that can be >4–5 cm for the thoracic aorta or >3 cm for the abdominal aorta (Figs. 8.3a, b).
- After performing the contrast-enhanced 3D FFE sequences, the evaluation of the MIP reconstruction alone highlights solely the internal surface of aneurysm (Fig. 8.3c), not taking into account the eventual presence of parietal thrombotic stratification, causing an underestimation of the vessel diameters (Fig. 8.3d).
- FSE T1, T2-weighted, and FSE IR sequences on the axial or coronal plane can be helpful for depiction of endoluminal parietal thrombosis. Homogeneous and stable thrombus will show an even hypo intense signal in all sequences (Fig. 8.3a) whereas

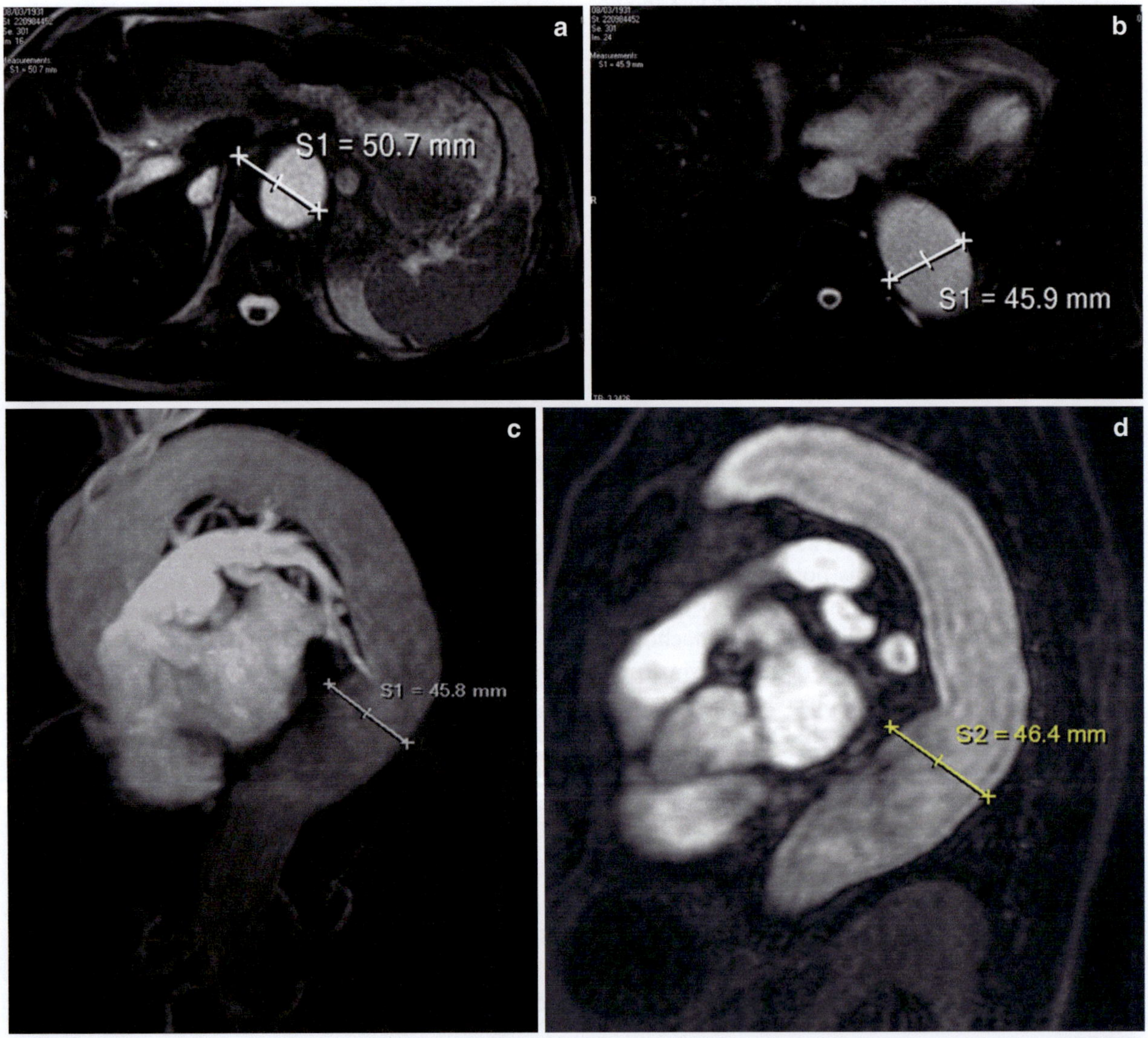

Fig. 8.3. MRA study of a patient with thoraco-abdominal aortic aneurysm (**a**) Axial BTFE T1-weighted sequence clearly showing the aortic diameter of the descending tract. Notice the homogeneously hypo-intense thrombus on its right lateral wall. (**b**) Axial BTFE T1-weighted sequence at a cranial level with respect to 1a, showing the aortic diameter without evidence of parietal thrombus; it is mandatory to measure the aortic diameter on multiple planes because of the possible twisting course of this vessel in advanced atherosclerosis. (**c**) CE-MRA MIP reconstruction: MIP stand-alone evaluation may cause a substantial underestimation of aortic diameters because such electronic reconstruction represents a "lumen" image not taking into account parietal thrombotic apposition or regions of slow flow, which are both quite peculiar to aneurysm or atherosclerotic vessels. (**d**) CE-MRA source images. Source images stand-alone evaluation in this case would have masked the thrombotic apposition seen in 3° because of its position, which is parallel to the acquisition plane

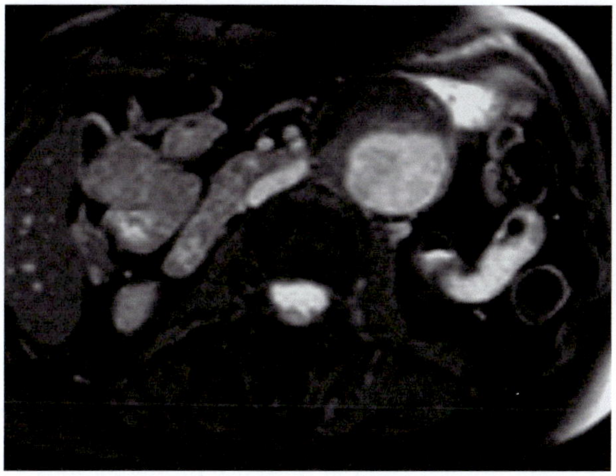

Fig. 8.4. Axial T2-weighted image showing a parietal thrombus in the descending aorta. Note the hypo-intense signal in the external layer, due to hemosiderin deposition and a homogeneous iso-intense internal layer representing stable thrombus

unstable thrombus or intra-plaque hemorrhage will show a hyper or hypo-intense signal on T2 and IR weighted sequences, depending on the phase of blood degradation: oxyhemoglobin (0–12 h, hypo intense on T1 and T2 weighted), deoxyhemoglobin (12–72 h, hypo intense on T1 and T2 weighted), early metahemoglobin (3–7 days, early sub-acute phase, sharply hyper intense signal on T1 weighted and hypo intense on T2 weighted), late metahemoglobin (7–15 days, hyper intense on both T1 and T2-weighted sequences) or hemosiderin (>15 days, hypo intense signal on both T1 and T2-weighted sequences) (Fig. 8.4).

- In case an infective or inflammatory cause is suspected, it is useful to acquire a GRE T1-weighted fat suppressed sequence (i.e., BFFE or true FISP) on the axial plane: in positive cases, the vessel wall will show a hyper intense signal (Fig. 8.5a) because of contrast media accumulating in the interstitial space, with or without an increase in arterial wall thickness (Choe et al. 1999; Nastri et al.2004).
- It is also necessary to complete the evaluation of peri-aortic tissues, in order to highlight oedema (hyper intense in FSE IR T2-weighted sequences) (Fig. 8.5b) and eventually a warning signal of instability or imminent rupture of the aneurysm itself (Hartnell GG 2001).

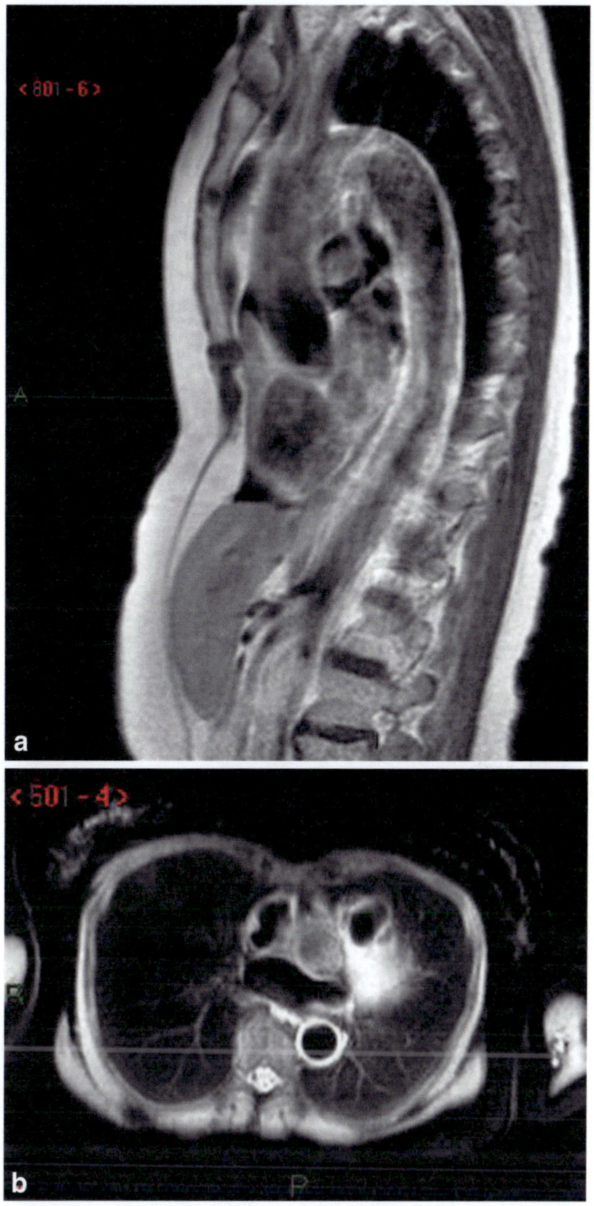

Fig. 8.5. (**a**) Black blood post contrast fat suppressed T1-weighted sagittal images showing diffusely hyper-intense signal from the aortic outer wall. (**b**) Black blood fat suppressed T2-weighted axial images showing hyper intense aortic wall (adventitia) signal due to inflammatory oedema in a case of aortitis (MR was requested after a positive FDG-PET had been performed)

8.5.1.4 Differential Diagnosis

- Posttraumatic Pseudo-aneurysm: Usually saccular, it is located mostly on the ventral aspect of the aortic arch, close to the arteriosus ligament of Botallo. Post-traumatic pseudo-aneurysms of the abdominal aorta are rare.
- Mycotic Aneurysm: usually saccular, focal, and limited in size, could eventually show hyper intense wall signal on IR and T2-weighted sequences in case of active disease.

- Retroperitoneal fibrosis: diffuse hyperintensity of the arterial wall and surrounding structures due to chronic inflammatory process.

The main consideration regarding an aneurysm is the maximum axial diameter. Four millimeter per year is the estimated average increase in size of an abdominal aortic aneurysm. The vessel wall has to oppose to the effect of the internal blood pressure with a circumferential force, regulated by Laplace law, on which depend the odds of rupture. The literature reports say that the rupture odds/year of aortic aneurysm based on his size are variable, depending on the multi-factorial aetiology and eventual presence of other co-morbidities (diabetes, hypertension, etc), ranging from 20% per year in aneurysm larger than 5 cm to 40% per year if larger than 6 cm to more than 50% per year if larger than 7 cm. Maximal axial diameters have to be evaluated on more planes or projections, taking into account the total longitudinal extension and side branches involved (supra aortic vessels, renal arteries). In some cases, using the 3D contrast enhanced sequences it is possible to distinguish between the origin of small lumbar or spinal cord feeding vessels and Adamkievicz artery (Fig. 8.6), which is of utmost importance, in case an endovascular aortic repair (EVAR) procedure is being planned.

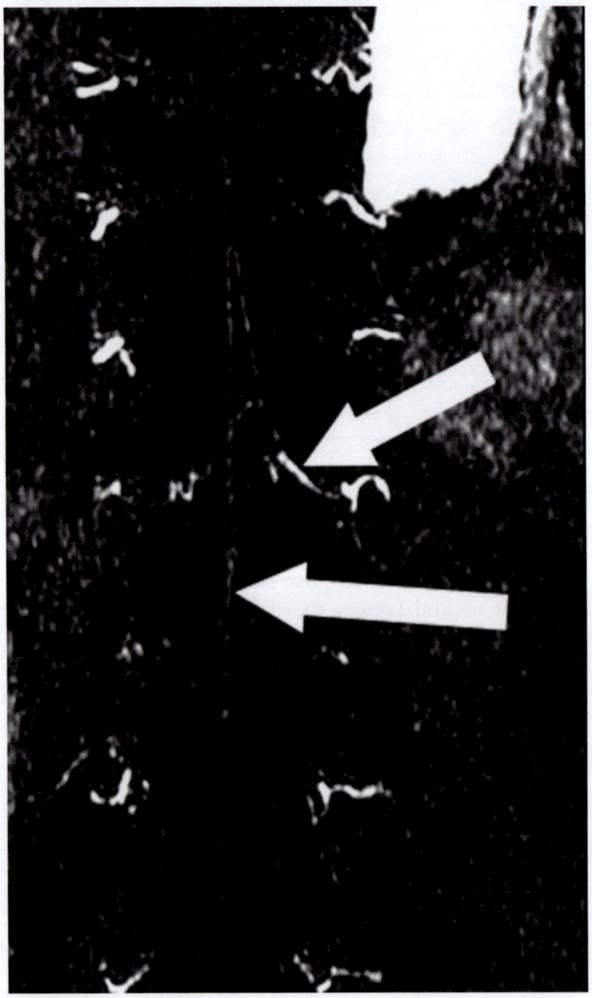

Fig. 8.6. Hi-resolution CE-MRA source images in a patient with thoraco-abdominal aortic aneurysm: the Adamkievicz spinal artery and a lumbar artery are clearly visible

In such cases it is important to correctly measure diameter and length of the aneurysm proximal neck, the maximal axial diameter, and longitudinal extension and to evaluate the presence of angulations or kinking. Measurements have to be performed on CE MRA images and on FSE, GRE, and 3D FFE T1-weighted images in order to correctly evacuate the external vessel size. The evaluation of the presence and extension of vessel wall calcification has to be demanded on CT examination.

8.5.2 Aortic Dissection and Intramural Hematoma

8.5.2.1 Clinical Findings

- Sudden onset of excruciating, stabbing-like thoracic or abdominal pain, sometime associated with hypotension or shock and asymmetric peripheral pulse, often in a patient with pre-existing aneurysm (dissecting aneurysm).
- Often associated to arterial hypertension (hypertensive fit being one of the cause of dissection).
- Very common in a patient with connective tissue pathology such as Marfan, Ehlers–Danlos, or cystic medionecrosis syndrome.
- When the ascending aorta is involved it is termed "type A" dissection as per Stanford classification (or previous De Bakey type I).
- When the aortic arch or descending/abdominal aorta is involved it is termed "type B" dissection as for Stanford classification (previous De Bakey type II and III).
- Dissection could extend to side branches (coronary arteries, supra-aortic, renal, splancnic or iliac vessels) causing ischemia or acute infarction of the respective vascular territories, or towards external rupture when the adventitial layer becomes disrupted.

8.5.2.2
Histopathology

- Aortic dissection is caused by a sudden tear in the intimal layer of the vessel wall, with high pressure arterial blood forming a false lumen into the medial layer. Entry site is often situated in correspondence to an atheroma or to an ulcerated claque. The "point-break" could be represented by an intra-plaque hemorrhage or by an intramural hematoma (often post traumatic). In the latter case the intramural bleeding with the increasing size of the hematoma ultimately causes the intimal tear. In connective tissue pathology such as Marfan syndrome, an eventual adventitial tear determinates an aortic fissuration, with active extravasation of blood in peri-aortic spaces (mediastinum, pericardium etc.).
- Intimal flap usually presents a spiral ascending or descending path.

8.5.2.3
MR Features

- In FSE T1-weighted images, the intimal flap appears as a thin hyper-intense line within the hypo intense aortic lumen (Khan and Nair 2002).
- In BTFE images the intimal flap will appear as a thin hypo-intense line within a hyper-intense aortic lumen (Figs. 8.7 and 8.8) (Pereles et al. 2002).
- In time resolved or cine GRE 3D contrast enhanced sequences, usually acquired into a para-sagittal plane, it is possible to distinguish the progressive enhancement of true and false lumen (Figs. 8.9 and 8.10) and sometimes to identify the dissection "entry site" (Fig. 8.11).
- Contrast enhanced MRA allows to identify and distinguish true and false lumen, both for morphological features (the true lumen is usually

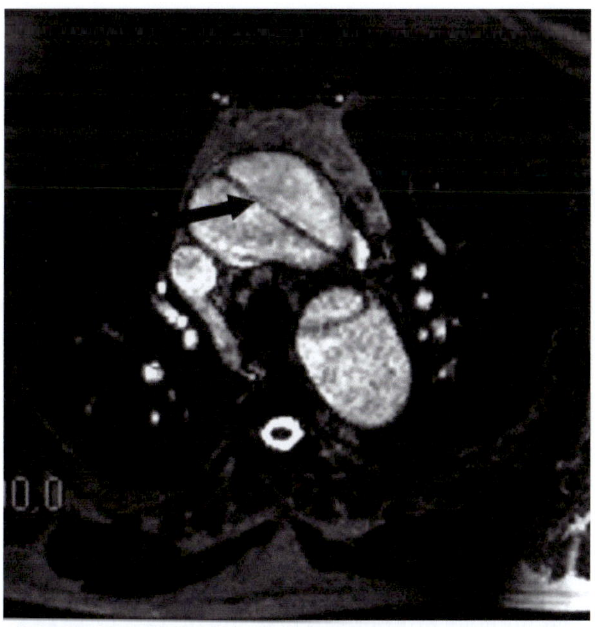

Fig. 8.7. Axial BTFE T1-weighted sequence at aortic arch level shows an intimal flap in a Stanford type A dissection. Intimal flap is represented by a thin hypo-intense linear structure within the hyper-intense lumen

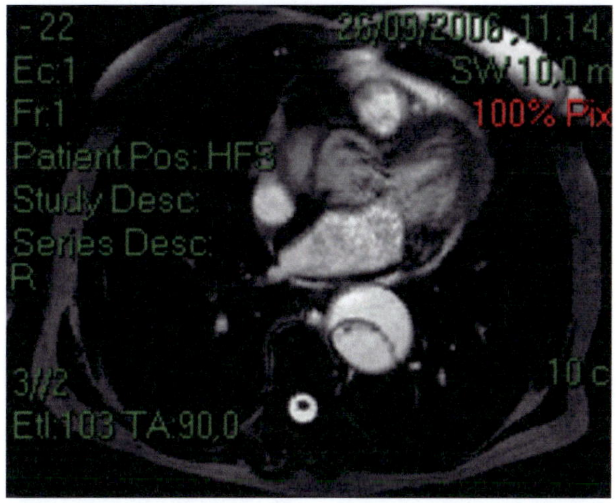

Fig. 8.8. Axial BTFE T1-weighted sequence at the level of descending aorta level shows the intimal flap in a Stanford type A dissection extending downstream. Intimal flap shows the same signal features as in Fig. 8.7

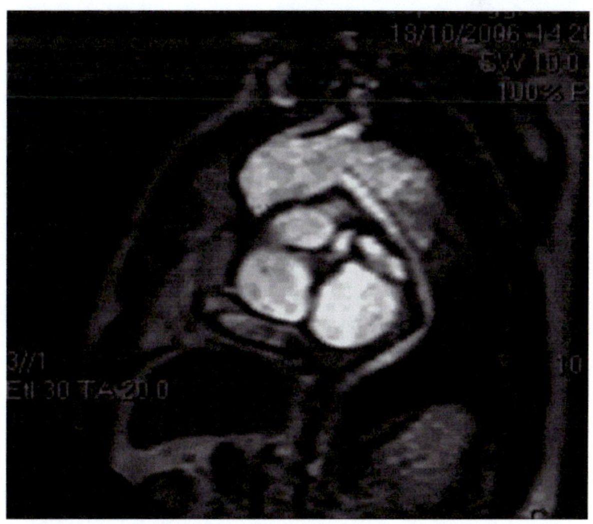

Fig. 8.9. Time resolved CE-MRA acquired on a para-sagittal plane: in early arterial phase the contrast medium is filling the true lumen-TL (anterior), where the blood flow is usually faster with respect to the false lumen-FL (posterior)

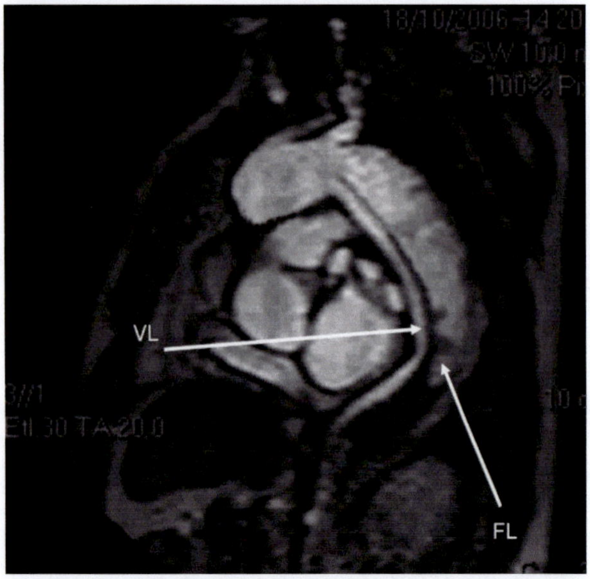

Fig. 8.10. Time Resolved CE-MRA acquired on a para-sagittal plane immediately after the early arterial phase shown in Fig. 8.9. True lumen has been completely filled whereas the false lumen filling is still incomplete

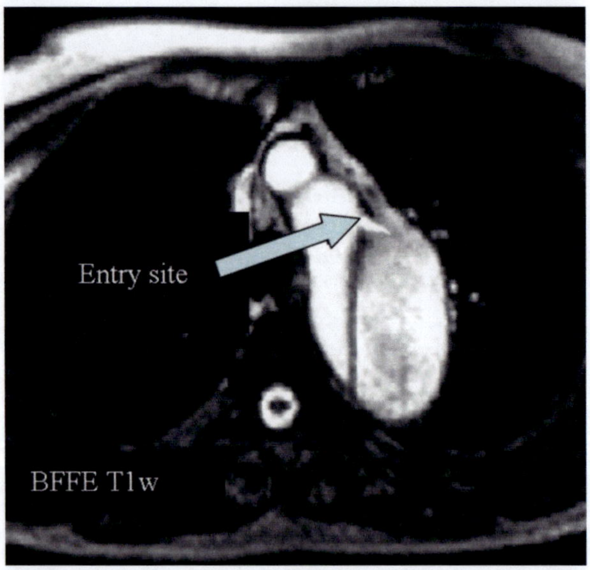

Fig. 8.11. Un-enhanced ECG-gated BFFE T1-weighted sequence acquired on the axial plane in correspondence to the supposed "entry site" of the aortic dissection. In cine mode it is possible to evaluate blood flow dynamics where high speed flow show hyper intense signal with respect to slower flow. In this case it was possible to highlight a small hyper intense high speed jet (*arrow*) indicating the dissection entry site

half-moon shaped with respect to the false lumen, which tends to be oval or circular in shape) and dynamic features of the vascular enhancement (flow in true lumen is usually faster then in false lumen). Moreover, CE MRA allows to correctly evaluate and confirm the origin of the side branches from true or false lumen.

- It is usually recommended to perform a multiphase contrast enhanced MRA acquisition in order to avoid missing features because of the slow flow in false lumen and side branches.

8.6 Trauma

- Taking into account that a trauma patient with suspected aortic lesion is frequently hemodynamically unstable, CT angiography and transoesophageal echo are still the preferred imaging methods.
- In a hemodynamically stable patient or in those in a sub-acute phase, MRA allows a comprehensive evaluation of main complication of aortic trauma, namely pseudo-aneurysm, dissection, and intramural hematoma, using the sequences that have been previously described for the evaluation of such pathologies.

8.7 Vascular Thoracic Outlet Syndrome

Thoracic outlet syndrome is a group of disorders that occur when the blood vessels or nerves in the thoracic outlet, that is, the space between the clavicle and the first rib become compressed. The vascular type of thoracic outlet syndrome occurs when one or more of the subclavian vessels are compressed. The diagnosis of this condition is usually clinical and eventually confirmed by color-doppler US. Multiphase (arterial and venous) CE-MRA performed with the arms first adducted and then abducted over the head is a very accurate technique for the study of this condition (Figs. 8.12a, b and 8.13a, b).

8.8 Post EVAR Evaluation

The number of Endovascular Aortic Repair (EVAR) procedures is still increasing worldwide, because of the introduction of new materials and better performance and durability of prosthesis, most of which

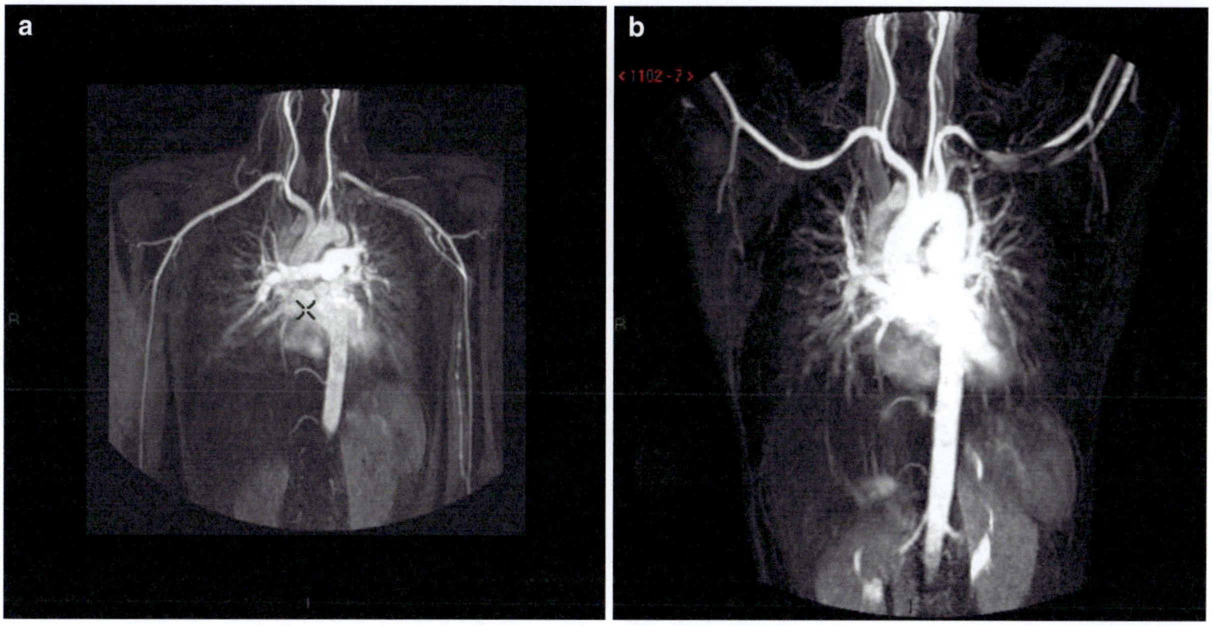

Fig. 8.12. CE-MRA acquired on a coronal plane for suspected thoracic outlet syndrome. (**a**) Arms in resting position. (**b**) Arms extended over the head. There is no evidence of any vascular compression

a b

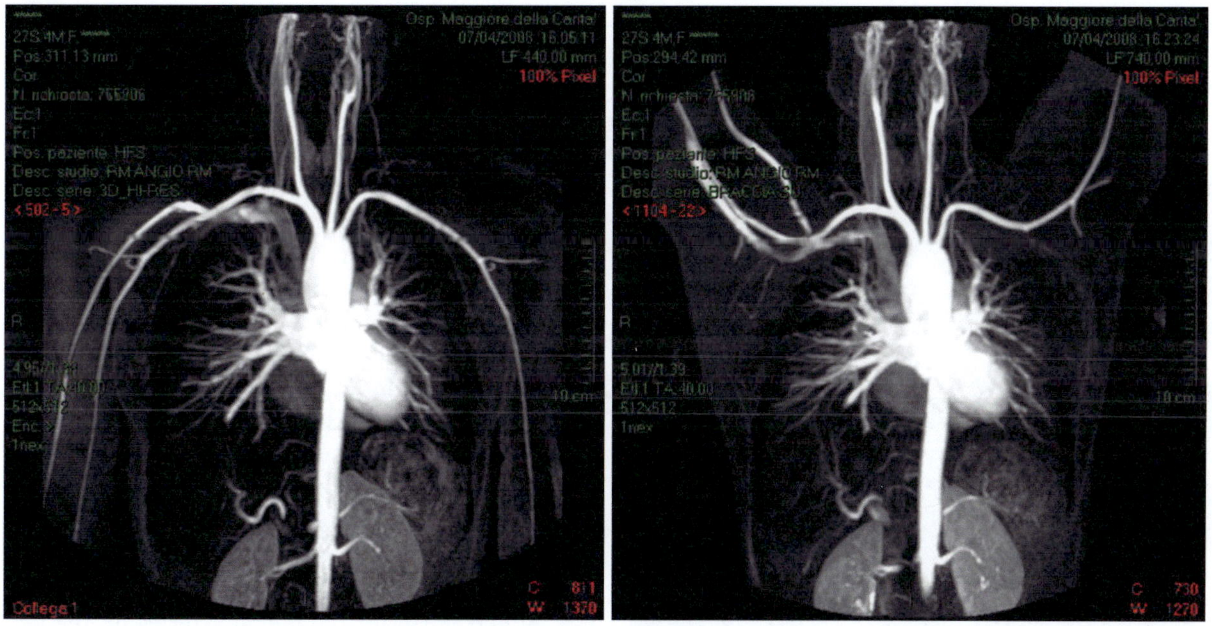

Fig. 8.13. CE-MRA acquired on a coronal plane for suspected thoracic outlet syndrome. (**a**) Arms in resting position. (**b**) Arms extended over the head. In (b) the presence of significant compression and stenosis of the proximal left subclavian artery becomes evident. The patient was subsequently successfully treated by means of surgical release

are actually MRI compatible. With MR it is now possible to perform multiple follow up evaluation in patients submitted to EVAR to gain control of prosthesis positioning and to search endoleaks (leakage of blood between the endoprosthesis and the vessel wall) using time resolved or cine GRE 3D contrast enhanced sequences as used for the evaluation of aortic dissection (Figs. 8.14 and 8.15)

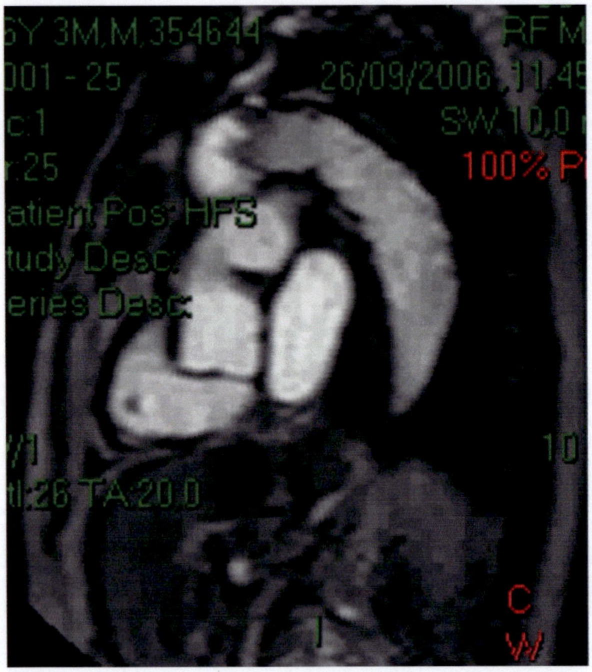

Fig. 8.14. Time Resolved CE-MRA acquired on a para-sagittal plane in a patient submitted to EVAR procedure for an aneurysm of the descending aorta. In the early phase the enhancement of the external para-prosthetic space, starting from its proximal end (so called type I endoleak) becomes evident

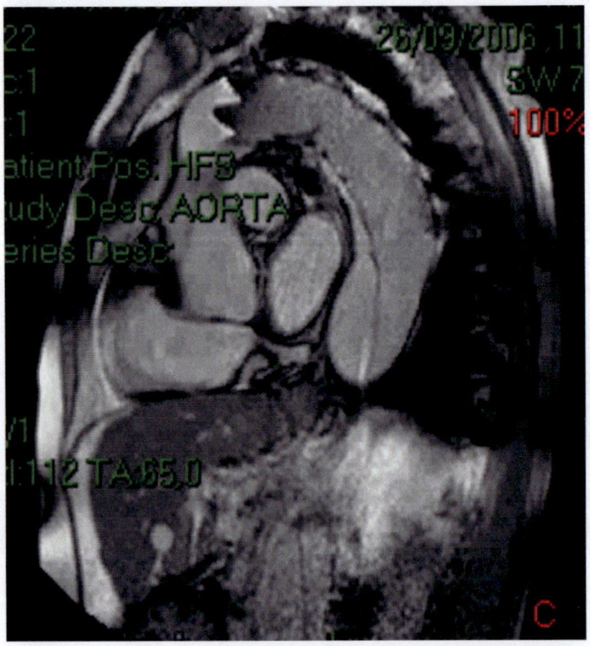

Fig. 8.15. Time Resolved CE-MRA acquired on a para-sagittal plane immediately after the early arterial phase shown in Fig. 8.14. Type I endoleak is clearly depicted because of the progressive enhancement of the para-prosthetic space

8.9 Pulmonary Vessels

Imaging of the pulmonary vasculature is a demanding MRA application, because a large volume has to be covered with short imaging times. The need for short imaging times is a consequence of the high susceptibility of the lung tissue. Dedicated MRA sequences use high readout bandwidths in order to achieve short echo times (TE) and to prevent increased intra-voxel spin dephasing. Correct contrast bolus timing is crucial to obtain a pure arterial image as the arterovenous transit times of the pulmonary vessels are in the order of 3–6 s (Schönberg et al. 1999).

8.9.1 Normal MR Pulmonary Vessels Anatomy

The pulmonary arteries, one on each side, arise from the bifurcation of the pulmonary trunk. The left main pulmonary artery goes posterior and it arches over the left main bronchus before dividing into upper and lower trunks. The right main pulmonary artery is longer than the left following a horizontal or slightly inferior course passing under the concavity of the aortic arch before dividing into a smaller upper and a larger lower trunk. The subsequent branching pattern of the pulmonary arteries then mirrors that of the segmental bronchi. The right gives rise to the apico-posterior, anterior, and posterior segmental arteries that feed the right upper lobe, a middle lobe artery that immediately divides into medial and lateral segmental arteries, and five segmental arteries to the right lower lobe (apical, anterior, posterior, medial, and lateral). On the left side, the arrangement is analogous presenting some variations—the lingular artery, a branch of the left upper lobe pulmonary artery, divides into superior and inferior branches. The apical and posterior arteries are shared on the left (the apico-posterior segmental artery) as are the medial and posterior segmental arteries on the left (the mediobasal segmental artery), thus giving a total of ten segmental arteries on the right and eight segmental arteries on the left.

Nomenclature of the pulmonary arteries on the basis of the distance from the right ventricle is widely used as follows: first order the branch and the pulmonary trunk; second order, the left and right main pulmonary arteries; third order, the upper and lower trunks and "lobar" arteries; fourth order, all ten segmental arteries; fifth order, sub-segmental arteries arising directly from a segmental artery; sixth order, arteries arising directly from the first division of a sub-segmental artery; and so on.

8.9.2 Bronchial Artery Anatomy

Bronchial artery anatomy is extremely variable, with most (>70%) arising from the descending thoracic aorta close but usually slightly below the carina (T5–T6 level). Anomalous bronchial arteries are defined as bronchial arteries that originate outside the T5–T6 range. Bronchial arteries typically run parallel to the bronco-vascular axes. A large published anatomic series described four classical branching patterns, as follows (52):

- Two arteries on the left and one on the right that arises as an intercostobronchial trunk (ICBT) (40.6%)
- One artery on the left and one ICBT on the right (21.3%)
- Two on the left and two on the right, one of which is an ICBT (20.6%)
- One on the left and two on the right, one of which is an ICBT (9.7%)

8.9.3 Magnetic Resonance Imaging and Magnetic Resonance Angiography

The presence of diaphragmatic and cardiac motion artifacts, long scan times, susceptibility artifacts at air-tissue interfaces, and poor contrast between flowing blood and emboli limited the early approaches to pulmonary vessel imaging. With the introduction of contrast-enhanced (CE) MRA techniques many of those limitations were overcome, enabling high-quality diagnostic images of the pulmonary arteries to be generated during one breath-hold (Fig. 8.16). Initial studies reported good success for detection of PE compared with catheter angiography up to the segmental level, but spatial resolution was still suboptimal for an accurate detection of PE in smaller sub-segmental arteries (Oudkerk et al. 2002). With the introduction of parallel imaging techniques, the improvements in gradient technology, and the optimization of acquisition protocols actual MRA techniques are almost comparable to CTA as for spatial and temporal resolution. The acquisition

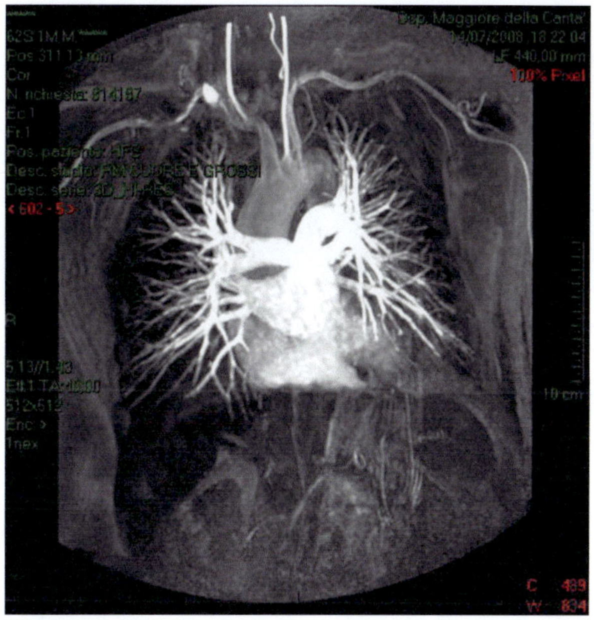

Fig. 8.16. CE-MRA MIP reconstruction in a healthy volunteer showing the pulmonary artery main trunk and its main branches up to the sub-segmental ones. Obviously non filling defect related to thrombus was detected

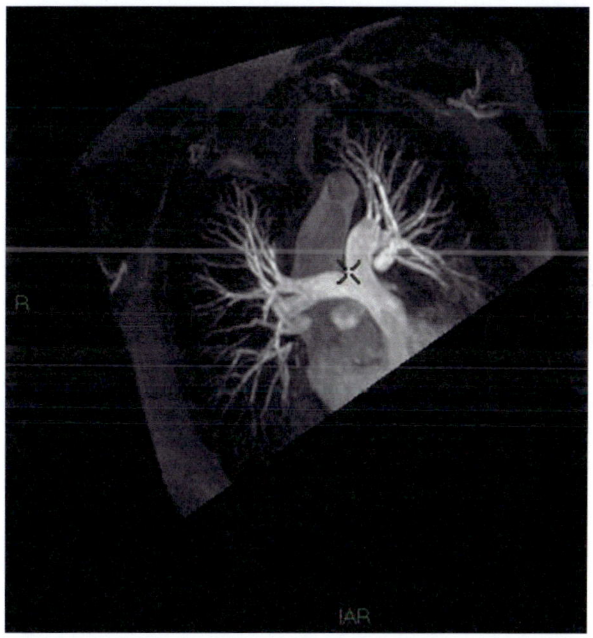

Fig. 8.17. CE-MRA MIP reconstruction selectively showing the right pulmonary artery with no peripheral or central perfusion defects

parameters for pulmonary MRA and MR perfusion imaging could be different among institutions, depending on scanner performance and clinical scenario. Acquisition time for one 3D data set ranges usually between 15 and 30 s. Acquisition is performed during a

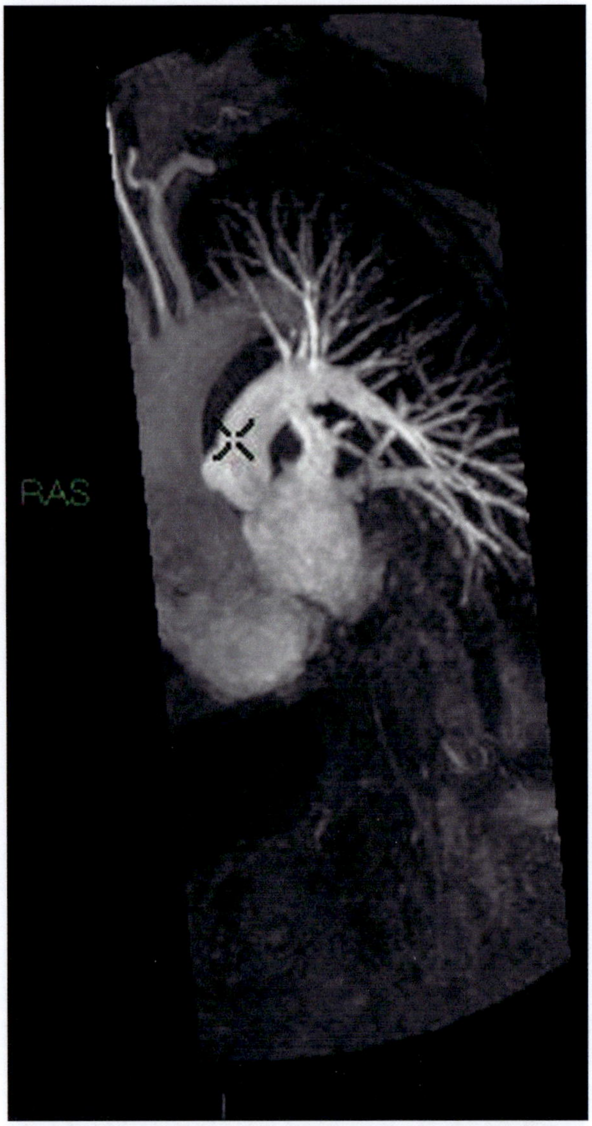

Fig. 8.18. CE-MRA MIP reconstruction selectively showing the left pulmonary artery with no peripheral or central perfusion defects

single breath-hold (in inspiration), and the k-space readout scheme is usually centrically reordered. Because of the asymmetric sequential k-space readout scheme, image acquisition has to start a few seconds before the estimated arrival of the contrast material bolus. Precontrast MR angiograms are then subtracted from the post-contrast images and then MPR and MIP reconstructions of the complete data are obtained for image analysis (Figs. 8.17 and 8.18). Fast imaging is essentially accomplished with short TR (typically ≤5 ms) and short TE (typically ≤2 ms) 3D sequences. With sufficiently strong gradients, extremely short TR/TE combinations can be achieved (i.e., 2.9/1.2 ms, respectively). A flip angle of 15–25° will be satisfactory for these sequences (Nikolaou et al. 2005).

8.9.3.1 Perfusion Imaging and Ventilation

Perfusion imaging with MRI offers substantial benefit in terms of spatial and temporal resolution.

Several studies compare scintigraphy and MR pulmonary perfusion in a variety of disorders. Although results with both techniques are similar, reproducibility is significantly greater with MR. Differential lung perfusion, which may be of value in assessing patients with chronic obstructive pulmonary disease prior to and after lung volume reduction surgery, is also well demonstrated with MR lung perfusion.

8.9.3.2 Time-Resolved MRA

Because of the short transit time from pulmonary artery to pulmonary vein, some researchers have implemented extremely rapid imaging of the pulmonary vasculature, in an effort to capture an arterial phase, similar to what is obtainable in the study of aortic dissections as shown previously. In order to achieve this goal, however, images must be acquired in 4 s or less per scan, thus requiring lower resolution than might otherwise have been employed (Korosec et al. 1996). With this approach, a "clean" arterial phase is acquired in favor of a higher resolution scan that can depict sub-segmental arteries. Often, in-plane resolution is sacrificed to achieve a sufficient combination of in-plane spatial resolution and rapid temporal resolution. That will limit the number of useful projections that can be generated.

8.9.3.3 BFFE Imaging

BFFE imaging is an acquisition sequence that depicts vessels as bright structures and held promises for the diagnosis of both deep venous thrombosis (DVT) and PE (Kluge et al. 2004). Images can be acquired in any plane and although breath-holding is essential, small imaging volumes tailored to the breath-hold capability can be acquired. However, recent evidences suggest that some small clots may go undetected with True-FISP imaging (Pedrosa et al. 2005).

8.9.3.4 Thrombus Imaging

Similar to the depiction of the intramural hematoma in case of aortic dissection previously reported (see Fig. 8.4), direct thrombus imaging is on the basis of the principle that changes in blood clots over time correspond to changes in MR signal intensity, offering a promising new tool for noninvasive diagnosis of PE without the need contrast material injection. As time-dependent changes in MR appearance reflect evolution of the clot, the high signal is seen in new clots only. Further advantages include the ability to scan in the coronal plane, thereby allowing time-efficient coverage of the lower extremity veins with two coronal overlapping acquisitions in a relatively short scan time.

8.9.3.5 Blood-Pool Imaging

Blood-pool agents play a promising role in pulmonary vascular diagnosis, especially in situations where higher spatial resolution is required (Kroft and de Roos 1999). To date, no prospective comparative study of blood-pool agent compared with DSA has been performed. Because of their long vascular persistence time, high-resolution breath-hold images of the entire pulmonary vasculature can be performed in multiple acquisitions tailored to the breath-hold capability of the patient. Another possibility that is attractive for patients with severe respiratory compromise is the use of navigator-echo techniques to eliminate the need for breath-holding. The long vascular persistence of blood-pool agents may offer the adjunctive potential to detect clots within the lower extremities (Perreault et al. 2003) in patients with suspected PE.

References

AHA, American Heart Association (2002)Heart and Stroke Statistical Update.

Baskaran V, Pereles FS, Nemcek AA, Jr., et al. (2002) Gadolinium-enhanced 3D MR angiography of renal artery stenosis: a pilot comparison of maximum intensity projection, multi-planar reformatting, and 3D volume-rendering postprocessing algorithms. Acad Radiol 9:50–59

Benachenhou K, Azarnouch K, Filaire M, et al. (2005) Evolution of healthy thoracic aortic segment diameter during follow-up of patients with aortic aneurysm or dissection: a magnetic resonance imaging study. Surg Radiol Anat 27:142–146

Choe YH, Kim DK, Koh EM (1999) Takayasu arteritis: diagnosis with MR imaging and MR angiography in acute and chronic active stages. J Magn Reson Imaging 10:751–757

Earls JP, DeSena S, Bluemke DA (1998) Gadolinium-enhanced three-dimensional MR angiography of the entire aorta and iliac arteries with dynamic manual table translation. Radiology 209:844–849

Grathwohl KW, Afifi AY, Dillard TA (1999) Vascular rings of the thoracic aorta in adults. Am Surg 65(11):1077–1083

Griswold MA, Jakob PM, Heidemann RM, et al. (2002) Generalized autocalibrating partially parallel acquisitions (GRAPPA). Magn Reson Med 47(6):1202–1210

Haddad JL, Rofsky NM, Ambrosino MM, et al. (1995) T2-weighted MR imaging of the chest: comparison of electrocardiograph-triggered conventional and turbo spin-echo and nontriggered turbo spin-echo sequences. J Magn Reson Imaging 5:325–329

Hartnell GG (2001) Imaging of aortic aneurysm and dissection: CT and MRI. J Thorac Imaging 16:35–46

Jara H, Yu BC, Caruthers SD, et al. (1999) Voxel sensitivity function description of flow-induced signal loss in MR imaging: implications for black-blood MR angiography with turbo spin- echo sequences. Magn Reson Med 41(3):575–590

Lee VS, Flyer MA, Weinreb JC, et al. (1996) Image subtraction in gadolinium-enhanced MR imaging. AJR Am J Roentgenol 167:1427–1432

Kersting-Sommerhoff BA, Sechtem UP, Fisher MR, et al. (1987) MR imaging of congenital anomalies of the aortic arch. AJR Am J Roentgenol 149:9–13

Khan IA, Nair CK (2002) Clinical, diagnostic, and management perspectives of aortic dissection. Chest 122:311–328

Kluge A, Muller C, Hansel J, et al. (2004) Real-time MR with TrueFISP for the detection of acute pulmonary embolism: initial clinical experience. Eur Radiol 14:709–718

Korosec FR, Fraine R, Grist TM, Mistretta CA (1996). Time Resolved Contrast Enhanced 3D MR Angiography. Magn. Res. Med. 36:345–361

Kroft LJ, de Roos A (1999) Blood pool contrast agents for cardiovascular MR imaging. J Magn Reson Imaging 10(3):3 95–403

Mallouhi A, Schocke M, Judmaier W, et al. (2002) 3D MR angiography of renal arteries: Comparison of volume rendering and maximum intensity projection algorithms. Radiology 23:509–516

Nastri MV, Baptista LP, Baroni RH, et al. (2004) Gadolinium-enhanced three-dimensional MR angiography of Takayasu arteritis. Radiographics 24:773–786

Nikolaou K, Schoenberg SO, Attenberger U, et al. (2005) Pulmonary arterial hypertension: diagnosis with fast perfusion MR imaging and high-spatial-resolution MRA – preliminary experience. Radiology 236:694–703

Olsson C, Thelin S, Stahle E, et al. (2006) Thoracic aortic aneurysm and dissection: increasing prevalence and improved outcomes reported in a nationwide population-based study of more than 14,000 cases from 1987 to 2002. Circulation 114:2611–2618

Oudkerk M, van Beek EJ, Wielopolski P, et al. (2002) Comparison of contrast-enhanced magnetic resonance angiography and conventional pulmonary angiography for the diagnosis of pulmonary embolism: a prospective study. Lancet 359:1643–1647

Pedrosa I, Morrin M, Oleaga L, et al. (2005) Is true FISP imaging reliable in the evaluation of venous thrombosis? AJR Am J Roentgenol 185(6):1632–1640

Pereles FS, McCarthy RM, Baskara V, et al. (2002) Thoracic aortic dissection and aneurysm: evaluation with nonenhanced true FISP MR angiography in less than 4 minutes. Radiology 223(1):270–274
Perreault P, Edelman MA, Baum RA, et al. (2003) MR angiography with gadofosveset trisodium for peripheral vascular disease: phase II trial. Radiology 229(3):811–820
Persson A, Dahlstrom N, Engellau L, Larsson EM, et al. (2004) Volume rendering compared with maximum intensity projection for magnetic resonance angiography measurements of the abdominal aorta. Acta Radiol 45:453–459
Pettigrew RI, Oshinski JN, Chatzimavroudis G, et al. (1999) MRI techniques for cardiovascular imaging. J Magn Reson Imaging 10:590–601
Plein S, Bloomer TN, Ridgway JP, et al. (2001) Steady-state free precession magnetic resonance imaging of the heart: comparison with segmented k-space gradient-echo imaging. J Magn Reson Imaging 14:230–236
Prince MR (1994) Gadolinium-enhanced MR aortography. Radiology 191:155–164
Prince MR, Chenevert TL, Foo TK, et al. (1997) Contrast-enhanced abdominal MR angiography: optimization of imaging delay time by automating the detection of contrast material arrival in the aorta. Radiology 203:109–114
Prince MR, Narasimham DL, Jacoby WT, et al. (1996) Three-dimensional gadolinium-enhanced MR angiography of the thoracic aorta. AJR Am J Roentgenol 166:1387–1397
Pruessmann KP, Weiger M, Scheidegger MB, et al. (1999) SENSE: sensitivity encoding for fast MRI. Magn Reson Med 42(5):952–962
Reeder SB, Atalar E, Faranesh AZ, et al. (1999) Multi-echo segmented k-space imaging: an optimized hybrid sequence for ultrafast cardiac imaging. Magn Reson Med 41:375–385
Riederer SJ, Fain SB, Kruger DG, et al. (1999) Real-time imaging and triggering of 3D contrast-enhanced MR angiograms using MR fluoroscopy. MAGMA 8:196–206
Sadowski EA, Bennett LK, Chan MR, et al. (2007) Nephrogenic systemic fibrosis: risk factors and incidence estimation. Radiology 243(1):148–157
Scheel AK, Meller J, Vosshenrich R, et al. (2004) Diagnosis and follow up of aortitis in the elderly. Ann Rheum Dis 63: 1507–1510
Schonberg SO, Bock M, Knopp MV, Essig M, Laub G, Hawighorst H, Zuna I, KallinowskiF, van Kaick (1999) Renal Arteries. Optimization of three dimensional Gadolinium enhanced MR angiography with bolus-timing-independent fast multiphase acquisition in a single breathold. Radiology 211:667–679
Semelka RC, Kelekis NL, Thomasson D, et al. (1996) HASTE MR imaging: description of technique and preliminary results in the abdomen. J Magn Reson Imaging 6: 698–699
Simonetti OP, Finn JP, White RD, et al. (1996) "Black blood" T2-weighted inversion-recovery MR imaging of the heart. Radiology 199(1):49–57
Sodickson DK, Manning WJ (1997) Simultaneous acquisition of spatial harmonics (SMASH): fast imaging with radiofrequency coil arrays. Magn Reson Med 38(4): 591–603
Stefens JC, Boume MW, Sarkuma H, et al. (1994) Quantification of collateral blood flow in coarctation of the aorta by velocity encoded cine magnetic resonance imaging. Circulation 90:937–943
Stemerman DH, Krinsky GA, Lee VS, et al. (1999) Thoracic aorta: rapid black-blood MR imaging with half-Fourier rapid acquisition with relaxation enhancement with or without electrocardiographic triggering. Radiology 213(1): 185–191
Van Dyke CW, White RD (1994) Congenital abnormalities of the thoracic aorta presenting in the adult. J Thorac Imaging 9:230–245
Vignaux OB, Augui J, Coste J, et al. (2001) Comparison of single-shot fast spin-echo and conventional spin-echo sequences for MR imaging of the heart: initial experience. Radiology 219:545–550
Weiger M, Pruessmann KP, Kassner A, et al. (2000) Contrast-enhanced 3D MRA using SENSE. J Magn Reson Imaging 12(5):671–677

Heart and Coronary Arteries

9

Massimo Lombardi and Matteo Milanesi

CONTENTS

9.1 Introduction *105*

9.2 Technological Requirements *106*

9.3 Pulse Sequence Aspects *107*
9.3.1 Cardiac and Respiratory Motion Suppression *107*
9.3.2 Making a Good Contrast *107*
9.3.3 From 2D to 3D: Whole Heart CMRA *108*
9.3.4 New Benefits from Parallel Imaging *109*
9.3.5 Contrast Agents to Improve Image Quality *110*

9.4 Clinical Applications *110*

9.5 MRA Coronary Stenting and Bypass *111*

9.6 Functional Evaluation of Coronary Arteries and Bypass *112*

9.7 Conclusion *113*

References *113*

Massimo Lombardi, MD
Matteo Milanesi, PhD
U.O.C. Cardiovascular RM, C.N.R./"G. Monasterio" Foundation, Via Moruzzi, 1, 56124 Pisa, Italy

ABSTRACT

The current status of technological development is such that coronary magnetic resonance angiography (MRA) has a marginal role in the field of noninvasive evaluation of this arterial district. Nevertheless, in several clinical applications, such as congenital abnormalities of coronaries or pathologies affecting young patients, coronary MRA is considered as the first choice tools and accredited with high diagnostic accuracy. New technical improvements such as high-field scanners (3 T) and multichannel surface coils may turn in favor of coronary MRA because of the intrinsic flexibility and lack of ionizing radiations. In this chapter, the main technological aspects as well the more relevant clinical applications and limitations are commented.

9.1 Introduction

The lack of ionizing radiations, the use of a noniodine contrast agent, and the possibility of obtaining reliable and clinically accepted information such as cardiac function, rest and stress myocardial perfusion as well as the presence and extension of necrotic and viable tissue within the myocardium make MRI (magnetic resonance imaging) a very flexible tool for the evaluation of coronary artery disease. However, there is still a weak point which is in someway reducing the clinical relevance in a large cohort of patients. Coronary artery angiography is in fact hard to be proposed systematically because of heavy technological limitations which confine the use of this approach within the research centers. However, there are already some

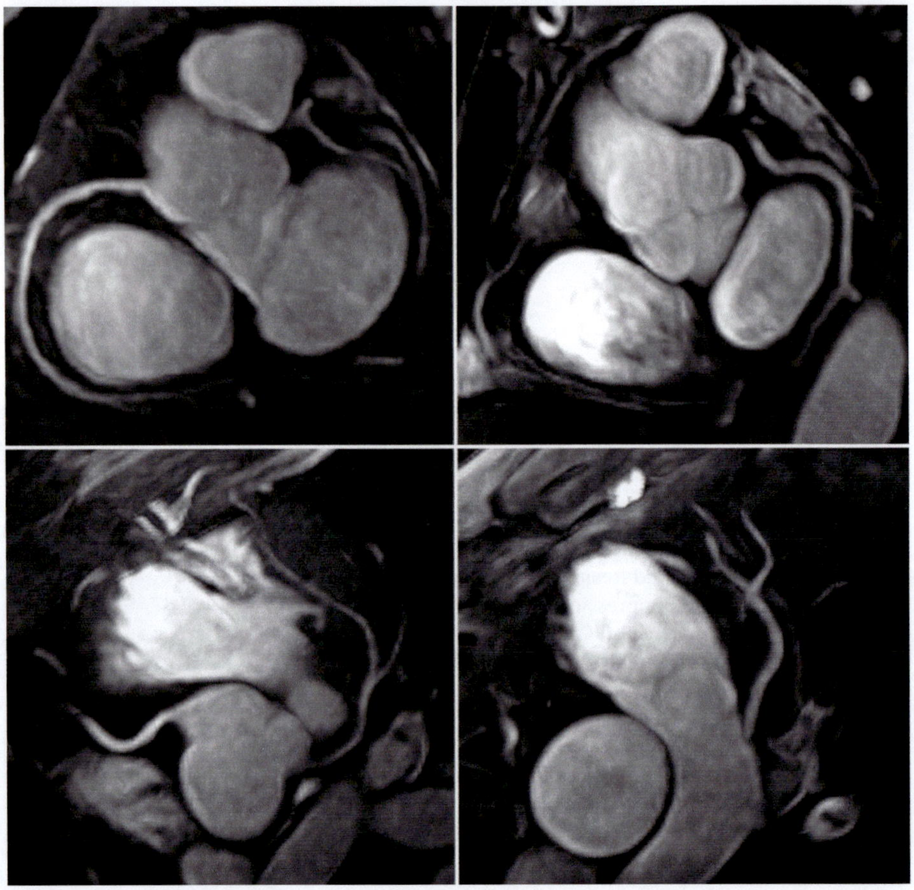

Fig. 9.1. Picture of a new designed multichannel coil for cardiac acquisition. Courtesy of Dr. Christopher J Hardy, GE Global Research, Niskayuna, NY

published results and some research lines which are addressing the main limitations of this technology such as the poor spatial resolution and the long acquisition time. While these results are still preliminary and far from a practical clinical application, they have the clear value to indicate that there are no intrinsic limits which cannot be overcome. The technological territories which are going to be explored are comprehensive of the whole aspects of the methodology such as the field strength (3 T at least), the receiving coil (32 channels at least) (Fig. 9.1), three-dimensional sequences of acquisition (the whole heart coverage in single breath hold or free breathing with multiple gating), acceleration factor (4 or more), etc.

Though biomedical companies and research laboratories are involved in this effort, the use of coronary MRA still remains limited to a small number of patients and within a few referral centers. Nevertheless, the present reduced use is still somehow useful because it avoids the uncontrolled application of a method for noninvasive coronary angiography which might be charged of unrealistic expectation as it is happening for multidetector computed tomography (Redberg and Walsh et al. 2008).

In the present chapter, besides the on-going developments, the few clinical indications and appropriateness criteria will be discussed.

9.2 Technological Requirements

The use of MRI is becoming progressively and extensively beneficial in the practice of cardiology. All the leading cardiology centers are routinely using MRI to answer the morphological as well as the functional aspects of congenital and adult cardiac disease. In the vast majority of clinical applications, the use of a field strength of 1.5 T is considered optimal as at this field the signal/noise ratio is reasonable and artifacts are usually negligible. There are a few and still preliminary indications on the advantages of performing cardiovascular examinations using a field of 3 T. Myocardial perfusion is one of the applications, which seems to have significant advantages, shifting toward a higher field strength in terms of contrast/noise ratio and temporal resolution (Araoz et al. 2005). In the case of coronary MRA, there are several advantages because of the increased signal available, which makes possible technical approaches that are not at 1.5 T such as the acquisition of the whole coronary tree in a single breath hold (SANTOS et al. 2006) or allowing the acquisition of images of coronary arteries during both the systolic and diastolic phase (GHARIB et al. 2007), or improving the spatial resolution up to 0.6 mm, which is one of the prerequisites for the clinical use (WITTLINGER et al. 2005).

9.3 Pulse Sequence Aspects

9.3.1 Cardiac and Respiratory Motion Suppression

In order to produce coronary arteries images not affected by cardiac motion artifacts, the acquisition must be ECG-gated and data must be collected over multiple cardiac cycles, hinging on a technique known as k-space segmentation. Typically, data acquisition is performed in a period of minimal myocardial motion that is found in mid-late diastole. Several methods have been proposed to determine such a period, like the standardized formula introduced in (STUBER et al. 1999a), based on the current subject heart rate, or more subject-specific techniques like those based on ECG-triggered M-mode navigator echo. However, the interval of minimal myocardial motion can be optimized also by visually inspecting cine MR images acquired prior to playing the actual CMRA sequence.

Another problem faced in CMRA is the respiratory motion, which can dramatically invalidate image quality. Two possible solutions are represented by breath-hold and free-breathing imaging. The former approach has the advantage of being fast and easy to implement but its success is strictly related to the ability of the patient to repeat sustained breath-holds. On the contrary, the development of respiratory navigator techniques have allowed free-breathing CMRA imaging by gating the acquisition to the movement of the right lung–diaphragm interface measured using a two-dimensional selective excitation pulse. Initially, in the retrospective gating approach data were oversampled regardless of respiratory motion, and are discarded a posteriori if they did not fit a certain gating window. But, almost all the current free-breathing pulse sequences have adopted the prospective gating control, where both navigator and CMRA data are acquired at each cardiac cycle (STUBER et al. 1999a; KIM et al. 2001). In this case, if the lung–diaphragm interface is found to be in a previously defined gating window (typically adjusted by the user at the end-expiratory period) the following k-space segment is accepted for data reconstruction, otherwise the acquisition is repeated in the next cardiac cycle. Prospective gating techniques also require adaptive motion correction, given that the diaphragm motion is definitely larger than coronary artery.

9.3.2 Making a Good Contrast

The pulse sequences developed for CMRA employ preparatory RF pulses to maximize the contrast between the coronary blood pool and the surrounding tissue.

The so-called "bright blood" gradient echo-based sequences are based on the suppression of the myocardium signal. Usually, in order to image the coronary arteries the natural difference between myocardium and blood T2 values, respectively 50 and 250 ms at 1.5 T. With a first preparation T2 pulse (T2prep), the longitudinal magnetization of both myocardium and blood is tipped to the transverse plane, where the myocardium experiences the T2 decay more quickly than the coronary blood. As a consequence, when the magnetization is turned back to longitudinal after a tip-up pulse, the signal would be higher for the arterial blood pool. Blood arteries will accordingly appear bright, with respect to the myocardium and other short T2 tissues like fat and venous blood, leading to an enhanced contrast. Fat signal can be also suppressed separately through a frequency-selective fat suppression.

"Bright blood" images can also be obtained taking advantage of the T1 shortening related to the paramagnetic contrast agent (CA), such as gadolinium chelated with diethylenetriaminepentaacetic acid (Gd-DTPA). An inversion recovery prepulse is usually applied and the imaging sequence is started when the longitudinal magnetization of the myocardium crosses zero. By this time, the coronary blood magnetization will be fully relaxed thanks to the presence of the contrast agent, which shortens the T1 form 1,200 ms to about 100 ms at 1.5 T, whereas the myocardium T1 still remains round 850 ms.

"Black Blood" CMRA is instead realized by a double-inversion prepulse in combination with a fast spin-echo imaging sequence. A first nonselective

180° pulse inverts the magnetization throughout the body and is immediately followed by a second selective pulse that reinverts the magnetization in the whole imaged slice without affecting the blood about to come in.

9.3.3
From 2D to 3D: Whole Heart CMRA

Initially, coronary artery images were obtained with 2D spin-echo-based pulse sequences. But a significant increase to CMRA was given in the early 1990s by the development of breath-hold 2D segmented k-space gradient echo, and most recently, in the late 1990s, by the steady-state free precession techniques (SSFP), also known as true fast imaging in steady-state precession (TrueFISP), balanced turbo field echo (FFE) or fast imaging employing steady-state acquisition (Fiesta). In SSFP, in fact, the high T2/T1 level of the blood acts as an intrinsic contrast agent producing high image quality in term of both SNR (signal-to-noise ratio) and CNR especially in combination with T2prep and fat saturation.

However, 2D imaging pulse sequences, though easy to implement, show a lot of drawbacks related to long scan time, low SNR because of thin slices and slice misgragation due to the diaphragmatic and cardiac position drifts. Furthermore, following the coronary arties, a rather tortuous path, the success of the exam is often bound up with the capability of the operator to prescribe a good set of slices.

For all these reasons, 3D pulse sequences were developed and have currently completely replaced 2D. Volume coronary angiography with targeted volumes was proposed for a 3D breath-hold technique to image each arterial segment. In spite of high resolution, this technique was still time-consuming and rather operator-dependent; so the effort is nowadays completely devoted to whole-heart 3D pulse sequence (Weber et al. 2004). By covering the entire myocardium, whole-heart imaging is made possible to reach not only major coronary arteries, like proximal and middle coronal segments, but also more distal coronary segments.

Currently, most whole-heart studies employ 3D SSFP pulse sequences, which can achieve higher blood signal intensity as compared to 3D gradient echo sequences (Fig. 9.2). The combination of 3D SSFP

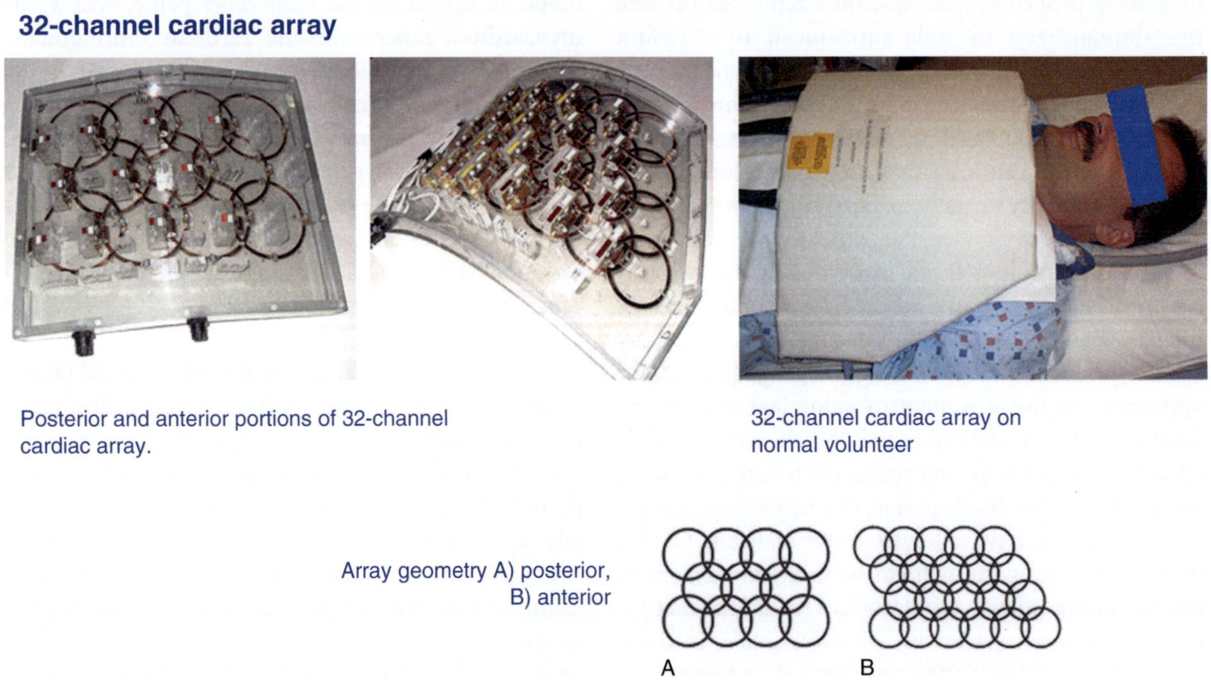

Fig. 9.2. Image obtained by a 3D SSFP sequence (FIESTA) and a 32-channel coil on a 1.5T scanner. TE/TR = 1.7/4.4 ms, ±125 kHz, 2.1 mm partitions, 256 × 224/0.5 NEX, 24 s breath-hold 12 partitions/slab, 24 s overall scan time. Courtesy of Dr. Christopher J Hardy, GE Global Research, Niskayuna, NY

with radial k-space sampling strategy has proven to attain superior vessel sharpness (Spuentrup et al. 2004). Though a radial approach is less sensitive to motion artifacts, it is also true that the image SNR is lower than with Cartesian k-space strategy. On the other hand, the combination of 3D Cartesian SSFP with the variable sampling in time (VAST) technique (Foo et al. 2005) has allowed to get an optimal trade-off between the degree of segmentation, i.e. the temporal resolution, and the spatial coverage while reducing cardiac motion-related blurring.

Three dimension breath-hold approaches have advantaged in term of time efficiency with respect to free-breathing 3D CMRA. This makes breath-hold the only approach suited for first-pass contrast enhancement studies that employ extracellular contrast agents like the aforementioned Gd-DTPA.

Since the imaging time can be lengthened without any constraints on patient breath-holds, 3D free-breathing CMRA can achieve larger heart coverage along with improved spatial resolution and SNR. Still, the scan time remains too long (up to 13–15 min). For instance, in (Stehning et al. 2004) an isotropic volumetric CMRA using free-breathing 3D radial SSFP was obtained, but the scan time could not be pushed down to 10–14 min. It is worth noting that in such a long acquisition period a drift of the diaphragm position over time might happen that can impair the diagnostic value of the images. Cardiac motion is not always entirely solved by correction factors like in prospective gating control. To overcome this problem, a self-navigated image reconstruction approach associated with 3D radial free-breathing has been proposed in (Stehning et al. 2005), where the myocardial motion information is extracted directly from the imaged data. Most recently, a 3D SSFP pulse sequence has been used in conjunction a cardiac fat navigator (Nguyen et al. 2006), proving better image quality and similar SNR and CNR when compared to conventional diaphragm navigator.

9.3.4 New Benefits from Parallel Imaging

An additional technological effort have been recently carried out also in the filed of phase-array coil design and parallel imaging technique development. Parallel imaging has definitely speeded up image acquisition, though at expense of reduced SNR. However, for higher fields like 3.0 T experience an increasing conventional, full gradient encoding SNR, parallel imaging techniques can provide an alternative means of treading SNR for acquisition speed. Another help might come also from phased array with an increased number of coils, which have been demonstrated to reduce the geometry factor for the benefit of SNR, while allowing also acceleration in two dimensions (2D acceleration), i.e. in both the view and slice direction. Such technological development has brought benefits not only for the breath-hold approach, where the whole-heart coverage in a single breath-hold became feasible, but also for the free-breathing approach.

For instance, in (Hardy et al. 2006) a FOV equal to $40 \times 40 \times 12\,cm^3$ with up to 120 slice partitioning was covered with a 32-channel coil yielding a nominal voxel of $1.5 \times 1.5 \times 1.0–2.0\,mm^3$ in a single breath-hold (acceleration factor up to $R = 4 \times 4 = 16$) with a 3D SSFP sequence. A high-density 128 system has been recently employed (Shankaranarayanan et al. 2008) for 3D CMRA in a single breath-hold, with an acceleration in both dimensions up to $R = 4 \times 4 = 16$, and nominal voxel size of $1.6 \times 1.6 \times 2.6\,mm^3$ By attaining a temporal resolution of 210 ms (data for each slice-encoding step was acquired in a single hearth cycle), this study demonstrated the importance of high acceleration factor to reduce both the breath-hold time and the imaging window in the cardiac cycle, which is responsible for motion blurring. Both in (Hardy et al. 2006; Shankaranarayanan et al. 2008) a SSFP sequence was used for image data, preceded by a fat suppression RF pulse.

Regarding free-breathing approach, the use of 2D acceleration strategies were compared with 1D in a recent study that employed a 10-element phased array coil (Okada et al. 2009). A reduction of the scan time up to 45% and a decrease in SNR and CNR (though not statistically significant) were observed for the 2D parallel imaging (Bornstedt et al. 2008). Besides, a 32-channel coil has attained an isotropic voxel of $1.3 \times 1.3 \times 1.3\,mm^3$ with 2D acceleration factor up to $R = 3 \times 3 = 9$ in a scan time between 90 and 450 s for a heart rate of 60 bpm.

Thus, although there are important and significant advances present high-channel MR systems, further research work is needed to perform a whole-heart isotropic 3D imaging with the same resolution and ease obtained by cardiac CT imaging in a reasonable scan time.

9.3.5 Contrast Agents to Improve Image Quality

The sequences suitable for coronary MRA can be provided with an inversion prepulse which usually precedes the acquisition part of the sequence; the purpose of the prepulse being to null the signal from the myocardium (Hoffman et al. 1999). Furthermore, the contrast between the blood and the surrounding tissue can be further enhanced by the combined use of contrast agents (Brenner et al. 1999; Goldfarb et al. 1998). As the extravascular contrast agent rapidly move toward the interstitium, the acquisition of the images should be performed during the first pass. This implies the use of breath-holding sequences (Brenner et al. 1999). This peculiar problem can also be overcome by using the injection of intravascular contrast agent (Hoffman et al. 1999; Li et al. 1998; Stuber et al. 1999b). Because these contrast agents remain confined within the intravascular compartment for hours, this can be used in combination with free-breathing coronary MRA.

9.4 Clinical Applications

Despite the fact that the first description of the coronaries by MRI had been published more than 20 years ago (Paulin et al. 1987), the current technological limitations are such that it is virtually impossible to obtain images constantly of good quality with a spatial resolution suitable for a routine clinical use. In the unique multicenter study which has been published so far, the comparison between MR coronary angiography and invasive angiography (positive and negative predictive values of 70 and 80%, respectively) has been shown. This diagnostic accuracy can be considered not satisfactory for the clinical use. However, it is to be mentioned that in the same study the patients with a left main or three vessel disease were detected with a sensitivity and specificity of 100 and 85%, and with positive and negative predictive values of 54 and 100%, respectively. However, the same authors correctly admitted that the false-positive was an unresolved issue (Kim et al. 2001).

In recent years, several methodological approaches have been proposed for the clinical use. The current hardware frontier is represented by the three-dimensional acquisition of the whole cardiac mass using a 3T scanner (Sakuma et al. 2005; Santos et al. 2006) employing a surface acquisition coil equipped with 32 channels (Niendorf et al. 2006). However, there is still no consensus about the acquisition methodology (single apnea, repeated apnea, free breathing, etc). The latter is obviously easier but affected by breathing artifacts and long acquisition time. These artifacts are only partially reduced by the use of techniques as the navigator approach, where not only ECG gating is employed for acquiring the images during the diastolic phase but also the position of the diaphragm is monitored allowing the acquisition only when the muscle is close to a predefined position. The double gating has the disadvantage of a prolonged acquisition time (5–10 min) and a relatively low success rate. For these reasons, the evaluation of coronary artery disease by MRA is still confined in few research laboratories and it should not be proposed for clinical purposes.

The guidelines of the scientific societies recognize to this method a position far from clinical use in coronary artery disease atherosclerotic in origin (Class III, level of evidence C) (Bluemke et al. 2008).

A different position can be attributed to the method with regard to the evaluation of congenital abnormalities (Figs. 9.3 and 9.4) and changes of the vessel caliber as in Kawasaky syndrome (Fig. 9.5). Such an application is reported by the guide lines as *Class IIa, evidence b.* where the weight of evidence/opinion is in favor of usefulness/efficacy. Interestingly, MDCT is recognized within the same class of appropriateness

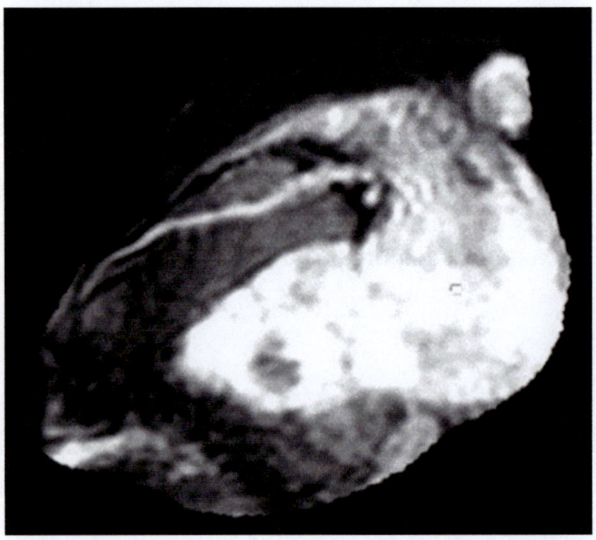

Fig. 9.3. Coronary MRA (CE-MRA technique) in a patient with an intramyocardial course of left descending artery. The young patient was referred for recurrent chest pain on effort

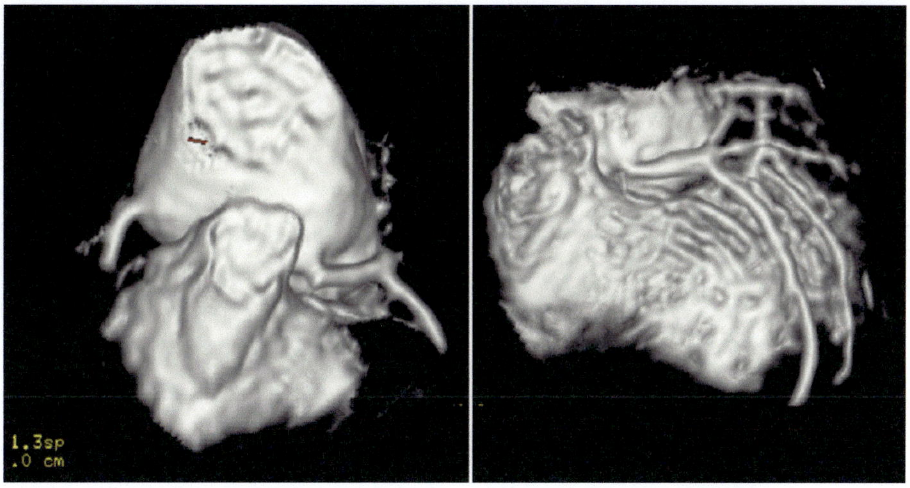

Fig. 9.4. Coronary MRA (CE-MRA technique) in a patient with a previous intervenction for transposition of great arteries. Arterial switch procedure. One of the main target of the follow-up remains the evaluation of the course and the exclusion of stretching of native coronary arteries which have been reimplanted during the surgical procedure. *Left panel*: the origin and the proximal course of the two reimplanted arteries are visible in the 3D reconstruction. *Right panel*: the proximal and mid-part of the left coronary artery and its main branches are visible in the 3D reconstruction

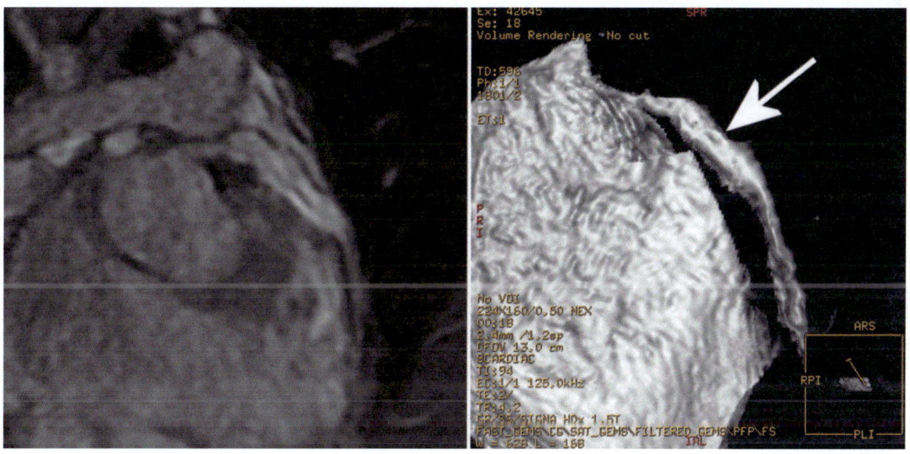

Fig. 9.5. Coronary MRA (CE-MRA technique) in a patient affected by Kawasaki disease. *Left panel*: 3D image of left anterior descending artery. *Right panel*: the same 3D image after volume rendering postprocessing. The aneurism affecting almost the whole proximal part of the artery is clearly visible (*arrow*)

in this group of patients and probably there is no significant difference between the two methodologies. However, these are usually rather young patients and in the case of Kawasaky disease there is the clinical neediness of repeated examinations to follow-up the evolutive nature of the disease. Thanks to the lack of ionizing radiations in such patients, MRA is the first choice of imaging technology to be used, if available. Furthermore, in several cases the association of a suspected course anomaly of the coronary artery with multiple congenital abnormalities, such as in the Fallot tetralogy, is in favor of MRI (Bluemke et al. 2008).

9.5 MRA Coronary Stenting and Bypass

The visualization of a coronary artery where a stent has been placed simply can be unfeasible because of the presence of signal artifacts induced by the metallic nature of the stent itself. MRI compatible and artifact free stents have been reported, but their use is still relatively limited (White et al. 1987).

It is well known that the visualization of bypass is feasible both in the case of venous graft and probably with better results and arterial bypass (Aurigemma

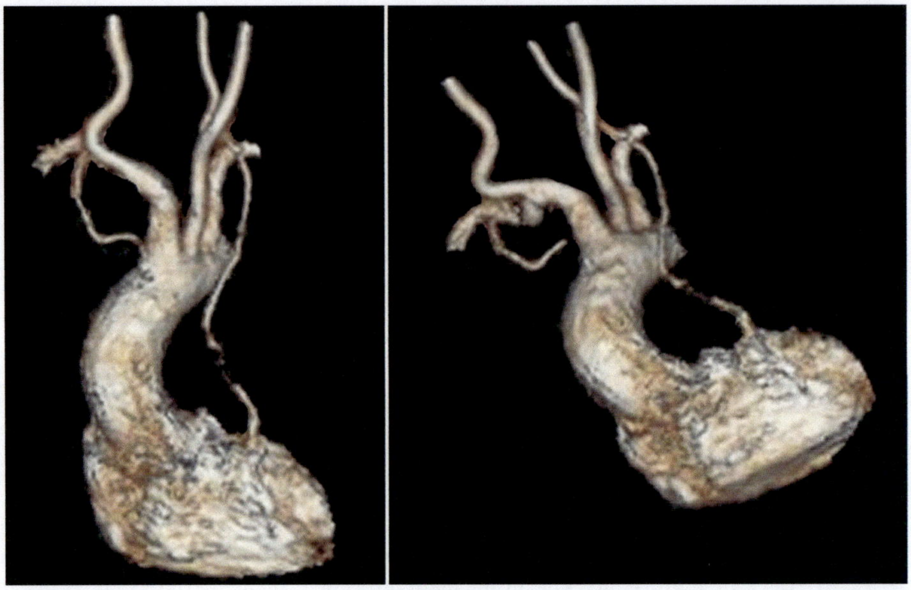

Fig. 9.6. Coronary MRA (CE-MRA technique) in a patient with two arteria bypass. The right internal mammarian artery which was supposed to support the right coronary artery is occluded after approximately 8 cm from the origin. The left internal mammarian artery connected to the left anterior descending artery shows a critical stenosis (subocclusion) in the distal part

et al. 1989; Von Smekal et al. 1997; Kessler et al. 1997; Langerak et al. 2002). The sensitivity in detecting occluded bypass is up to 100%; sensitivity in correctly describing a patent graft is slightly lower, 90% (Van Rossum et al. 1999) (Fig. 9.6).

In this field also, the different methods used to detect the vessel can give some variability in terms of results and in a recent paper a sensitivity of 83% and a specificity of 98% has been shown in the diagnosis of occluded bypass (Spuentrup et al. 2005). Furthermore, no multicenter data are available and no methodolocally consistent approach can be proposed. Finally, as for other non invasive techniques, the evaluation of existing bypass has a reduced value "per se" unless also the native coronary arteries are evaluated. For these reasons, although the capability of MRA in accurately describing the morphology of bypass has been proven, it can be considered still clinically non suitable.

9.6 Functional Evaluation of Coronary Arteries and Bypass

The residual function of the coronary artery can be effectively evaluated measuring the flow distal to the stent using phase-contrast velocity quantification at rest and during adenosine-stimulated hyperemia. Using this method and adopting a threshold of 1.2, a sensitivity of 83% with a specificity of 94% has been achieved for ≥75% stenoses (Nagel et al. 2003).

Flow velocity and volume flow in bypass grafts can be measured by applying velocity-encoded phase contrast cine MRI sequences, thus allowing assessment of graft function in addition to a morphologic evaluation (Sakuma et al. 1999; Ishida et al. 2001; Kreitner et al. 2000; Walpoth et al. 1999). Galjee et al. (1996) demonstrated that adequate velocity profiles throughout the cardiac cycle could be obtained in 85% of angiographically patent vein grafts, using non-breath-hold MR velocity mapping. Graft flow velocity was characterized by a biphasic pattern, with one peak in systole and a second peak in diastole.

Breath-hold segmented k-space sequences have been applied to measure flow velocity and volume flow at rest and after pharmacological stress. A preliminary report by Voigtländer et al. (2001) demonstrates the feasibility of measuring the flow reserve in bypass grafts and its potential to differentiate between patent and stenotic grafts. In 21 grafts without stenoses compared with 6 grafts with >75% stenoses, a flow velocity reserve was found to be 2.6 ± 1.5 vs. 0.8 ± 0.4 (p, 0.005) and a flow reserve to be 2.9 ± 1.9 vs. 1.2 ± 0.5 (p, 0.05). Using non-breath-hold and breath-hold techniques, functional results were also obtained in IMA grafts, despite the imaging artifacts due to metallic clips (Voigtländer et al. 2001; Debatin et al. 1993; Sakuma et al. 1996; Kawada et al. 1999). The diastolic/systolic peak velocity ratio was found to be higher in IMA grafts than in native IMA (Kawada et al. 1999). Findings obtained in a small cohort of patients with internal mammary artery grafts by (18 patients without and 5 patients with >75% stenosis) indicate a higher sensitivity and specificity for the detection of

IMA graft stenosis using measurements of the mean flow rate and diastolic/systolic peak flow ratio at rest, than using the flow reserve after administration of dipyridamole (Kawada et al. 1999).

9.7 Conclusion

While in other aspects of cardiology the use of MRI is nowadays considered an unavoidable step in the diagnostic process, in the field of coronary artery angiography the methodology still suffers unresolved technological limitations and cannot be proposed as a clinical tool to be used systematically for the diagnosis of atherosclerotic coronary artery disease. The current developments are very promising but still they are confined within a few research centers. The use of high-field scanners (3T or more) together with sophisticated multichannel surface coils and new acquisition sequences are expected to give a substantial improvement toward this direction. However, at present the only accepted use in the clinical setting is the detection of course anomalies and non atherosclerotic morphological disorders. Limited to this application, coronary MRA is competitive with regards to MDCT (Von Smekal et al. 1997). It is also worth noting that there are no substantial theoretical limitations in the physical domain to be overcome but a tenacious effort to solve the many problems still on the way is required.

References

Araoz PA, Glockner JF, McGee KP, et al. (2005) 3 Tesla MR imaging provides improved contrast in first-pass myocardial perfusion imaging over a range of gadolinium doses. J Cardiovasc Magn Reson 7(3):559–564

Aurigemma GP, Reichek N, Axel L, et al. (1989) Noninvasive determination of coronary artery bypass graft patency by cine magnetic resonance imaging. Circulation. 80(6): 1595–1602

Bluemke DA, Achenbach S, Budoff M, et al. (2008) Noninvasive coronary artery imaging: magnetic resonance angiography and multidetector computed tomography angiography: a scientific statement from the American heart association committee on cardiovascular imaging and intervention of the council on cardiovascular radiology and intervention, and the councils on clinical cardiology and cardiovascular disease in the young. Circulation 118:586–606

Bornstedt A, Hombach V, Kouwenhoven M, et al. (2008) Whole-heart coronary angiography at isotropic spatial resolution: high sense acceleration at 3T utilizing a 32 element cardiac receive coil. In: Proceedings of the 16th Annual Meeting of ISMRM, Toronto, Canada, p 941

Brenner P, Wintersperger B, von Smekal A, et al. (1999) Detection of coronary artery bypass graft patency by contrast enhanced magneticresonance angiography. Eur J Cardiothorac Surg 15(4):389–393

Debatin JF, Strong JA, Sostman HD, et al. (1993) MR characterization of blood flow in native and grafted internal mammary arteries. J Magn Reson Imaging 3:443–450

Foo TK, Ho VB, Saranathan M, et al. (2005) Feasibility of integrating high-spatial-resolution 3D breath-hold coronary MR angiography with myocardial perfusion and viability examinations. Radiology 235:1025–1030

Galjee MA, van Rossum AC, Doesburg T, et al. (1996) Value of magnetic resonance imaging in assessing patency and function of coronary artery bypass grafts: an angiographically controlled study. Circulation 93:660–666.

Gharib AM, Herzka DA, Ustun AO, et al. (2007 Oct) Coronary MR angiography at 3T during diastole and systole. J Magn Reson Imaging 26(4):921–926

Goldfarb JW, Edelman RR. (1998) Coronary arteries: breath-hold, gadolinium-enhanced, three-dimensional MR angriography. Radiology 206:830–834

Hardy CJ, Cline HE, Giaquinto RO, et al. (2006) 32-Element receiver-coil array for cardiac imaging. Magn Reson Med 55:1142–1149

Hoffman MBM, Henson RE, Kovaks SJ, et al. (1999) Blood poll agent strongly improves 3D magnetic resonance coronary angiography using an inversion pre-pulse. Magn Reson Med 41(2):360–367

Ishida N, Sakuma H, Cruz BP, et al. (2001) Mr flow measurement in the internal mammary artery-to-coronary artery bypass graft: comparison with graft stenosis at radiographic angiography. Radiology 220(2):441–447

Kawada N, Sakuma H, Cruz BC, et al. (1999) Noninvasive detection of significant stenosis in the coronary artery bypass grafts using fast velocity-encoded cine MRI. In: Book of Abstracts, 2nd Annual Meeting of the Society for Cardiovascular Magnetic Resonance, p 82

Kessler W, Achenbach S, Moshage W, et al (1997) Usefulness of respiratory gated magnetic resonance coronary angiography in assessing narrowings > or = 50% in diameter in native coronary arteries and in aortocoronary bypass conduits. Am J Cardiol 80(8): 989–993

Kim WY, Danias PG, Stuber M, et al. (2001) Coronary magnetic resonance angiography for the detection of coronary stenoses. N Engl J Med 345:1863–1869

Kreitner KF, Voigtländer T, Wittlinger T, et al. (2000) Flow quantification in coronary and bypass vessels with MR phase contrast technique Radiologe 40(2):143–149

Langerak SE, Vliegen HW, de Roos A, et al. (2002) Detection of vein graft disease using high-resolution magnetic resonance angiography. Circulation 105(3):328–333

Li D, Dolan RP, Walkovitch RC, et al. (1998) Three-dimensional MRI of coronary arteries using an intravascular contrast agent. Magn Reson Med 39:1014–1018

Nagel E, Thouet T, Klein C, et al. (2003) Noninvasive determination of coronary blood flow velocity with cardiovascular magnetic resonance in patients after stent deployment. Circulation 107:1738–1743

Nguyen TD, Spincemaille P, Prince MR, et al. (2006) Cardiac fat navigator-gated steady-state free precession 3D magnetic resonance angiography of coronary arteries. Magn Reson Med 56:210–215

Niendorf T, Hardy C., Giaquinto RO, et al. (2006) Toward single breath-hold whole-heart coverage coronary MRA using highly accelerated parallel imaging with a 32-channel MR system. Magn Reson Med 56:167–176

Okada T, Kanao S, Ninomiya A, et al. (2009) Whole-heart coronary magnetic resonance angiography with parallel imaging: comparison of acceleration in one-dimension vs. two-dimensions. Eur J Radiol doi:10.1016/j.ejrad.2008.06.005

Paulin S, von Schulthess GK, Fossel E, et al. (1987) MR imaging of the aortic root and proximal coronary arteries. Am J Roentgenol 148(4):665–670

Redberg RF, Walsh J (2008) Pay now, benefits may follow–the case of cardiac computed tomographic angiography. N Engl J Med 359(22):2309–2311

van Rossum AC, Bedaux WLF, Hofman MBM (1999) Morphologic and functional evaluation of coronary artery bypass conduits. J Magn Reson Imaging 10:734–740

Sakuma H, Globits S, O'Sullivan M, et al. (1996) Breath-hold MR measurements of blood flow velocitiy in internal mammary arteries and coronary artery bypass grafts. J Magn Reson Imaging 6:219–222

Sakuma H, Ichikawa Y, Suzawa N, et al. (2005) Assessment of coronary arteries with total study time of less than 30 minutes by using whole-heart coronary MR angiography. Radiology 237(1):316–321

Sakuma H, Kawada N, Takeda K, et al. (1999) MR measurement of coronary blood flow. J Magn Reson Imaging 10(5): 728–733

Santos JM, Cunningham CH, Lustig M, et al. (2006) Single breath-hold whole-heart MRA using variable-density spirals at 3T. Magn Reson Med 55(2):371–379

Shankaranarayanan A, Fung M, Beatty P, et al. (2008) 128-Channel highly-acceerated breah-held 3D coronary MR imaging. In: Proceedings of the 16th Annual Meeting of ISMRM, Toronto, Canada, p 314

von Smekal A, Knez A, Seelos KC, et al. (1997) A comparison of ultrafast computed tomography, magnetic resonance angiography and selective angiography for the detection of coronary bypass patency. Rofo Fortschr Geb Rontgenstr Neuen Bildgeb Verfahr 166(3):185–191

Spuentrup E, Katoh M, Buecker A, et al. (2004) Free-breathing 3D steady-state free precession coronary MR angiography with radial k-space sampling: comparison with cartesian k-space sampling and cartesian gradient-echo coronary MR angiography-pilot study. Radiology 231:581–586

Spuentrup E, Ruebben A, Mahnken A, et al. (2005) Artifact-free coronary magnetic resonance angiography and coronary vessel wall imaging in the presence of a new, metallic, coronary magnetic resonance imaging stent. Circulation 111: 1019–1026

Stehning C, Bornert P, Nehrke K, et al. (2004) Fast isotropic volumetric coronary MR angiography using free-breathing 3D radial balanced FFE acquisition. Magn Reson Med 52(1):197–203

Stehning C, Bornert P, Nehrke K, et al. (2005) Free-breathing whole-heart coronary MRA with 3D radial SSFP and self-navigated image reconstruction. Magn Reson Med 54: 476–480

Stuber M, Botnar RM, Danias PG, et al. (1999a) Submillimiter three-dimensional coronary MR angiography with real-time navigator correction: comparison of navigator locations. Radiology 212:579–587

Stuber M, Botnar RM, Danias PG, et al. (1999b) Contrast agent-enhanced free breathing, three dimensional coronary magnetic resonance angiography. J Magn Reson Imaging 10(5): 790–799

Voigtländer T, Kreitner KF, Wittlinger T, et al. (2001) MR angiography and flow measurement in coronary arteries and coronary bypass grafts Z Kardiol 90(12):929–938

Walpoth BH, Müller MF, Genyk I, et al. (1999) Evaluation of coronary bypass flow with color-Doppler and magnetic resonance imaging techniques: comparison with intraoperative flow measurements. Eur J Cardiothorac Surg 15(6): 795–802

Weber OM, Martin AJ, Higgins CB (2004) Whole-heart steady-state free precession coronary artery magnetic resonance angiography. Magn Reson Med 50(6):1223–1228

Wittlinger T, Martinovic I, Noeske R, et al. (2005) High-field MR angiography on an in vitro stenosis model determination of the spatial resolution on 1.5 and 3T in correlation to flow velocity and contrast medium concentration. J Cardiovasc Magn Reson. 7(4):623–630

White RD, Caputo GR, Mark AS, et al. (1987) Coronary artery bypass graft patency: noninvasive evaluation with MR imaging. Radiology 164(3):681–686

Abdominal Aorta and Renal Arteries

10

Virna Zampa, Francesca Turini, Lorenzo Faggioni, and Gabriele Caproni

CONTENTS

10.1 Introduction *115*

10.2 Technical Approaches *116*
10.2.1 CEMRA *116*
10.2.2 Phase Contrast *116*
10.2.3 Perfusion *117*
10.2.4 Balanced Fast Gradient Echo *117*
10.2.5 Arterial Spin Labeling (ASL) *117*

10.3 Clinical Applications *117*
10.3.1 Abdominal Aorta *117*
10.3.1.1 Atherosclerotic and Inflammatory Aneurysms *117*
10.3.1.2 Infected Aneurysm *118*
10.3.1.3 Dissection *119*
10.3.1.4 Aortic Occlusion and Leriche's Syndrome *119*
10.3.1.5 Control of Vascular Stents and Endoleak *119*
10.3.1.6 Retroperitoneal Fibrosis and Erdheim–Chester Disease *120*
10.3.2 Renal Arteries *121*

References *125*

Virna Zampa, MD
Gabriele Caproni, MD
Francesca Turini, MD
Lorenzo Faggioni, MD
Diagnostic and Interventional Radiology, Department of Oncology, Transplants and Advanced Technologies in Medicine, University of Pisa, Via Roma 67, 56100 Pisa, Italy

ABSTRACT

The recent advent of newer multidetector computer tomography (MDCT) scanners has dramatically changed the role of magnetic resonance imaging (MRI) in the evaluation of the abdominal aorta and its peripheral branches, due to its faster acquisition times, higher spatial resolution allowing to obtain submillimetric sections with voxel isotropy, greater technical simplicity, wider availability, and lower overall cost. However, lack of ionizing radiation and of intravenous administration of iodinated contrast material remain substantial advantages of MRI over MDCT, that make it the technique of choice for young patients and individuals with high serum creatinine values. Moreover, for the assessment of renal arteries MRI offers not only adequate anatomical detail but, and more importantly, is also able to provide quantitative data relative to renal blood flow, as well as to yield functional information about the parenchymal effect of a vascular stenosis by means of phase contrast and perfusion techniques, respectively. In this particular concern, MRI may be successfully proposed as a novel noninvasive tool for functional evaluation of the kidneys.

10.1 Introduction

While the widespread use of multislice computed tomography (MSCT) has dramatically changed the role of magnetic resonance imaging (MRI) in the assessment of abdominal aorta, the evaluation of renal arteries still remains an important application of MRI.

The higher speed of MSCT, its excellent spatial resolution with the possibility of achieving voxel isotropy, and its wider volume coverage make it more attractive than MRI for studying the abdominal aorta and its pathology, allowing the assessment of peripheral visceral branches with greater accuracy than contrast-enhanced MR angiography (CEMRA). This is particularly true for preinterventional evaluation of aortic aneurysms, in which morphological data must be extremely accurate and the depiction of calcified plaques of the aortic wall is a crucial diagnostic element. Therefore CT angiography (CTA) has taken over the prime role in imaging of the abdominal aorta, while CEMRA has become a second-choice imaging modality in this clinical setting. Nevertheless, the lack of radiation and of administration of iodinated contrast agent associated with MRA make it the technique of choice for young patients and individuals with high serum creatinine values (Böckler et al. 2007; Shih and Hagspiel 2007).

Conversely, for the assessment of renal arteries MRI offers not only acceptable anatomical details of vessels but, and more importantly, is also able to provide quantitative data relative to renal flow and is able to assess the parenchymal effect of a vessel stenosis by means of phase contrast and perfusion techniques. The feature of functional assessment of the kidneys makes MRI preferable to CT.

10.2 Technical Approaches

10.2.1 CEMRA

CEMRA still represents the elective, widely used technique for studying the aorta and its main branches, providing images with high diagnostic value.

Technical advances, such as parallel imaging and k-space sharing methods (time-resolved imaging of contrast kinetics, or TRICKS), have allowed a further increase in the spatial and temporal resolution of the technique.

When studying the aorta or the renal arteries, images are acquired on the coronal plane and CEMRA is usually associated with fat saturation to increase vessel conspicuity and to avoid the problems connected with image subtraction, such as imperfect overlap between unenhanced and contrast-enhanced series caused by patient motion.

Proper bolus timing is crucial for obtaining high quality CEMRA images, as it is essential to synchronize the sampling of the centre of the k-space (which retains information about contrast resolution) with the peak intravascular concentration of gadolinium. Indeed, too early sampling of the k-space centre leads to poor enhancement of the vascular lumen and ringing artefacts (also called the Maki effect), while a late acquisition will result in suboptimal arterial enhancement with venous overlap. Therefore, contrast material administration for CEMRA now relies on the test bolus technique or real-time bolus detection tools (such as fluoroscopic triggering or automatic signal intensity detection system [Smart Prep]).

Contrast material is usually administered at a flow-rate of 2 mL/s by means of an automatic power injector, which provides more precise control over injection parameters compared to manual injection.

The sampling scheme of the k-space also plays an important role in maximizing contrast resolution while reducing imaging time, thereby optimizing usage of contrast material. This goal is accomplished by using centric or elliptical k-space filling schemes, in which the centre of the k-space is sampled at the beginning of the acquisition, corresponding to the peak intravascular concentration of contrast material.

Currently, the adoption of 3 T MR equipment is increasing and angiographic exams will especially benefit directly from the higher field strength. The theoretically doubled signal-to-noise ratio at 3 T allows for abdominal MRA exams with sub-millimetre spatial resolution and very short acquisition time. Furthermore, as longitudinal relaxation time is longer at 3 T compared with lower field strength, MRA exams can be performed with a significantly reduced amount of contrast material (Michaely et al. 2007).

10.2.2 Phase Contrast

The information provided by CEMRA can be further enriched by Phase-Contrast (PC) technique.

PC exploits the spin-induced phase shift to differentiate flowing blood from stationary tissue. Based on the phase accumulation of the spins, it is possible to calculate the speed of spins and flow direction. For PC-MRA, signal intensity reflects the absolute velocity of voxels, and it is always positive and independent from the direction of flow.

The three-dimensional (3D) PC-MRA has the advantage of high sensitivity to turbulent flow that can be associated with vascular stenoses. It is mainly used to study renal vessels, because it immediately reveals the hemodynamic significance of a stenosis caused by the flow dropout related to turbulent flow.

The bidimensional (2D) no-breath cine PC technique has been used to calculate the flow value obtaining flow curves. The limited spatial and temporal resolution of the cine PC technique has been overcome by the fast implementation of the sequence with k-space segmentation, in which more phase encodings per cardiac cycle (views per segment) are acquired. This sequence is performed on breath-hold on a plane perpendicular to the target vessel. Images are reconstructed on an independent workstation by means of automatic detection of the area of the vessel under investigation, thus obtaining graphs that display the velocity and the flow as a function of time.

PC sequences are a useful complement to MRA morphological data in assessing the hemodynamic significance of a stenosis.

10.2.3 Perfusion

The perfusion technique has been widely used for studying different anatomical districts, especially the heart and the brain. Nowadays, its application has been extended to other anatomical districts, with a high interest in evaluating renal function.

In the literature, different technological approaches have been proposed; an important technical requirement is high spatial and temporal resolution. In particular, a temporal resolution of one acquisition per second is recommended for detecting the first pass of contrast material.

Perfusional images have to be analyzed by dedicated software aimed at providing quantitative data.

10.2.4 Balanced Fast Gradient Echo

[Steady state free precession (SSFP) – FIESTA – true FISP]

Balanced SSFP sequences have enjoyed a recent resurgence of interest across a wide range of applications. The technique can be used for angiographic studies because image contrast is determined by the T2/T1 ratio, yielding bright blood imaging without reliance on inflow.

Some limitations are the susceptibility to field inhomogeneities, such as at air-tissue interfaces or in the presence of metallic implants; moreover, the sequence is not specific for arteries or veins.

10.2.5 Arterial Spin Labeling (ASL)

The ASL technique relies on the replacement of blood in the imaging volume with tagged blood. Protons are "tagged" as they experience an inversion pulse in a target region, typically in a blood vessel upstream of the imaging field of view. After a fixed interval or delay time, when tagged protons enter the field of view, imaging is performed. The time delay depends on the distance between tagged and imaged areas and the rate of blood flow. This process is repeated, with alternate acquisition of tagged and untagged images, and the two sets are then subtracted from each other, thus resulting in a bright-blood angiogram with no background signal.

The combination of ASL with tagged and untagged balanced SSFP provides bright-blood, venous-free angiographic images with high signal-to-noise ratio. This approach is particularly well suited to evaluating renal vessels, because the sequence is flow-compensated in all three spatial directions and, therefore, the complex orientation of the aorta, as relative to the renal arteries, can be well depicted.

Because the ASL technique is based on continuous refilling of blood in the imaging volume, the method requires reasonably high arterial velocity to be successful (Miyazaki and Lee 2008).

10.3 Clinical Applications

10.3.1 Abdominal Aorta

10.3.1.1 Atherosclerotic and Inflammatory Aneurysms

Most aneurysms in the abdominal aorta are atherosclerotic and are usually located below the origin of renal arteries. They usually have a fusiform shape and, in rare cases, a saccular morphology.

The radiological findings of a vessel aneurysm are well-known and crucial for a correct presurgical evaluation. This requires the exact measurement of the maximum diameter of the aneurysm and of its superior

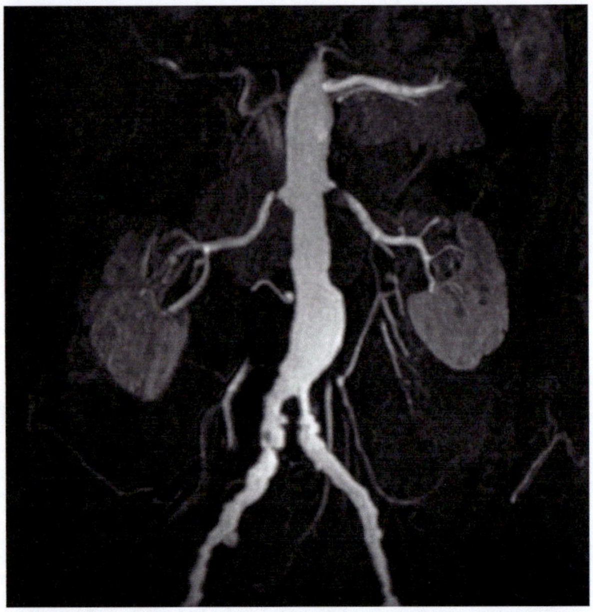

Fig. 10.1. Aortic aneurysm. The MIP coronal reconstruction shows an infrarenal aortic aneurysm as well as stenoses of both the renal arteries and diffuse atherosclerosis of the iliac arteries

neck, the precise definition of its longitudinal extension, the caliber of the patent lumen in case of thrombotic wall deposit, the spatial relationship with the renal and visceral arteries, as well as the evaluation of iliac vessels. These measurements must be taken along the plane perpendicular to the vessel's greater axis.

In addition, evidence of rupture, complications of the aortic wall, supernumerary renal arteries, obstructive disease of renal, celiac, or mesenteric vessels, or an anatomical anomaly (such as a horseshoe kidney) may significantly modify the surgical plan (Ho and Corse 2003). MIP and MPR reconstructions are particularly helpful for the identification of stenotic branch vessels, which should preferably be revascularized at the time of the aortic aneurysm repair (Fig. 10.1).

Gadolinium-enhanced 3D MRA has been shown to be accurate for the detection of significant occlusive celiac, renal, mesenteric, or iliac arterial disease (94% sensitivity and 98% specificity) (Ho and Corse 2003). The critical diagnostic point is represented by the assessment of the presence and extension of wall calcification that requires a CT exam, since this element may condition the therapeutic choice. Calcification can actually hinder clamping and suturing and impair the mechanical support to be provided to an endoprosthesis.

In view of a percutaneous treatment with endoprosthesis placement, the overall evaluation of the iliac-femoral vessels is required, since tortuous courses or stenoses may jeopardize the technical outcome of the procedure.

Five to twenty-three percent of all aneurysms of the abdominal aorta are inflammatory. By the term "inflammatory" we refer to several histopathological findings: thickening and fragmentation of the elastic lamina, loss of smooth muscle cells from the media, presence of dense neovascularized connective tissue surrounding the aortic wall and seeping of lymphocytes, plasma cells, and histiocytes; furthermore, there may be adherences to visceral organs (duodenum, ureters) (Nitecki et al. 1996).

In MR images, the aneurysmatic inflammatory reaction can be recognized by high signal intensity on T2-weighted images, which can be better appreciated with fat suppression, and by marked enhancement following the administration of a contrast agent (Fig. 10.2).

10.3.1.2
Infected Aneurysm

An infected or mycotic aneurysm is defined as an infectious break in the wall of an artery with the formation of a blind, saccular outpouching that is contiguous with the arterial lumen (Kaufman et al. 1978) and usually occurs in portions of the aorta that are not commonly affected by atherosclerosis.

Early changes of aortitis preceding aneurysm formation include an irregular arterial wall, periaortic edema, periaortic soft-tissue mass, and periaortic gas.

On MRI, periaortic edema is characterized by high signal intensity on T2-weighted images that are better appreciable when fat saturation is applied. Concentric or eccentric periaortic inflammatory soft-tissue may develop and usually exhibits uniform enhancement (Lee et al. 2008).

Necrosis may develop from the inflammatory mass and shows heterogeneous signal intensity on MRI, together with a rim of poor enhancement after administration of contrast material. Periaortic mass or stranding is the most common imaging finding with infected aortic aneurysms and is found in 48% of cases (Macedo et al. 2004), while periaortic gas is an uncommon feature.

Lack of or delayed treatment of infected aneurysms may lead to fulminant sepsis, spontaneous arterial rupture, and death. Early detection of infected aneurysms is critical for timely treatment aimed at optimizing patient outcome (Lee et al. 2008).

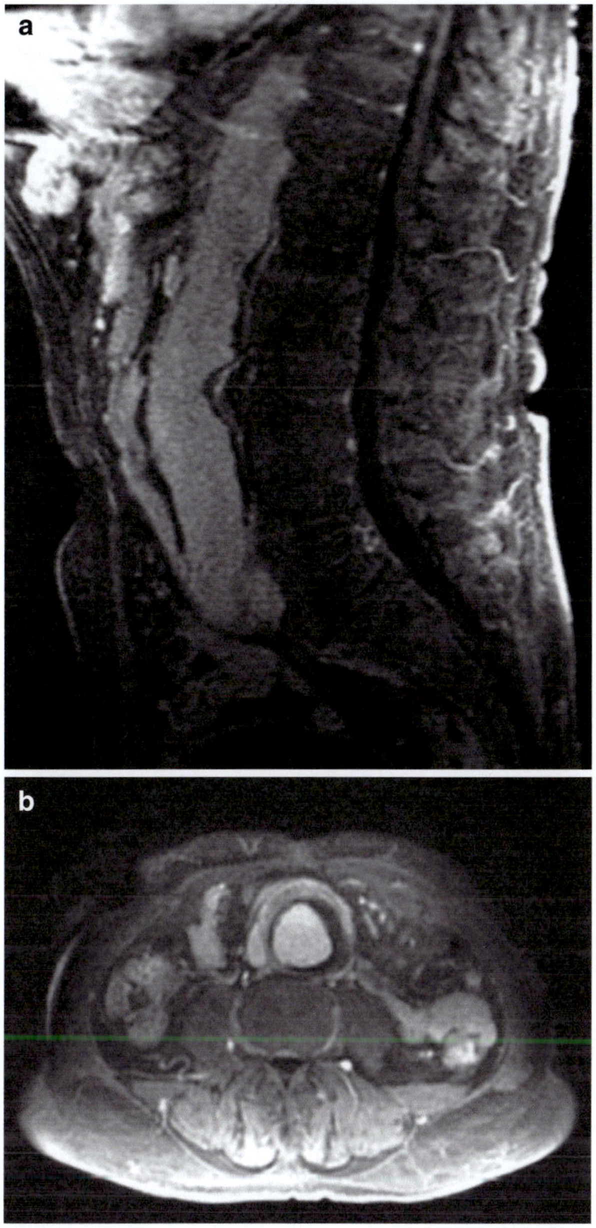

Fig. 10.2. Inflammatory aortic aneurysm with the typical feature of periaortic tissue showing contrast enhancement on gadolinium-enhanced LAVA images on the sagittal and axial planes (**a**, **b**)

10.3.1.3
Dissection

Dissections arise much more frequently in the thoracic aorta and often extend inferiorly into the abdominal aorta involving the renal, celiac, mesenteric, and iliac arteries. The main concern with dissections is the involvement of visceral branch vessels, which may result in their obstruction.

CEMRA can display the extent of the dissection and the involvement of branch vessels with exquisite clarity. At least two delayed phase contrast-enhanced acquisitions are advisable to guarantee optimal visualization of the false channel, where blood flow is much slower than in the true lumen (McGuigan 2005).

MPR of the 3D datasets enables selective view of individual aortic branch vessels and the identification of their blood supply (i.e., true vs. false channel). The extension of an intimal tear into the abdominal aorta typically spirals posterior laterally about the arch with the false channel coursing to the left of the aorta, potentially to involve the left renal artery and possibly the celiac and superior mesenteric arteries. Delayed-phase imaging is recommended because the flow within the false channel may be slow and not adequately mixed with contrast medium during the initial acquisition (Ho and Corse 2003).

10.3.1.4
Aortic Occlusion and Leriche's Syndrome

Abdominal aortic occlusion may occur as a result of several causes, but in the acute setting, it most commonly results from embolism. Thrombosis superimposed on severe atherosclerotic involvement of the distal abdominal aorta and common iliac arteries is the main cause of chronic occlusion. Chronic occlusion may produce Leriche's syndrome, which refers to the clinical syndrome that results from occlusion of the infrarenal aorta. The syndrome is characterized by the absence of femoral pulses, gluteal claudication, impotence, and, when symptoms persist over time, by hypotrophy of the muscular masses of the lower limbs.

Invariably, patients with chronic occlusion develop a rich collateral circulation.

In these individuals arterial access is limited and CEMRA is sufficient for the primary assessment of the occlusion (Ho and Corse 2003).

10.3.1.5
Control of Vascular Stents and Endoleak

The degree of safety and reliability of CEMRA in evaluating the positioning of aortic prostheses has been demonstrated since the introduction of prostheses made of MR-compatible materials. Stents made of nitinol are generally more suitable for MR imaging; elgiloy stents may obscure the lumen, and

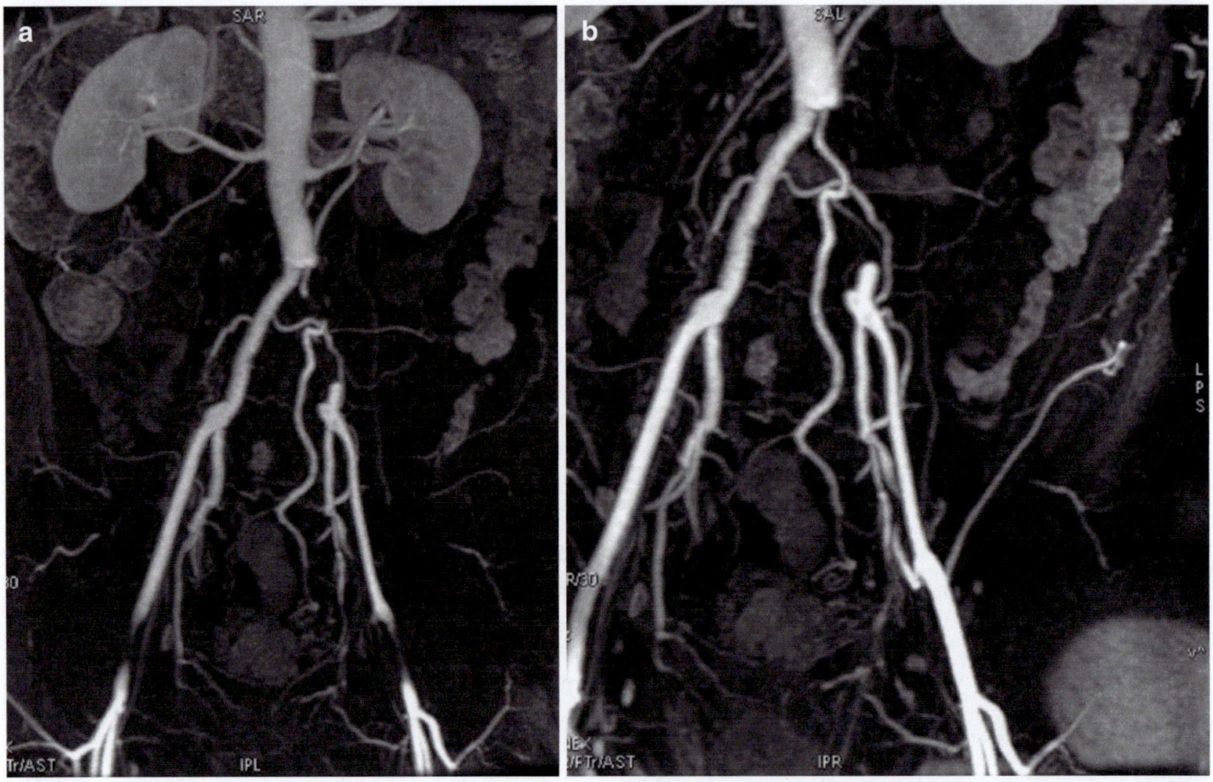

Fig. 10.3. (**a**, **b**) Obstruction of left iliac branch of aorto-bisiliac prosthesis, as depicted from CEMRA images (MIP reconstructions). Notice the presence of retroperitoneal and inferior epigastric collateral pathways

stainless steel stents cause extensive artefacts that render the study nondiagnostic (Stavropoulos and Charagundla 2007). In several studies involving small numbers of patients with predominantly nitinol stents, MRA was at least as sensitive as CTA, and in some cases, highlighted endoleaks that were not detected by CTA.

The protocol for the MRI study of prostheses and stents positioned at the level of the abdominal aorta requires the use of axial sequences performed before and after administration of contrast material. Such a procedure allows the evaluation of retroperitoneal peri-aortic tissues, and the correct diagnostic assessment of potential complications (such as periprosthesic hematoma, abscess, thrombosis, or pseudoaneurysm). CEMRA examinations feature high sensitivity as required for the evaluation of graft patency and for the detection of prosthetic deformation. In addition, they can provide a complete visualization of the entire aorto-iliac axes and their collaterals (Fig. 10.3). CEMRA is also accurate in the depiction of endoleaks, and new techniques, such as time-resolved MR angiography, may allow a better characterization of the endoleak type by demonstrating the gradual passage of contrast material into the aneurysm sac, analogous to a conventional angiogram (Stavropoulos and Charagundla 2007). Moreover, the availability of blood pool contrast media has opened new perspectives for studying endoleaks; in particular, type IV endoleaks, which are caused by stent-graft porosity and may take advantage of blood pool contrast media. By using such contrast agents, which remain intravascularly longer than ordinary intravascular-interstitial contrast agents, the temporal window between contrast medium injection and imaging can be extended. Such a long delay allows the accumulation of a higher amount of contrast material in the endoleak while maintaining adequate contrast resolution between the blood and the surrounding tissues (Cornelissen et al. 2008).

10.3.1.6 Retroperitoneal Fibrosis and Erdheim-Chester Disease

Retroperitoneal fibrosis is characterized by a proliferation of fibrous tissue extending along the aorta through a plaque-like infiltrative soft-tissue process that is typically localized to the distal abdominal (infrarenal) aorta and the common iliac arteries; involvement of the pelvis is uncommon. About two-thirds of cases are idiopathic.

This entity may be considered as an isolated disease or as part of a systemic syndrome known as multifocal fibrosclerosis, which may also include autoimmune pancreatitis, sclerosing cholangitis, scleroderma, Riedel thyroiditis, fibrotic pseudotumor of the orbit, and fibrosis involving multiple organ systems. Typical imaging manifestations of perirenal fibrosis include a soft-tissue mass that envelops the kidneys without displacing them.

Three stages of the disease have been described, ranging from chronic active inflammation to fibrous scarring. On MRI signal intensity on T2-weighted images and contrast enhancement on dynamic acquisitions depend on the disease stage. While areas affected by active inflammation demonstrate high signal intensity on T2-weighted images and early contrast enhancement, areas of fibrosis show low signal intensity on T2-weighted images and delayed contrast enhancement.

Erdheim-Chester disease is a rare form of systemic non-Langerhans cell histiocytosis of unknown etiology. Histologically, this disease is defined by a mononuclear infiltrate consisting of lipid-laden, foamy histiocytes. It affects middle-aged individuals without any specific sex predilection; lower extremity bone pain is a typical symptom.

Retroperitoneal involvement has been noted in approximately a third of patients. The fibrotic involvement of the retroperitoneum demonstrates decreased signal intensity in T1-weighted, T2-weighted, fast inversion recovery, gradient-echo chemical shift in-phase/out-of-phase, and gadolinium-enhanced T1-weighted images (Fig. 10.4) (Gottlieb and Chen 2002).

Perirenal involvement in Erdheim-Chester disease is characteristic and shows as rind-like soft-tissue lesions surrounding the kidneys and ureters (Murray et al. 2001). Severe compression of the renal parenchyma and ureters as a consequence of fibrous perinephritis may lead to progressive renal failure (Surabhi et al. 2008).

Skeletal involvement, characterized by symmetric sclerosis of the meta- and diaphyses of long tubular bones, is a typical sign. Other extraskeletal manifestations, including central nervous system, lungs, skin, and heart involvement, are seen in about half of patients (Surabhi et al. 2008).

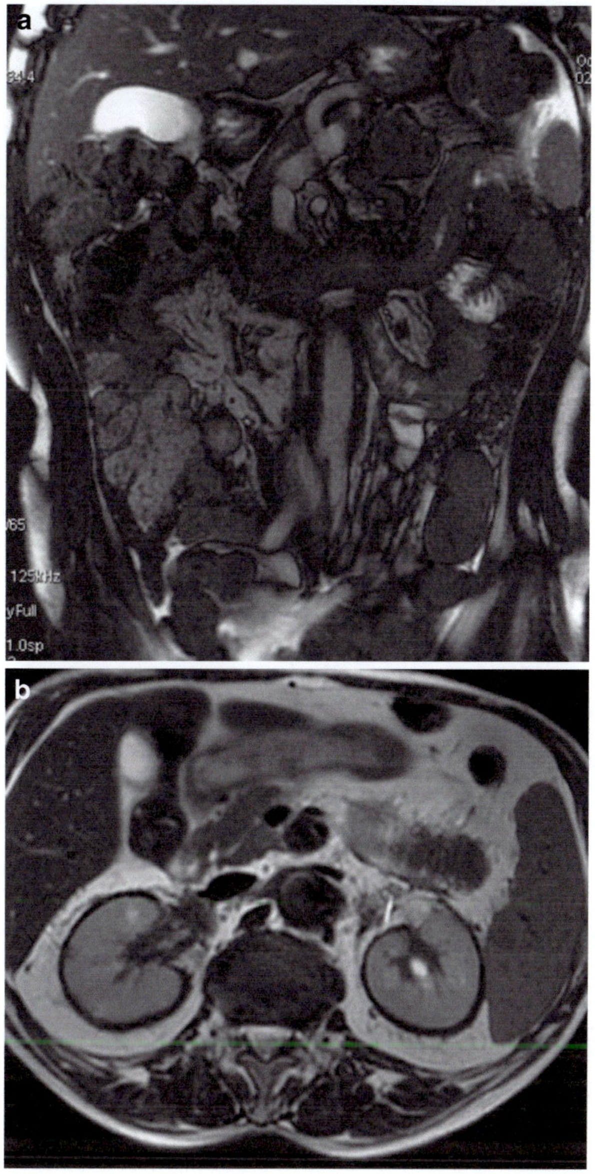

Fig. 10.4. Erdheim-Chester disease: (**a**) coronal FIESTA image shows periaortic tissue with low signal intensity. (**b**) Axial FSE T2-weighted image shows the typical perirenal low band. Also notice periaortic tissue surrounding the aorta

10.3.2 Renal Arteries

Renal artery stenosis (RAS) frequently manifests clinically as systemic hypertension (70%), which can often be reversed by renal revascularization using balloon angioplasty, stenting, or vascular surgery. Patients with RAS may progress to end-stage renal disease if left untreated (Ho and Corse 2003).

Atherosclerotic disease is the most common pathologic condition of RAS, which typically involves the

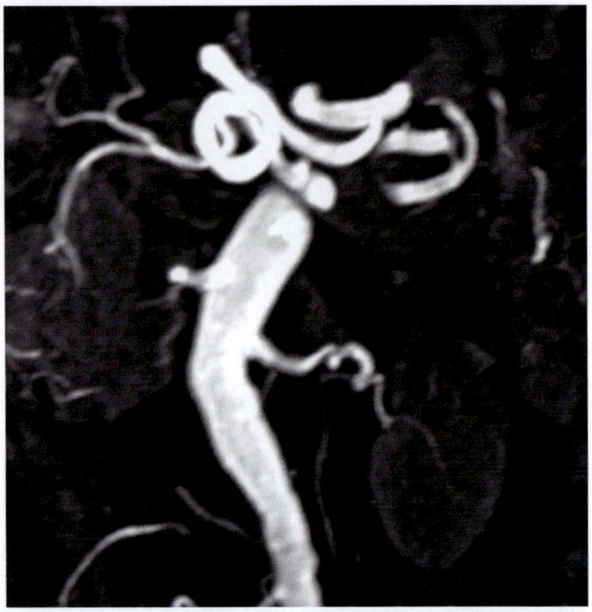

Fig. 10.5. Maximum Intensity Projection (MIP) reconstruction shows involvement of the middle tract of the left renal artery and a small aneurysm, both typical signs of fibromuscular dysplasia

ostium or the proximal 1–2 cm of the renal arteries, and usually also affects the abdominal aorta. Fibromuscular dysplasia is the second most common cause of RAS. This condition typically involves the distal two-thirds of the main renal artery with bead-like stenosis alternating with small fusiform or saccular aneurysms (Fig. 10.5).

Like atherosclerotic RAS, fibromuscular dysplasia may result in reversible systemic hypertension. Patients with fibromuscular dysplasia, however, are more typically young women, whereas atherosclerosis tends to occur in older men.

The preoperative identification of RAS is important and may augment or change the surgical plan (Ho and Corse 2003).

The MRI protocol for the examination of the renal arteries includes the 3D PC and 3D CEMRA techniques. The first sequence gives diagnostic elements relative to the hemodynamic significance of a stenosis, while the second one is useful for morphological depiction and grading of vessel stenosis.

The MRA assessment of hemodynamic significance of RAS can be further refined by considering additional factors, such as arterial stop of signal, post-stenotic dilatation, delayed renal enhancement, and functional changes in the renal parenchyma (i.e., reduced kidney longitudinal length and parenchymal thickness, and loss of corticomedullary differentiation).

A precise evaluation of the degree of stenosis requires the use of dedicated software.

3D PC images are acquired in the axial plane, with the slab positioned on a coronal plane to cover the renal arteries. An inferior saturation band is applied to null signal from the venous flow.

On PC MRA, spin dephasing is invariably present in hemodynamically significant RAS (Ho and Corse 2003).

As with digital angiography, CEMRA is a luminography, and as such provides exclusively morphological information. Data from the literature indicate that CEMRA has 97% sensitivity and 93% specificity in the grading of RAS. However, CEMRA alone has a tendency to overestimate stenosis, as has been documented in several other vascular districts. This overestimation is caused by the limited spatial resolution of CEMRA as compared to digital angiography (which remains the diagnostic gold standard), as well as by artefacts related to bad contrast bolus timing (blurring and ringing artifacts), and an intravoxel dephasing component that, albeit to a lesser extent than for PC sequences, also affects CEMRA (Fig. 10.6).

CEMRA images are usually acquired on a coronal plane in order to minimize acquisition time by reducing the number of phase encoding steps. Usage of a phased array body coil with a wide cranio-caudal coverage (such as a torso or a cardiac coil) is recommended so as to ensure adequate panoramicity, potentially allowing to assess accessory renal arteries stemming from the distal aorta or the iliac arteries, as well as to keep imaging time down to a reasonable level. The CEMRA acquisition is carried out in a single breath-hold with a high matrix in slice and frequency direction, with at least 160 phase-encoded steps as a compromise between sufficiently short imaging time and adequate spatial resolution. A slice thickness of usually 2–3 mm with partial overlap (typically 50%) yields good spatial resolution while maintaining adequate signal-to-noise ratio and acquisition time. The k-space is usually sampled with a centric filling scheme.

In addition, MRI-based flow measurements can be performed to evaluate the hemodynamic significance of RAS. The new ultrafast ECG-triggered PC sequence is able to provide quantitative measurements of the renal flow for each renal artery by

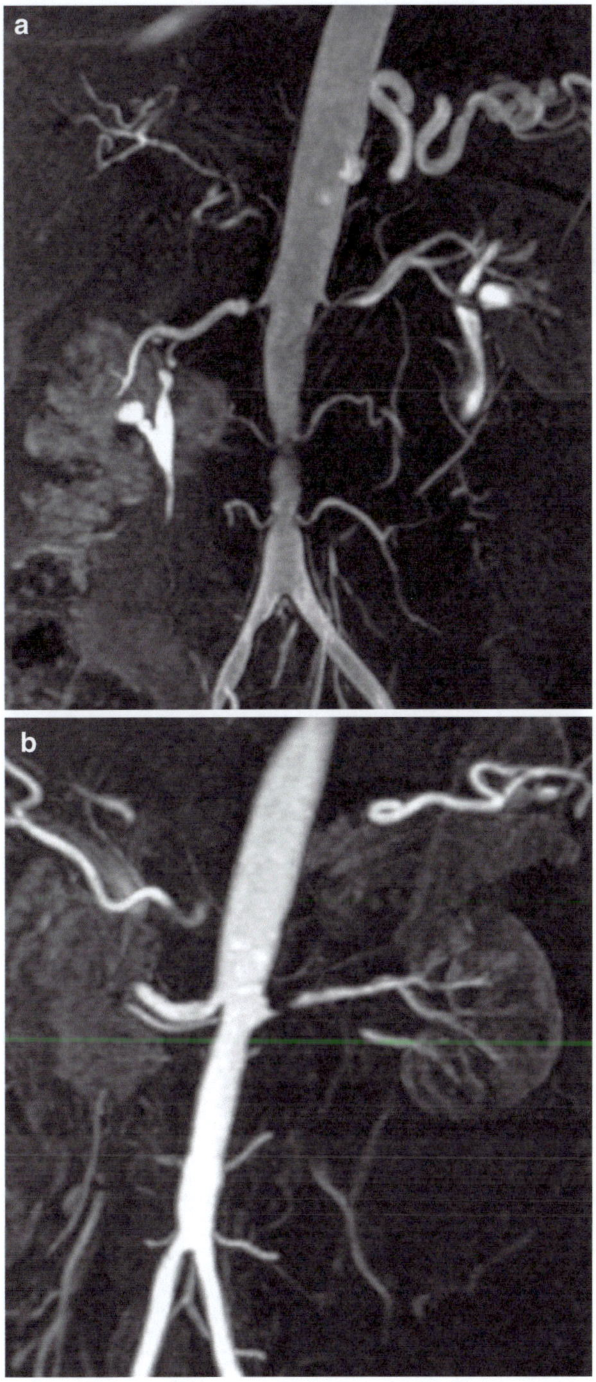

Fig. 10.6. RAS depicted with CEMRA. (**a**) Severe stenosis with post-stenotic dilation of right renal artery, and severe stenosis of the left renal artery. (**b**) Severe stenosis of the left renal artery overestimated because of the intravoxel dephasing. Both dephasing and post-stenotic dilation are indicative of severe, hemodynamically significant vascular stenosis

calculating renal blood flow as a function of time (Fig. 10.6). For the assessment of the renal artery, the acquisition plane is placed 1 cm past the ostium or downstream of a stenosis; the calculation of the flow value is obtained by multiplying the average velocity by the vessel surface area. In this respect, it is noteworthy that the renal volume and flow ratio seem to have a prognostic value in predicting the outcome of angiographic percutaneous procedures.

One of the most recent and promising implementations of MRI for functional evaluation of the kidneys is MR perfusion imaging (PWI). Its applications include the evaluation of segmental RAS, the assessment of in-stent restenosis, or the detection of parenchymal disease in the absence of RAS.

Imaging is performed in the oblique coronal plane, with a coverage including the abdominal aorta and both kidneys. Signal intensity (SI)-time curves can be obtained from a region of interest over the renal cortex, consisting of two parts that represent the first pass of the contrast agent through the renal circulation and its excretion phase, respectively. A normal SI-curve is characterized by three distinct morphological features: a steep upslope, an early SI peak, and a slow signal decline.

A gamma-variate fit is required to take interstitial dispersion of contrast material into account when an intravascular-interstitial contrast agent is used, as usual in clinical practice. The gamma-variate function has useful analytical properties that allow to derive semiquantitative perfusion parameters from signal intensity-time curves, such as time to peak (TTP), mean transit time (MTT), maximum upslope (MUS), and maximum signal intensity (MSI). It has been shown that a modification of these parameters reflects an alteration of renal perfusion, such as in patients with hemodynamically significant RAS. For instance, in patients with hemodynamically significant RAS an increase of TTP and MTT occurs (indicating that longer time is needed for the contrast bolus to pass through the renal vasculature), together with lower MUS and MSI values (which reflect decreased parenchymal perfusion). This results in an altered shape of the perfusion curve in high grade RAS, characterized by a slower upslope, flattened peak, and delayed signal decay (Fig. 10.7).

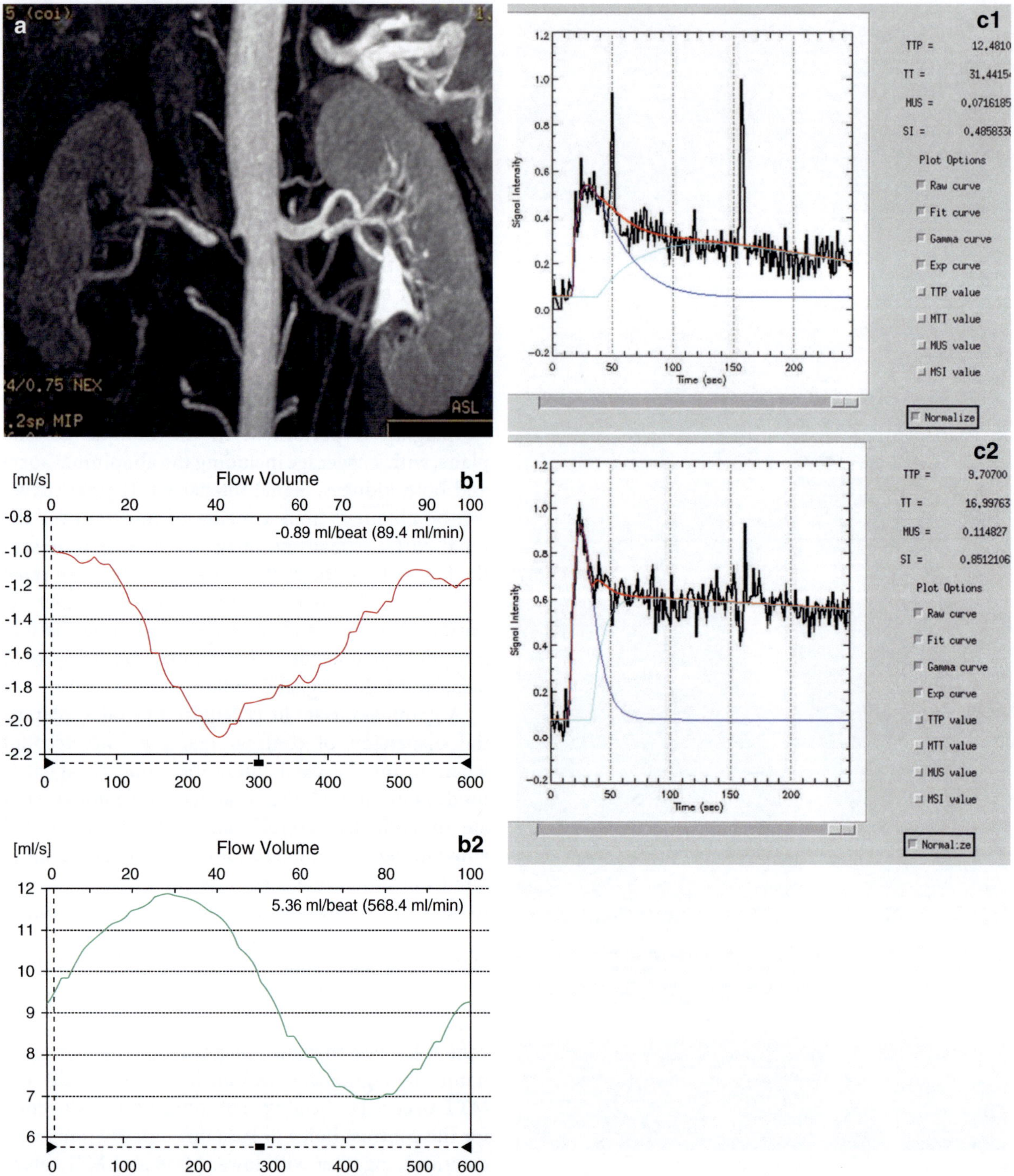

Fig. 10.7. On CEMRA MIP (**a**) a severe stenosis of the right renal artery. Flow analysis of renal arteries with ECG-gated PC technique reveals a substantial decrease of flow values (89 mL/min, normal >350 mL/min) of the right renal artery compared with the left one (568 mL/min) (**b**). Note the different morphology of the curves (**c**) of renal perfusion as well as the values of quantitative perfusion parameters (increased MTT and reduced MUS and MSI at the stenosis site) reported on the right top of the graphs (**c1**) right renal artery (**c2**) left renal artery

References

Böckler D, Hylik-Dürr A, von Tengg-Kobligk H, et al. (2007) Clinical requirements of aortic imaging. Radiologe 47(11): 962–973

Cornelissen SA, Verhagen HJ, Prokop M, et al. (2008) Visualizing type IV endoleak using magnetic resonance imaging with a blood pool contrast agent). J Vasc Surg 47(4):861–864

Gottlieb R, Chen A (2002) Pictorial essay MR findings of Erdheim-Chester disease. J Comput Assist Tomogr 26(2): 257–261

Ho VB, Corse WR (2003) MR angiography of the abdominal aorta and peripheral vessels. Radiol Clin N Am 41(1): 115–144

Kaufman SL, White RI Jr, Harrington DP, et al. (1978) Protean manifestations of mycotic aneurysms. AJR Am J Roentgenol 131(6):1019–1025

Lee WK, Mossop PJ, Little AF, et al. (2008) Infected (mycotic) aneurysms: spectrum of imaging appearances and management. Radiographics 28(7):1853–1868

Macedo TA, Stanson AW, Oderich GS, et al. (2004) Infected aortic aneurysms: imaging findings. Radiology 231(1):250–257. Erratum in: Radiology 2006;238(3):1078

McGuigan EA, Sears ST, Corse WR, et al. (2005) MR angiography of the abdominal aorta. Magn Reson Imaging Clin N Am 13(1):65–89

Michaely HJ, Attenberger UI, Kramer H, et al. (2007) Abdominal and pelvic MR angiography. Magn Reson Imaging Clin N Am 15(3):301–314

Miyazaki M, Lee VS (2008) Nonenhanced MR angiography. Radiology 248(1):20–43

Murray D, Marshall M, England E, et al. (2001) Erdheim-Chester disease. Clin Radiol 56:481–484

Nitecki SS, Hallet JW Jr, Stanson AW, et al. (1996) Inflammatory abdominal aortic aneurisms: a casa control study. J Vasc Surg 23:860–868

Shih MC, Hagspiel KD (2007) CTA and MRA in mesenteric ischemia: part 1, Role in diagnosis and differential diagnosis. AJR 188(2):452–461

Stavropoulos SW, Charagundla SR (2007) Imaging techniques for detection and management of endoleaks after endovascular aortic aneurysm repair. Radiology 243(3):641–655

Surabhi VR, Menis C, Prasad SR, et al. (2008) Neoplastic and non-neoplastic proliferative disorders of the perirenal space: cross-sectional imaging findings. Radiographics 28(4):1005–1017

Peripheral Vessels

11

FRANCESCO DE COBELLI, ELENA BELLONI, and ALESSANDRO DEL MASCHIO

CONTENTS

11.1 Introduction *127*

11.2 Lower Limb *128*
11.2.1 Steno-Occlusive Disease *129*
11.2.1.1 Leriche's Syndrome *131*
11.2.2 Aneurysms *131*
11.2.2.1 Popliteal Aneurysms *132*
11.2.2.2 Femoral Aneurysms *132*
11.2.3 Popliteal Entrapment *132*

11.3 Upper Limb *133*
11.3.1 Steno-Occlusive Disease *135*
11.3.1.1 Subclavian Steal Syndrome *136*
11.3.2 Aneurysms *136*
11.3.2.1 Subclavian Aneurysms *136*
11.3.3 Takayasu's Arteritis *136*
11.3.4 Thoracic Outlet Syndrome *139*

References *139*

ABSTRACT

Magnetic resonance angiography (MRA) has become a fundamental imaging modality in the assessment of peripheral arterial disease. Three-dimensional contrast-enhanced MRA (3D CE-MRA) provides a luminographic study of the arteries, which resembles digital subtraction angiography (DSA). In the literature, 3D CE-MRA has been compared to DSA, which is the standard of reference, and it has shown its superiority in terms of sensitivity and specificity. For this reason and for its high accuracy, nowadays, 3D CE-MRA represents the preferred technique for MRA of the peripheral vessels. The chapter will focus on the clinical applications of MRA in the evaluation of arterial disease of upper and lower limbs, with particular attention on the steno-occlusive disease and on aneurysms. Specific topics will also be addressed, which are the popliteal entrapment, the Takayasu's arteritis, and the thoracic outlet syndrome. The recent introduction of the blood pool contrast agents in the clinical routine will also be discussed.

11.1 Introduction

In recent years, because of significant technical advances, magnetic resonance angiography (MRA) has become a fundamental imaging modality in the assessment of peripheral arterial disease. Among non-contrast-enhanced MRA sequences, phase contrast angiography has been progressively dropped for peripheral vascular studies, while time-of-flight (TOF)

FRANCESCO DE COBELLI, MD
ELENA BELLONI, MD
ALESSANDRO DEL MASCHIO, MD
Radiology Department, Vita-Salute University, San Raffaele Scientific Institute, Via Olgettina, 60, 20132 Milan, Italy

angiography has still some specific clinical applications in the distal runoff vessels (Ho et al. 1997; McDermott et al. 1995; Unger et al. 1995). "Black blood" sequences are spin-echo derived sequences, which allow a better depiction of mural or intraluminal abnormalities, such as thickening of walls, thrombi, atherosclerotic plaques, penetrating atherosclerotic ulcers, and intramural hematomas. In fact, the signal produced by flowing blood is intentionally suppressed and this allows an excellent delineation of the vessel wall (Dellegrottaglie et al. 2007; Tatli et al. 2003). Three-dimensional contrast-enhanced MRA (3D CE-MRA) provides a luminographic study of the arteries, which resembles digital subtraction angiography (DSA). For the first time, 3D CE-MRA was described by Prince et al. (1993) and nowadays it represents the preferred technique for MRA of the peripheral vessels (Ho et al. 1999). In fact, 3D CE-MRA is highly accurate in the assessment of limbs arteries, as demonstrated in a large meta-analysis performed for lower extremity vasculature. In the same article, 3D CE-MRA proved to offer a better diagnostic performance compared to that offered by two-dimensional MRA (2D MRA), with higher sensitivity and specificity (Koelemay et al. 2001). Moreover, 3D CE-MRA has been compared to DSA and again it showed its superiority also because of the possibility of extensive postprocessing of the images (Loewe et al. 2004; Vavrik et al. 2004; Winterer et al. 2002; Wyttenbach et al. 2003). Duplex ultrasound (DUS) may play a role for a fast, widely available and less expensive evaluation of the peripheral vessels, but its use is limited to specific clinical settings and selected vessels because the technique cannot guarantee the panoramic assessment of long segments of vasculature, as DSA and CE-MRA do. Moreover, CE-MRA is more sensitive and specific for the diagnosis and presurgical assessment in patients with peripheral arterial occlusive disease (Visser and Hunink 2000; Gjonnaess et al. 2006; Leiner et al. 2005). The chapter will focus on the clinical applications of MRA in the evaluation of upper and lower limb arterial disease, with particular attention on the latter. The recent introduction of the blood pool contrast agents in the clinical routine will also be discussed.

11.2 Lower Limb

Assessment of the lower limb arterial disease with CE-MRA requires the representation of long segments of vessels, from the abdomen to the ankle, whose lumen is enhanced by a gadolinium-based contrast agent: for this reason, the contrast agent bolus should be rapidly followed distally. This is obtained by means of a fast field-echo (FFE) T1-weighted 3D coronal sequence, which requires state-of-the-art 1.5 T scanners with high performance gradients providing very short TEs and TRs for dynamic acquisition (Ersoy and Rybicki 2008). MRA of peripheral vessels requires the imaging of 3–4 stations: aorto-iliac, thigh, calf and foot, depending on the patient's height. The introduction of the "moving-table" technology allowed the chasing of a single bolus of contrast agent in its distal progression (Carriero et al. 2002; Ho et al. 1998; Meaney et al. 1999). The patient is placed in the magnet, feet first, making sure that all the vascular structures to be examined are in the same anterior–posterior level. Scout and mask images (before contrast administration) are acquired for three locations, typically the pelvis, the thigh and the calf; low resolution TOF images can also be added for optimal planning. Usually, 0.2 mmol/kg of a gadolinium-based contrast agent is administered with an automatic injector, with a flow rate of 0.8 mL/s, followed by 20 mL of saline flush with a flow rate of 1.5 mL/s. The slow flow rates allow to obtain a "long" bolus which enhances the vascular structures for a longer period of time. Immediately after the injection starts, a fast, low resolution coronal T1-weighted sequence is performed repeatedly to visualize the influx of the contrast in the iliac arteries. As soon as the contrast enhancement is visible in the iliac arteries, the standard FFE T1-weighted 3D diagnostic imaging is acquired, first of all on the aorto-iliac region; then the table moves automatically and thigh imaging is performed; finally the table again moves and the calf imaging is obtained. The table moves at a speed rate of 10 cm/s. Another method to obtain an optimal timing of the imaging, after contrast infusion, is to estimate the contrast agent travel time with the "test-bolus" technique (Dellegrottaglie et al. 2007; Earls et al. 1996). A small amount of contrast (1–2 mL) and 20 mL of saline solution are injected at the same flow rate, which will be used for the diagnostic imaging and a fast, low resolution coronal T1-weighted sequence is performed repeatedly to visualize the influx of the contrast in the arteries of interest and to choose the best acquisition time for the following diagnostic imaging with the proper amount of contrast agent. Imaging of the lower limb arterial vasculature can be performed either with the combined use of a body coil and a surface coil, or with a dedicated lower extremity reception coil. The latter has been introduced recently in the clinical practice and it allows to obtain an optimized, more homogeneous signal reception. It is also characterized by a higher signal-to-noise ratio than the body coil and this in turn can be exploited to provide higher resolution images, to shorten the acquisition time or to increase the anatomic coverage (Huber et al. 2003; Leiner et al. 2004). Even if the "moving-table" technique is robust and accurate, some problems can be associated with multiple stations imaging. The nonlinearity of gradients can be associated with distortions of the images, no standard protocols can be applied to all scanners, venous superimposition in the calf can lead to suboptimal imaging, and finally the imaging is not isotropic and there is the tendency towards overestimation of the stenosis. However, 3D MRA is considered an accurate, sensitive and specific imaging technique for the assessment of lower extremity vascular disease, when compared to DSA and 2D MRA (Deutschmann et al. 2006; Ho et al. 1998; Koelemay et al. 2001; Lapeyre et al. 2005; Meaney et al. 1999; Winterer et al. 2002). A new contrast agent has been introduced in the clinical use since 2005: Vasovist® (Gadofosveset, Bayer Schering Pharma, Germany) is the first intravascular contrast agent approved for the use in CE-MRA in Europe. Vasovist® is a small-molecule gadolinium-based contrast agent which binds reversibly to the albumin in the human plasma and provides extended vascular enhancement compared to conventional extracellular contrast agents. In fact, its main characteristic is a 5–7 times higher relaxivity at 1.5 T if compared, for example, to Magnevist® (GD-DTPA, Bayer Schering Pharma, Germany) (Lauffer et al. 1998). The administered dose of Vasovist® is lower than that of the conventional contrast agents: in fact, 0.03 mmol/kg is the standard to obtain good quality diagnostic images, as stated in the preliminary studies that recently appeared in the literature (Goyen et al. 2005; Vogt et al. 2007). After the injection of Vasovist®, two acquisition phases can be distinguished. The arterial first-pass acquisition is similar to the conventional arterial angiograms obtained with extracellular contrast agents. Following this dynamic phase, the intravascular contrast agent distributes to all the body vasculature, both arterial and venous, and reaches an equilibrium phase in about 5 min. In this steady state phase, high spatial resolution images can be obtained, as lower extremities can be imaged for several minutes: in fact, Vasovist® has a half-life of about 15 h and allows to obtain a good signal-to-noise ratio for about 1 h after intravenous administration (Nikolaou et al. 2006). However, some limitations exist for steady-state imaging of lower extremities. First of all, the venous overlay may be a potential diagnostic problem, because in lower limbs the venous vascular structures are in close vicinity to arterial vascular structures. Moreover, because of long imaging times, limb movement can occur, with well-evident motion artefacts. Finally, the patient can experience some discomfort as a result of the prolonged permanence in the magnet bore (Nikolaou 2006). For these reasons, at the present time steady-state imaging of lower limb vasculature with Vasovist® is not considered a routine clinical application.

11.2.1 Steno-Occlusive Disease

Atherosclerosis is a systemic disease affecting the arterial bed of the body and represents the leading cause of steno-occlusive arterial disease in the population older than 40 years of age. In particular, the vasculature of the lower extremities is a frequent target of the atherosclerotic disease (Wasser 2003). Clinical manifestation of peripheral arterial occlusive disease (PAOD) is either intermittent claudication (pain during exercise which is relieved by rest) or critical ischemia (pain also at rest; possible tissue loss due to low oxygen levels and nutrient supply in the affected tissues) (Meaney 2003). In patients with acute limb-threatening ischemia, DSA should be the technique of choice because it allows to perform both imaging and endovascular therapeutic procedures (Wasser 2003). On the other hand, CE-MRA can be considered a safe and accurate diagnostic tool in patients with intermittent claudication. The current role of CE-MRA in patients affected by PAOD is to depict the location, the length and the severity of vascular stenosis (Fig. 11.1). This is particularly important in patients who will undergo endovascular or surgical procedures, as it helps to plan the execution of such procedures. Careful evaluation of infrapopliteal vessels is essential for treatment planning in diabetic patients, who often have long segments of arterial occlusion in the lower limbs. In this kind of patients, information on distal pedal vessels is necessary and 3D MRA may reveal patent distal arteries that are not visible in DSA (Drescher et al. 2006; Lapeyre et al. 2005; Reid et al. 2001) Moreover, TOF angiography proved to be a useful complementary study in patients with peripheral

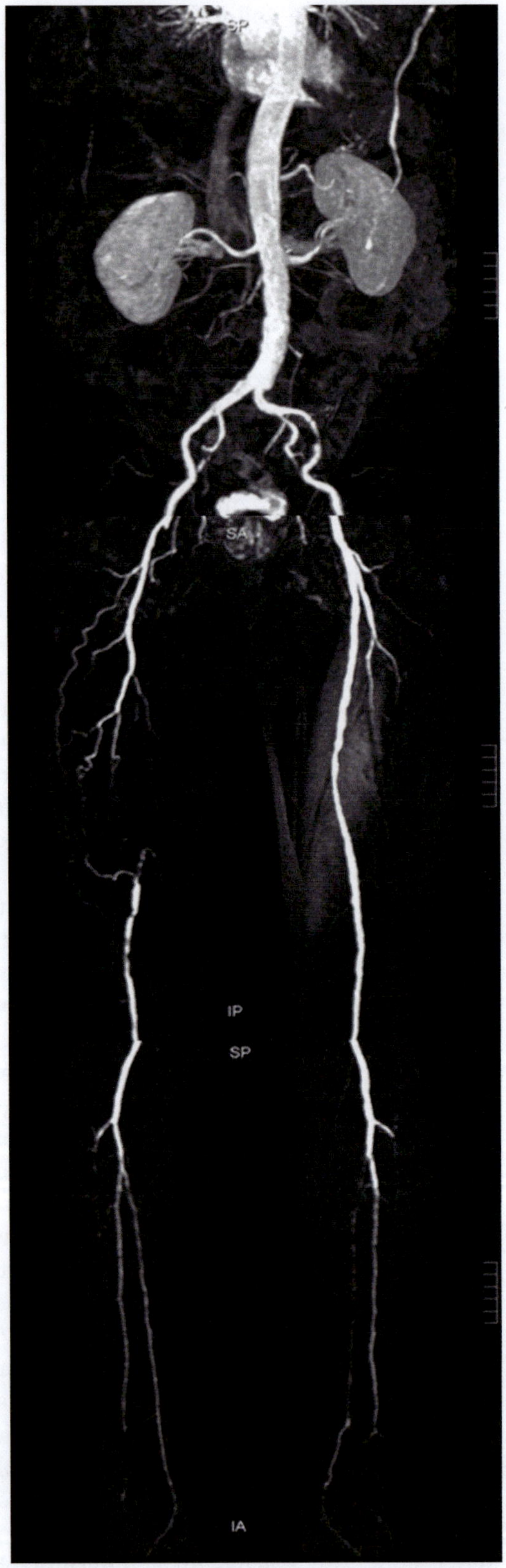

Fig. 11.1. Steno-occlusive disease in a 74-year-old diabetic patient. Coronal MIP (maximum intensity projection) image showing diffuse atherosclerotic parietal irregularities of the common iliac arteries and of the right external iliac artery, without evidence of significant stenotic lesions. Focal stenosis (60–70%) of the left external iliac artery. *Right lower limb*: occlusion of the origin of the superficial femoral artery, with distal revascularization due to collateral vessels from the deep femoral artery. Distally to the occlusion, the superficial femoral artery has a focal significant stenosis (>50%) and wall irregularities. The anterior tibial artery is nearly occluded 3 cm after its origin, it has multiple focal stenoses in the medium third and is completely occluded distally. Occlusion of the distal third of the interosseous artery. *Left lower limb*: diffuse atherosclerotic lesions of the superficial femoral artery, without significant stenotic disease. The anterior tibial artery is occluded 2 cm after its origin. The posterior tibial artery is nearly occluded proximally for 2 cm (with distal revascularization from the interosseous artery) and it has focal stenoses in the medium and distal third. The interosseous artery is occluded distally, with evidence of small collateral vessels

vascular disease of the foot and the ankle (Unger et al. 1995). It is well known that CE-MRA tends to overestimate stenosis (Ho et al. 1999; Lee et al. 2000; Winterer et al. 2002) and this may be an important aspect in evaluating the cases of borderline stenosis (40–60%), when its hemodynamic significance should be assessed carefully because it can be an important element when deciding if invasive therapies are needed. However, the same problems may be encountered with DSA, which still represents the reference standard imaging technique but is characterized by significant pitfalls (Wasser 2003). In fact, recent studies have assessed that DSA is not adequate in evaluating runoff vessels; CE-MRA seems to demonstrate more patent vessels in this case (Owen et al. 1992). MRA may be utilized even in the evaluation of patients after surgical therapy. The patients affected by PAOD, in fact, can undergo different surgical procedures involving the lower limbs, in particular aorto-bi-iliac, aorto-bi-femoral, axillo-bi-femoral bypasses proximally and popliteal and tibial bypasses distally (Tatli et al. 2003). In these cases, CE-MRA can assess graft location, patency and stenosis, even if the bypass grafts are very small and distal; the proximal and distal anastomoses of the bypasses can be evaluated. It is also accurate in the demonstration of the complications of the surgical therapy such as anastomotic pseudoaneurysms (Insko and Carpenter 2004). Unfortunately, long-term failure of the grafts is a common event, because of the progression of the disease in the native outflow vessels. Periodic screening is implemented with DUS, but MRA may be requested prior to graft revision. In these cases, special care should be taken to avoid misinterpretation of the images because of the presence of the magnetic susceptibility artefacts created by metallic clips. Source images, as well as 3D data sets, have to be evaluated in particular in these patients (Tatli et al. 2003). If endovascular treatment has been performed and metallic stents have been positioned in the vessels, susceptibility artefacts may mimic stenosis or occlusion; therefore, the clinical history and the source images should be evaluated carefully (Lee et al. 2000).

11.2.1.1 Leriche's Syndrome

The complete obstruction of the aortic bifurcation was originally described by Leriche, and now the term Leriche's syndrome refers to the chronic occlusion of the infrarenal aorta involving the iliac arteries. Common symptoms include intermittent claudication of the whole lower limb starting from the gluteus muscles, impotence in men, absence of the femoral pulses and symmetric atrophy of the lower limbs. The symptoms may vary according to the formation of collateral vessels (Chan et al. 2005). The disease can be well tolerated for years, but finally gangrene is observed (Sabiston 1997a). In the clinical setting of Leriche's syndrome, CE-MRA can depict the level of aortic occlusion, the most proximal arterial segments which may possibly undergo a surgical bypass procedure, as well as the anatomy and the extent of the collateral arterial vessels (Ruehm et al. 2000). Moreover, CE-MRA seems to be more accurate than DSA in the assessment of the distal runoff vessels (Link et al. 1998).

11.2.2 Aneurysms

An aneurysm is a localized dilatation of an artery. A true aneurysm is characterized by the involvement of all the three layers of the vessel wall: intima, media and adventitia. A false aneurysm, or pseudoaneurysm, is characterized by the destruction of the vessel wall because of trauma, infections and invasive vascular procedures, leading to dilatation of the adventitial layer only.

Aneurysms may be congenital but the majority of them are acquired. The most common cause of an aneurysm is atherosclerosis, followed by traumas, micotic infections, adventitial cystic medial necrosis and congenital diseases such as Ehlers–Danlos or Marphan's syndrome (Sabiston 1997b). Aneurysms can affect all the lower limb arteries, but the most commonly affected arteries are the popliteal and the femoral arteries. Aims of the MRA evaluation in aneurysmatic disease of the lower limb arterial vasculature are to assess the presence, location, morphology and dimensions of the aneurysms. Their patency should be evaluated as well and in this case particular attention is to be paid to the imaging timing after the contrast bolus injection. In fact, in patients with diffuse atherosclerotic disease, the blood flow can be very slow because of the extension of the aneuryms in the lower limbs, because of the possible concomitant presence of aneurysms in central arteries, as well as the possible concomitant heart failure, which is frequent in patients with extensive atherosclerotic disease and leads to low cardiac output and consequently altered blood flow dynamics all over the body vasculature. For these reasons, in this kind of patients the timing of CE-MRA sequence acquisition is crucial to avoid a false diagnosis of nonpatent aneurysmal lumen. Apart from the two methods described in the Sect. 11.2 (rapid low resolution sequence to chase the bolus and test-bolus), there are other methods to ensure reliable and accurate timing of the bolus. In particular, rapid temporally resolved acquisitions and automated techniques to start image acquisition after bolus arrival have been proposed (Foo et al. 1997; Korosec et al. 1996; Lee et al. 2000; Prince et al. 1997). It is also possible to study the characteristics of the vessel wall with MR imaging, evaluating the presence and the signal intensity of mural thrombi, hematomas and wall thickening. Atherosclerotic plaques can be depicted as well, even when they contain a considerable amount of calcium, which would be a disturbing element in multidetector computed tomography.

11.2.2.1
Popliteal Aneurysms

The popliteal artery is the peripheral artery most commonly affected by aneurysmatic disease (Tatli et al. 2003). The leading cause of popliteal aneurysms is the atherosclerotic disease (95%) (Dawson et al. 1991), but the compartment syndrome, bacterial infections, collagen disorders and traumas also may play a role. This kind of aneurysm is typically present in the seventh decade of life and is much more common in male than in female patients. It is very often associated with aneurysms of other arteries (85%), such as the aorta, the iliac arteries or the femoral arteries, and it can be bilateral (40%) (Makhoul 1997a). The majority of popliteal aneurysms present as pulsatile masses, or as nonpulsatile masses if they are thrombosed. They are usually symptomatic and the patient complains of ischemic claudication, gangrene, local compressive symptoms such as venous obstruction or nervous pain and paresthesia. It is fundamental to recognize and to treat surgically popliteal aneurysms because up to one third of the patients have distal embolization or limb ischemia caused by an unrecognized and nonpalpable aneurysm (Carpenter et al. 1994; Ramesh et al. 1993). The MRA study in popliteal aneurysms may depict the location, morphology and dimensions of the aneurysms, as well as the possible presence of mural thrombi and their signal intensity characteristics (Fig. 11.2).

11.2.2.2
Femoral Aneurysms

Even for femoral arteries, the most common cause of aneurysms is atherosclerosis. They are often associated with aortic and politeal aneurysms and are bilateral in 50% of the cases (Makhoul 1997b). Pseudoaneurysms of the femoral arteries are mostly iatrogenic, following percutaneous access during vascular procedures (McCann 1991). Anastomotic pseudoaneurysms are also fairly common, following surgical procedures such as aorto-bifemoral bypass (Makhoul 1997b).

In comparison to popliteal aneurysms, femoral aneurysms are characterized by less morbidity if untreated. In fact, while symptomatic aneurysms should be treated surgically, the surgical treatment of small, asymptomatic aneurysms is still controversial and they and can be treated conservatively (Graham et al. 1980). Once again, the MRA study in femoral aneurysms may depict the location, morphology and dimensions of the aneurysms, as well as the possible presence of mural thrombi and their signal intensity characteristics.

11.2.3
Popliteal Entrapment

Popliteal artery entrapment syndrome is a rare entity associated with an anomalous relationship between the medial head of the gastrocnemius muscle and the popliteal artery (Wasser 2003). This embryological abnormality causes compression of the popliteal artery; the popliteal vein can be involved as well. The syndrome may be due either to four main anatomic abnormalities or to a functional entrapment, as stated by the Popliteal Vascular Entrapment Forum in 1998. The disease is much more common in young male subjects (Turnipseed 2002; Utsunomiya and Sawamura 2007; Elias et al. 2003; Holden et al. 2008; Macedo et al. 2003). This anomalous anatomical condition results in paraesthesias and coldness of the foot and in intermittent claudication of the calf; ischemia is rare (Turnipseed 2002; Ersoy and Rybicki 2008; Macedo et al. 2003). Surgical treatment is indicated whenever anatomic popliteal artery entrapment syndrome is demonstrated because, if left untreated, it can lead to stenosis because of repeated trauma and fibrosis development, occlusion and embolization. Poststenotic aneurysms can also occur. In functional entrapment, surgical treatment should be reserved to

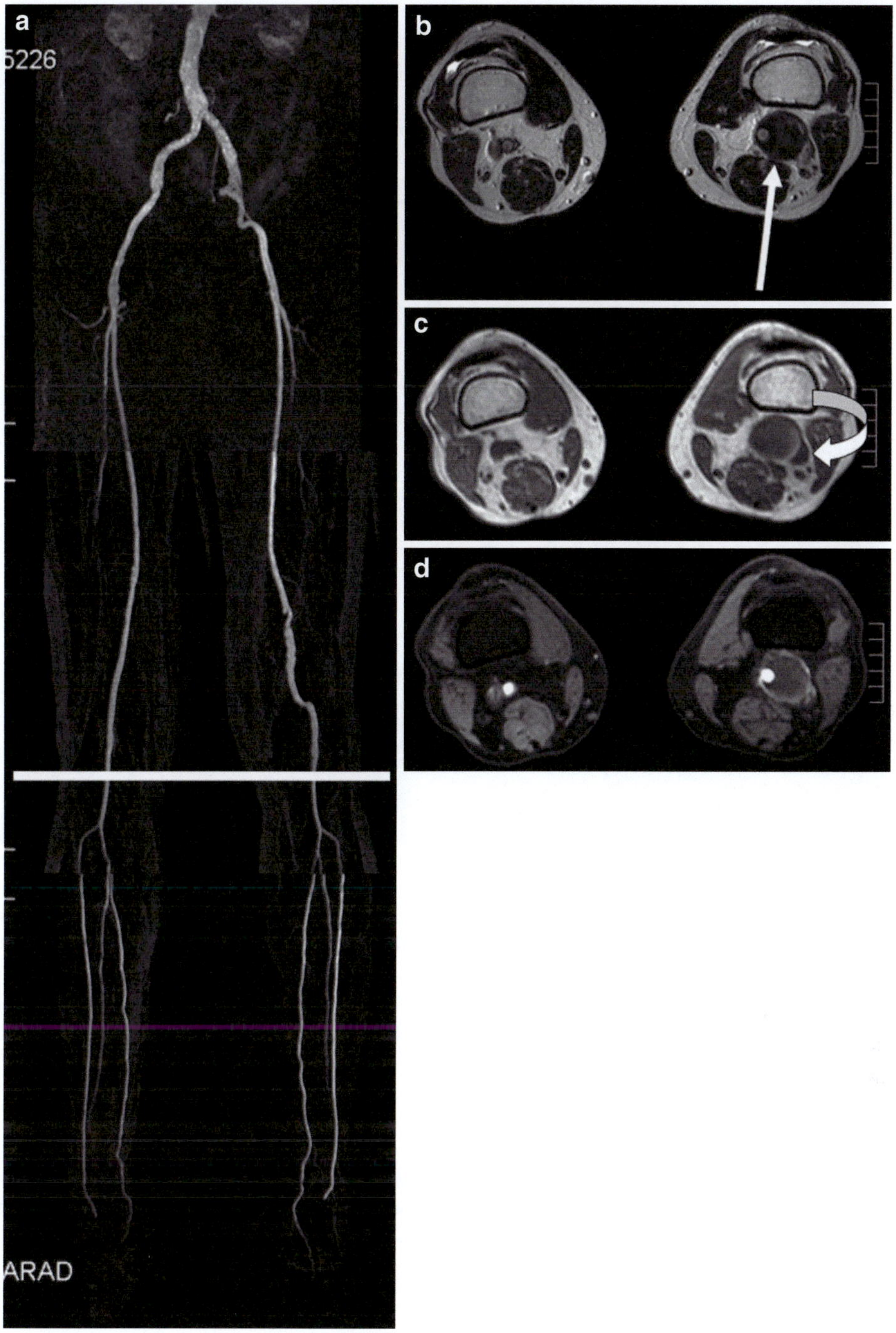

Fig. 11.2. Presurgical evaluation of a popliteal artery aneurysm in a 73-year-old man. (**a**) Coronal MIP image showing that the fusiform aneurysm of the left popliteal artery is extended longitudinally for 7 cm. (**b, c**) Morphologic T2-weighted (**b**) and T1-weighted (**c**) axial images at the level of the aneurysm shown in (**a**). The aneurysm (3.5 × 3 cm) is partially occluded by a 2-cm thick eccentric thrombus (see *straight arrow*); the residual lumen is 9 mm. The popliteal vein is compressed and dislocated by the aneurysm (see *curved arrow*). (**d**) Postcontrast axial T1-weighted fat-saturated image at the same level shown in (**b**) and in (**c**)

symptomatic patients (Turnipseed 2002; Elias et al. 2003). MRA has recently become a fundamental imaging technique in the assessment of the popliteal artery entrapment syndrome. In fact, it can depict both the vasculature and the muscle anatomy, and thereby help in distinguishing between anatomic and functional forms of the disease (Turnipseed 2002; Holden et al. 2008; Macedo et al. 2003) (Fig. 11.3).

11.3 Upper Limb

MRA assessment of the upper limb arterial disease is less diffuse than the assessment of the lower limb arterial disease. This should be possibly related to the lower incidence of vascular diseases such as

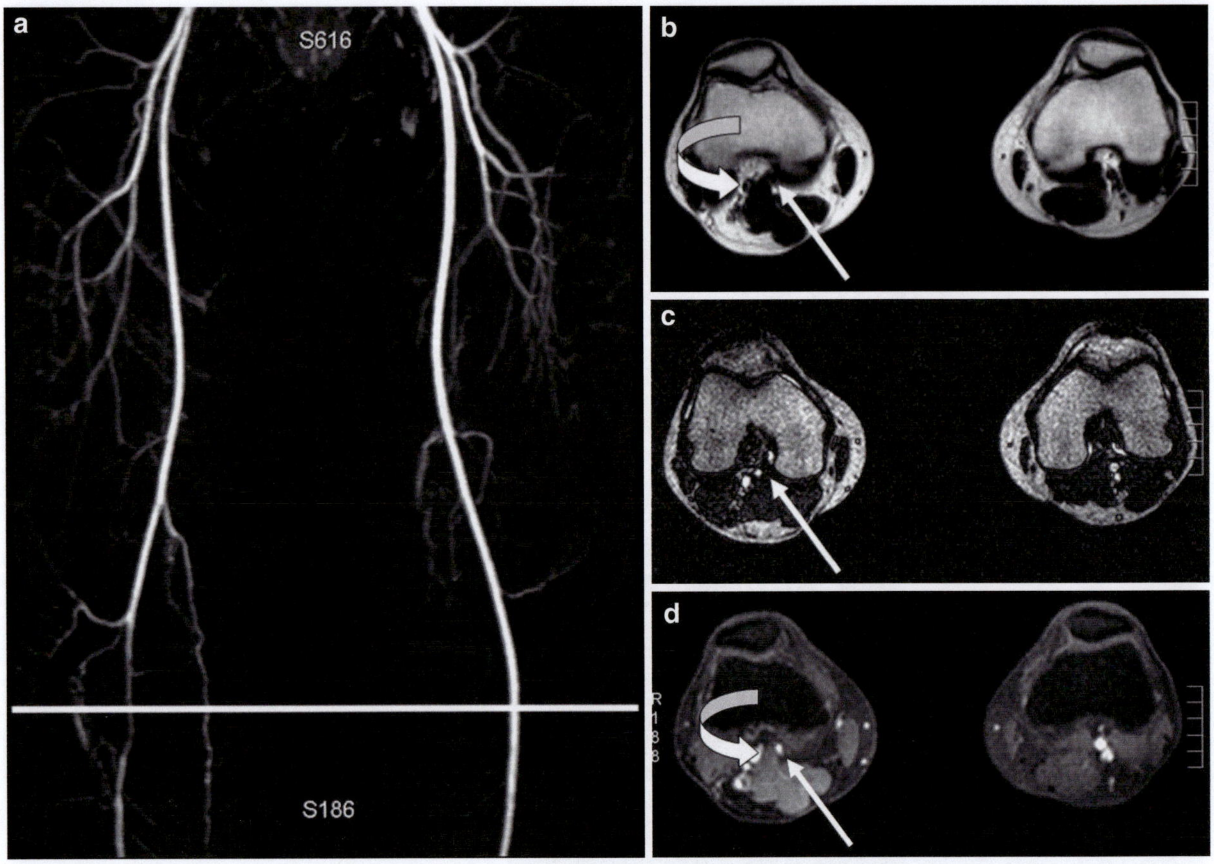

Fig. 11.3. Popliteal entrapment in a 21-year-old man with calf claudication during training. (**a**) Coronal MIP image showing the significant stenosis of the right popliteal artery, with the concomitant development of collateral vessels originating from the superficial femoral artery. (**b, c**) Axial T2-weighted (**b**) and balanced-FFE (**c**) images at the level showed in (**a**), demonstrating the presence of a tissue component of the medial head of the gastrocnemius muscle (see *curved arrow*) which divaricates the vein and the artery, and compresses the latter (see *straight arrow*). (**d**) Postcontrast axial T1-weighted fat-saturated image at the same level shown in (**b**) and in (**c**): the divarication and the compression of the left popliteal artery are even more evident than in the precontrast images (see *arrows*)

aneurysms and steno-occlusions. On the contrary, inflammatory arteritis and vasospastic disease are more common in the upper limbs, and they usually involve medium and small caliber arteries. In addition to the lower incidence of aneurysms and occlusions, underestimation of vascular disease in the upper limbs is frequent because patients are less commonly symptomatic. In fact, only 1% of patients with diffuse peripheral vascular disease complains of ischemic symptoms also to the upper limbs. The lower incidence of the symptoms is related to the extensive development of collateral vessels, to the smaller amount of muscular tissue and to its intermittent use in the normal life activities (Slocum and Silver 1997). Nowadays, DSA still represents the reference standard for the imaging of upper extremity vasculature. However, recent developments in hardware and software of state-of-the-art 1.5 T scanners have made MRA, in particular CE-MRA, an interesting imaging modality for the depiction of upper extremity vessel anatomy and disease because of its rapid and three-dimensional imaging over a large field of view (Planken et al. 2008). Some studies compared the diagnostic performances of CE-MRA and DSA and concluded that CE-MRA shows high accuracy in the detection and grading of lesions involving the aortic arch vessels. For this reason, its use can be recommended for screening, treatment planning, and follow-up in known or suspected arterial disease in aortic arch arteries (Carpenter et al. 1997; Loewe et al. 2004; Willinek et al. 2005). In imaging of the subclavian and brachial arteries, precontrast and postcontrast images are obtained, both with the breath-hold technique. Usually, 0.1 mmol/kg of a gadolinium-based contrast agent is administered with an automatic injector, with a flow rate of 2 mL/s, followed by 20 mL of saline flush at the same flow rate. Care should be taken to avoid venous overlap, and this can be managed by injecting the contrast agent in the contralateral arm. A body coil should be used to gain a higher signal-to-noise ratio. MRA can not be considered the imaging modality of choice in small distal vessels (typically in the hands) because some problems still exist. First of all, high resolution scanning with dedicated surface coils is necessary to obtain a good spatial resolution and the availability of this state-of-the-art equipment is still limited. Moreover, the arterial transit time of the contrast bolus in the hand is very short (approximately 12 s) and this requires a very accurate timing of the imaging (Wasser 2003; Winterer et al. 2000). DUS still can play a role for the imaging of the vessels in the upper arm, forearm and hand but, as stated in the introduction to the chapter, this is limited to particular clinical settings. Assessment of the upper extremity vasculature prior to hemodialysis vascular access creation in the terminal-stage renal disease patients may be an example, considering also the emerging problem of systemic nephrogenic fibrosis as a late adverse event following gadolinium-based contrast agent administration in this kind of patients (Thomsen 2006; Tordoir and Mickley 2003; Bongartz et al. 2008; Planken et al. 2008). As in the case of the lower limbs, Vasovist® can be used for CE-MRA of the upper limbs also, even though the literature in this field is very limited up to now. The arterial first-pass acquisition is similar to the conventional arterial angiograms obtained with extracellular contrast agents and is now implemented in the routine clinical setting, while the steady-state imaging of upper extremities with Vasovist® is not.

11.3.1 Steno-Occlusive Disease

Atherosclerosis is a systemic disease affecting the arterial bed of the body and represents the leading cause of steno-occlusive arterial disease in the population older than 40 years of age. The second most important cause of occlusive disease of the aortic arch vessels, subclavian arteries included, is the Takayasu's arteritis, which will be discussed further in the Sect. 11.3.4 (Krinsky et al. 1996). In the upper limb, atherosclerosis is less common than in the abdomen and the lower limb arteries; it mainly involves the innominate artery and the origin of the subclavian artery (Wasser 2003; Shadman et al. 2004). The stenosis or occlusion of the two arteries may be asymptomatic because of the development of collateral vessels. If symptomatic, they can lead to upper extremity claudication or ischemia and to vertebrobasilar insufficiency (Krinsky et al. 1996; Woo et al. 2006). The current role of CE-MRA in patients affected by steno-occlusive disease of the upper limb is to depict the location, the length and the severity of vascular stenosis. This is particularly important in the work-up prior to endovascular or surgical procedures, in order to plan their execution. It is well known that CE-MRA tends to overestimate stenosis (Ho et al. 1999; Lee et al. 2000; Winterer et al. 2002). This may be an important matter in evaluating the cases of borderline stenosis (40–60%), when its hemodynamic significance should be assessed carefully because it can be an important element when deciding if invasive therapies are needed. However, the same problems may be encountered with DSA, which still represents the reference standard imaging technique but is characterized by significant pitfalls (Wasser 2003). MRA may be utilized even in the evaluation of patients following surgical treatment. The symptomatic patients can undergo different surgical procedures involving the upper limbs, in particular transposition of the subclavian artery, carotid-to-subclavian artery bypass, in situ or extra-anatomic bypass for innominate artery disease, or even axillo-axillary bypass (Krinsky et al. 1998; Woo et al. 2006). In these cases, CE-MRA can assess graft location, patency and stenosis; the proximal and distal anastomoses of the bypasses can be evaluated as well. CE-MRA is also accurate in the demonstration of the complications of the surgical therapy, such as anastomotic pseudoaneurysms (Insko and Carpenter 2004). Particular care should be taken to avoid misinterpretation of the CE-MRA images because of the presence of the magnetic susceptibility artefacts created by metallic clips. Source images, as well as 3D data sets, have to be evaluated in particular in these patients (Tatli et al. 2003). If endovascular treatment has been performed and metallic stents have been positioned in the vessels, susceptibility artefacts may mimic stenosis or occlusion, so the clinical history and the source images should be evaluated carefully (Lee et al. 2000).

11.3.1.1
Subclavian Steal Syndrome

The subclavian steal syndrome is secondary to obstruction of the proximal subclavian artery. In this setting, the blood supply to the arm is maintained by the retrograde flow in the ipsilateral vertebral artery distal to the obstruction, which is obtained at the expense of the cerebral circulation by stealing from the contralateral vertebral and/or basilar arteries. Therefore, vertebrobasilar hypoperfusion may develop when the arm is exercised (Wasser 2003). The most common cause of the subclavian steal syndrome is the presence of an atherosclerotic lesion in the proximal subclavian artery. The treatment of the syndrome is usually the surgical bypass of the stenosis. Percutaneous transluminal angioplasty of the subclavian artery may be implemented as well (Van Grimberge et al. 2000). CE-MRA may depict the obstruction of the subclavian artery; the retrograde flow in the vertebral artery can be demonstrated with flow-encoded MR sequences (Wasser 2003; Flynn et al. 1993; Van Grimberge et al. 2000).

11.3.2
Aneurysms

An aneurysm is a localized dilatation of an artery. A true aneurysm is characterized by the involvement of all the three layers of the vessel wall: intima, media and adventitia. A false aneurysm, or pseudoaneurysm, is characterized by the destruction of the vessel wall because of trauma, infections and invasive vascular procedures, leading to dilatation of the adventitial layer only.

Aneurysms may be congenital but the majority of them are acquired. The most common cause of an aneurysm is atherosclerosis, followed by traumas, micotic infections, adventitial cystic medial necrosis and congenital diseases such as Ehlers-Danlos or Marphan's syndrome (Sabiston 1997a). Aneurysms of the upper limbs are less common than those of the lower limbs. In the upper limb, an aneurysm is most frequently located in the subclavian artery. Aims of the MRA evaluation in aneurysmatic disease of the upper limb arterial vasculature are to assess the presence, location, morphology and dimensions of the aneurysms, as in lower limb disease. Their patency should be evaluated as well, and also in this case particular attention to the imaging time after the contrast bolus injection is necessary. The timing of CE-MRA sequence acquisition is crucial to avoid a false diagnosis of nonpatent aneurysmal lumen. Apart from the two methods described in the Sect. 11.2 (rapid low resolution sequence to chase the bolus and test-bolus), there are other methods to ensure reliable and accurate timing of the bolus. In particular, rapid temporally resolved acquisitions and automated techniques to start image acquisition after bolus arrival have been proposed (Foo et al. 1997; Korosec et al. 1996; Lee et al. 2000; Prince et al. 1997). MR imaging also allows the assessment of the characteristics of the vessel walls, as well as evaluation of the presence and the signal intensity of mural thrombi, hematomas and wall thickening. Atherosclerotic plaques can be depicted as well, even when they contain a considerable amount of calcium, which would be a disturbing element in multidetector computed tomography.

11.3.2.1
Subclavian Aneurysms

The aneurysms of the subclavian artery are the most common in the upper limb. They are usually secondary to atherosclerotic disease, but also to traumas. A poststenotic dilatation may cause an aneurysm. The lesion can have either an intra-thoracic or supraclavicular development. In the latter case, they can present as pulsatile masses. The aneurysm of the subclavian artery may be bilateral.

The treatment of this disease is usually surgical (Sabiston 1997c; Halldorsson et al. 2007).

Aims of the MRA evaluation in aneurysmatic disease of the subclavian aneurysms are to assess the location, morphology and dimensions of the aneurysms, as in lower limb disease (Halldorsson et al. 2007). Their patency can be evaluated as well. MR imaging also allows the assessment of the characteristics of the vessel walls, and the evaluation of the presence and the signal intensity of mural thrombi, atherosclerotic lesions, hematomas and wall thickening.

11.3.3
Takayasu's Arteritis

Takayasu's arteritis is an inflammatory disease of unknown origin, which involves mainly the aorta, its principal branches and the pulmonary arteries. It is more common in young female subjects. The acute phase of the disease is characterized by inflammation of the vasa vasorum in the adventitia of the vessel walls. In the chronic phase of the disease, the vessel walls are thickened because of fibrosis of their three layers. Typical symptoms of the disease are fever, malaise, arthralgia, thoracic pain and tachycardia. Cerebral and upper limb ischemias are more common in the later stage of the disease (SABISTON 1997d).

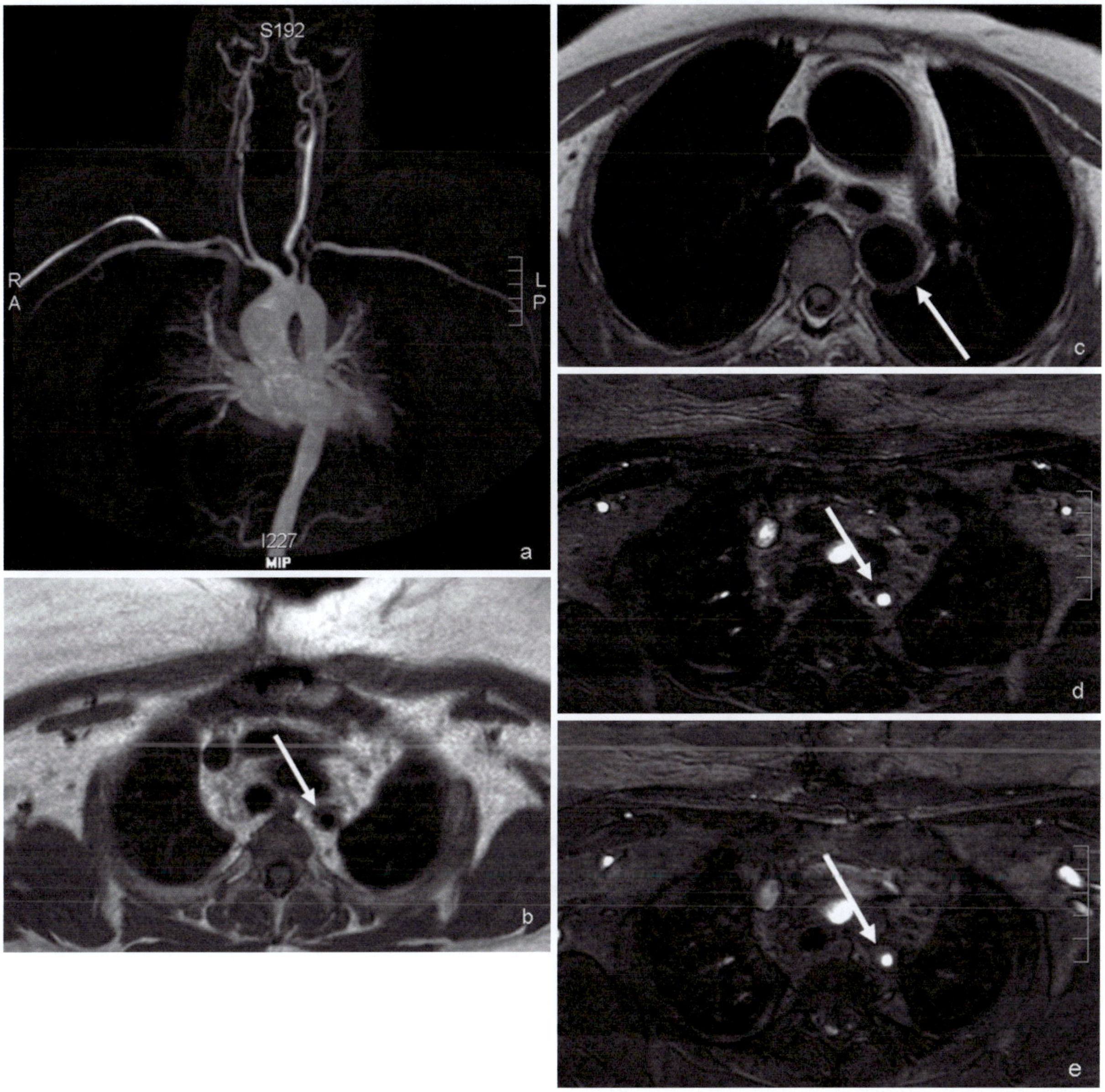

Fig. 11.4. Takayasu's arteritis in a 30-year-old woman, (CE-MRA performed with Vasovist®). (**a**) Coronal MIP image showing a focal significant stenosis of the origin of the right subclavian artery. A diffuse reduction of the caliber of the right common carotid artery is also visible. Significant stenosis of the origin of the left subclavian artery, extended for 2 cm. Also the origin of the left common carotid artery is stenotic for 2 cm (<50%). (**b**) Morphologic axial black-blood proton density (PD)-weighted image showing the concentric thickening of the walls of the left subclavian artery, which causes the significant stenosis described in (**a**) (see *arrow*). (**c**) Morphologic axial black-blood PD-weighted image showing the concentric thickening of the walls of the descending aorta (see *arrow*). (**d**) Axial 3D FFE T1-weighted high resolution image before contrast administration, showing the thickening of the wall of the left subclavian artery (see *arrow*). (**e**) Axial 3D FFE T1-weighted high resolution image obtained 15 min after Vasovist® intravenous administration, which shows the enhancement of the thickened wall of the left subclavian artery (see *arrow*). This can be considered a marker of active disease in Takayasu's arteritis

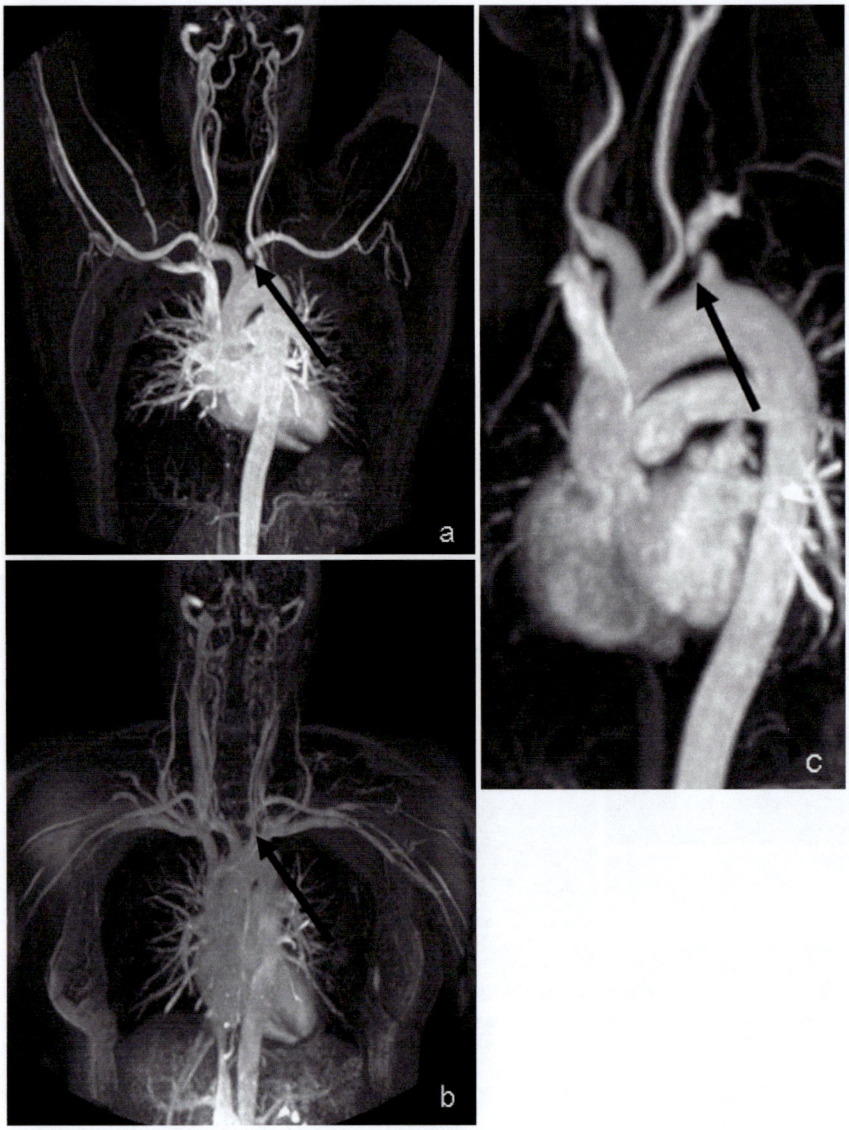

Fig. 11.5. Suspected thoracic outlet syndrome in a 52-year-old woman with pain during elevation and exercise of the left upper limb. (**a**) Coronal MIP image with arms in abduction showing the occlusion of the origin of the left subclavian artery (see *arrow*). (**b**) The occlusion of the left subclavian artery is present even in the coronal MIP image with arms in adduction, thus the final diagnosis was of an occlusive lesion of the artery (see *arrow*). Please note the superimposition of the venous vessels with respect to (**a**). (**c**) Closer view of the MIP image showing the occlusion of the origin of the left subclavian artery (see *arrow*)

The diagnosis of Takayasu's arteritis and the identification of the activity of the disease are fundamental in order to implement an adequate therapy. Histopathologic specimens are rarely available, so the diagnosis is on the basis of the American College of Rheumathology criteria proposed in 1990 (Arend et al. 1990) and the evaluation of activity is on the basis of the Kerr's criteria (Kerr et al. 1994). In this setting, the information derived from the diagnostic imaging may also be important. DSA has been considered for years the most important imaging technique in Takayasu's arteritis patients, but MRA may now be considered the technique of choice. First of all, the lack of ionizing radiations is fundamental when repeated follow-up studies must be performed on young women. Moreover, MRA allows both luminographic and vessel wall evaluation (Yamada et al. 2000). At MR imaging, Takayasu's arteritis involving the upper limbs is characterized by mural thrombi, arterial wall thickening and enhancement (typically in the active phase) (Choe et al. 2000). CE-MRA may depict the location, degree, length of the stenoses and dilatations, even in pulmonary arteries

(Gotway et al. 2005) (Fig. 11.4). The finding of the disease involving the pulmonary arteries is fairly typical of Takayasu's arteritis (70% of patients) and is rarely seen in atherosclerosis or other arteritides; therefore, it enables a definitive diagnosis of the disease (Yamada et al. 1993). CE-MRA is also useful in evaluating the patency of the collateral arteries or of the surgical bypasses (Yamada et al. 2000; Gotway et al. 2005), even if the therapeutic approach in these patients is usually conservative (Kerr et al. 1994).

11.3.4 Thoracic Outlet Syndrome

The thoracic outlet syndrome is due to the compression of either the subclavian vessels or the brachial plexus by anomalous osseous structures or by a broad insertion of the scalenus muscle anteriorly over the clavicle. For these reasons, the syndrome is characterized by vascular and neurological symptoms. The patients may experience pain during elevation of the arms and loss of sensation during exercise. The clinical examination may show a palpable thrill over the subclavian artery, diminished radial pulses and lowered brachial blood pressure (Wasser 2003). There are many provocative maneuvers that may be implemented to diagnose the thoracic outlet syndrome (Rayan 1998). The compression of the brachial plexus is more common than the stenosis of the subclavian artery but, if the latter is present, it may be clearly depicted by CE-MRA. In particular, CE-MRA should be performed with the arms in the usual position inside the magnet bore, which is with arms in adduction and also with arms in hyperabduction, which is one of the provocative maneuvers described for the clinical diagnosis of the syndrome. In the latter position, CE-MRA may show the narrowing or occlusion of the subclavian artery. Moreover, morphologic MR imaging may depict the cause of arterial compression (Dymarkowski et al. 1999) (Fig. 11.5).

References

Arend WP, Michel BA, Bloch DA, et al. (1990) The American College of Rheumathology 1990 criteria for the classification of Takayasu arteritis. Arthritis Rheum 33:1129–1134

Bongartz G, Mayr M, Bilecen D (2008) Magnetic resonance angiography (MRA) in renally impaired patients: when and how. Eur J Radiol 66:213–219

Carpenter JP, Barker CF, Roberts B, et al. (1994) Popliteal artery aneurysms: Current management and outcome. J Vasc Surg 19:65–72

Carpenter JP, Holland GA, Golden MA, et al. (1997) Magnetic resonance angiography of the aortic arch. J Vasc Surg 25:145–151

Carriero A, Maggialetti A, Pinto D, et al. (2002) Contrast-enhanced magnetic resonance angiography MoBI-track in the study of peripheral vascular disease. Cardiovasc Intervent Radiol 25:42–47

Chan HHL, Tai KS, Yip LKC (2005) Patient with Leriche's syndrome and concomitant superior mesenteric aneurysm: evaluation with contrast-enhanced three-dimensional magnetic resonance angiography, computed tomography angiography and digital subtraction angiography. Australas Radiol 49:233–237

Choe YH, Han B, Koh E, et al. (2000) Takayasu's arteritis: assessment of disease activity with contrast-enhanced MR imaging. AJR Am J Roentgenol 175:505–511

Dawson I, van Bockel JH, Brand R, et al. (1991) Popliteal artery aneurysms: Long-term follow-up of aneurysmal disease and results of surgical treatment. J Vasc Surg 13:398–407

Dellegrottaglie S, Sanz J, Macaluso F, et al. (2007) Technology insight: magnetic resonance angiography for the evaluation of patients with peripheral arterial disease. Nat Clin Pract Cardiovasc Med 4(12):677–687

Deutschmann HA, Schoellnast H, Portugaller HR, et al. (2006) Routine use of three-dimensional contrast-enhanced moving-table MR angiography in patients with peripheral arterial occlusive disease: comparison with selective digital subtraction angiography. Cardiovasc Intervent Radiol 29:762–770

Drescher R, Haller S, Köster O, et al. (2006) Standard-protocol moving-table magnetic resonance angiography for planning of interventional procedures in patients with peripheral vascular occlusive disease. Clin Imaging 30:382–387

Dymarkowski S, Bosmans H, Marchal G, et al. (1999) Three-dimensional MR angiography in the evaluation of thoracic outlet syndrome. AJR Am J Roentgenol 173:1005–1008

Earls JP, Rofsky NM, DeCorato DR, et al. (1996) Breath-hold single-dose gadolinium-enhanced three-dimensional MR aortography: usefulness of a timing examination and MR power injector. Radiology 201:705–710

Elias DA, White LM, Rubinstein JD, et al. (2003) Clinical evaluation MR imaging features of popliteal artery entrapment and cystic adventitial disease. AJR Am J Roentgenol 180:627–632

Ersoy H, Rybicki FJ (2008) MR angiography of the lower extremities. AJR Am J Roentgenol 190:1675–1684

Flynn PD, Delany DJ, Gray HH (1993) Magnetic resonance angiography in subclavian steal syndrome. Br Heart J 70:193–194

Foo TKF, Saranathan M, Prince MR, et al. (1997) Automated detection of bolus arrival and initiation of data acquisition in fast, three-dimensional, gadolinium-enhanced MR angiography. Radiology 203:275–280

Gjonnaess E, Morken B, Sandbaek G, et al. (2006) Gadolinium-enhanced magnetic resonance angiography, color duplex and digital subtraction angiography of the lower limb arteries from the aorta to the tibio-peroneal trunk in patients with intermittent claudication. Eur J Vasc Endovasc Surg 31:53–58

Gotway MB, Araoz PA, Macedo TA, et al. (2005) Imaging findings in Takayasu's arteritis. AJR Am J Roentgenol 184: 1945–1950

Goyen M, Edelman M, Perreault P, et al. (2005) MR angiography of aortoiliac occlusive disease: a phase III study of the safety and effectiveness of the blood-pool contrast agent MS-325. Radiology 236:825–833

Graham LM, Zelenock GB, Whitehouse WM, et al. (1980) Clinical significance of arteriosclerotic femoral artery aneurysms. Arch Surg 115:502–507

Halldorsson A, Ramsey J, Gallagher C, et al. (2007) Proximal left subclavian artery aneurysms: a case report and review of the literature. Angiology 58:367–371

Ho KY, de Haan MV, Oei TK, et al. (1997) MR angiography of the iliac and upper femoral arteries using four different inflow techniques. AJR Am J Roentgenol 169:45–53

Ho KY, Leiner T, de Haan MV, et al. (1998) Peripheral vascular tree stenosis: evaluation with moving bed infusion tracking MR angiography. Radiology 206:683–692

Ho KY, Leiner T, de Haan MV, et al. (1999) Peripheral MR angiography. Eur Radiol 9(9):1765–1774

Holden A, Merrilees S, Mitchell N, et al. (2008) Magnetic resonance imaging of popliteal artery pathologies. Eur J Radiol 67:159–168

Huber A, Scheidler J, Wintersperger B, et al. (2003) Moving-table MR angiography of the peripheral runoff vessels: comparison of body coil and dedicated phased array coil systems. AJR Am J Roentgenol 180:1365–1373

Insko EK, Carpenter JP (2004) Magnetic resonance angiography. Semin Vasc Surg 17(2):83–101

Kerr GS, Hallahan CW, Giordano J, et al. (1994) Takayasu arteritis. Ann Intern Med 120:919–929

Koelemay M, Lijmer J, Stoker J, et al. (2001) Magnetic resonance angiography for the evaluation of lower extremity arterial disease: a meta-analysis. JAMA 285: 1338–1345

Korosec FR, Frayne R, Grist TM, et al. (1996) Time-resolved contrast-enhanced 3D MR angiography. Magn Reson Med 36:345–351

Krinsky G, Rofsky N, Flyer M, et al. (1996) Gadolinium-enhanced three-dimensional MR angiography of acquired arch vessel disease. AJR Am J Roentgenol 167:981–987

Krinsky G, Jacobowitz G, Rofsky G (1998) Gadolinium-enhanced MR angiography of extraanatomic arterial bypass grafts. AJR Am J Roentgenol 170:735–741

Lapeyre M, Kobeiter H, Desgranges P, et al. (2005) Assessment of critical limb ischemia in patients with diabetes: comparison of MR angiography and digital subtraction angiography. AJR Am J Roentgenol 185:1641–1650

Lauffer RB, Parmelee DJ, Dunham SU, et al. (1998) MS-325: albumin-targeted contrast agent for MR angiography. Radiology 207:529–538

Lee VS, Martin DJ, Krinsky GA, et al. (2000) Gadolinium-enhanced MR angiography: artifacts and pitfalls. AJR Am J Roentgenol 175:197–205

Leiner T, Kessels AGH, Nelemans PJ, et al. (2005) Peripheral arterial disease: comparison of color duplex US and contrast-enhanced MR angiography for diagnosis. Radiology 235:699–708

Leiner T, Nijenhuis RJ, Maki JH, et al. (2004) Use of a three-station phased array coil to improve peripheral contrast-enhanced magnetic resonance angiography. J Magn Reson Imaging 20:417–425

Link J, Steffens JC, Brossmann J, et al. (1998) Contrast-enhanced MR angiography in Leriche's syndrome. Rofo 169(1):22–26

Loewe C, Schillinger M, Haumer M, et al. (2004) MRA versus DSA in the assessment of occlusive disease in the aortic arch vessels: accuracy in detecting the severity, number, and length of stenoses. J Endovasc Ther 11:152–160

Macedo TA, Johnson CM, Hallet JW, et al. (2003) Popliteal artery entrapment syndrome: role of imaging in the diagnosis. AJR Am J Roentgenol 181:1259–1265

Makhoul RG (1997a) Popliteal artery aneurysms. In: Sabiston JC (ed) Textbook of surgery: the biological basis of modern surgical practice, 15th edn. W.B. Saunders, Philadelphia, PA, pp 1675–1678 (italian version)

Makhoul RG (1997b) Femoral artery aneurysms. In: Sabiston JC (ed) Textbook of surgery: the biological basis of modern surgical practice, 15th edn. W.B. Saunders, Philadelphia, PA, pp 1673–1675 (italian version)

McCann RL, Schwartz LB, Pieper KS (1991) Vascular complications of cardiac catheterization. J Vasc Surg 14:375–381

McDermott VG, Meakem TJ, Carpenter JP, et al. (1995) Magnetic resonance angiography of the distal lower extremity. Clin Radiol 50:741–746

Meaney JFM (2003) Magnetic resonance angiography of the peripheral arteries: current status. Eur Radiol 13:836–852

Meaney JFM, Ridgway JP, Chakraverty S, et al. (1999) Stepping-table gadolinium-enhanced digital subtraction MR angiography of the aorta and lower extremities arteries: preliminary experience. Radiology 211:59–67

Nikolaou K (2006) Whole-body MR angiography using the intravascular contrast agent Vasovist®. In: Goyen M (ed) Real whole-body MRI. ABW Wissenschaftsverlag GmbH, Berlin, pp 38–46

Nikolaou K, Kramer H, Grosse C, et al. (2006) High-spatial-resolution multistation MR angiography with parallel imaging and blood pool contrast agent: initial experience. Radiology 241:861–872

Owen RS, Carpenter JP, Baum RA, et al. (1992) Magnetic resonance imaging of angiographically occult runoff vessels in peripheral arterial occlusive disease. N Engl J Med 326:1577–1581

Planken NR, Tordoir JH, Duijm LE, et al. (2008) Magnetic resonance angiographic assessment of upper extremity vessels prior to vascular access surgery: feasibility and accuracy. Eur Radiol 18:158–167

Prince MR, Chenevert TL, Foo TKF, et al. (1997) Contrast-enhanced abdominal MR angiography: optimization of imaging delay time by automating the detection of contrast material arrival in the aorta. Radiology 203:109–114

Prince MR, Yucel EK, Kaufman JA, et al. (1993) Dynamic gadolinium-enhanced three-dimensional abdominal MR arteriography. J Magn Reson Imaging 3:877–881

Ramesh S, Michaels JA, Galland RB (1993) Popliteal aneyrysm: Morphology and management. Br J Surg 80:1531–1553

Rayan GM (1998) Thoracic outlet syndrome. J Shoulder Elbow Surg 7:440–451

Reid SK, Pagan-Marin HR, Menzoian JO, et al. (2001) Contrast-enhanced moving-table MR angiography: prospective comparison to cathether ateriography for treatment planning peripheral arterial occlusive disease. J Vasc Interv Radiol 12(1):45–53

Ruehm SG, Weishaupt D, Debatin JF (2000) Contrast-enhanced MR angiography in patients with aortic occlusion (Leriche syndrome). J Magn Reson Imaging 11:401–410
Sabiston DC (1997b) Aneurysms. In: Sabiston JC (ed) Textbook of surgery: the biological basis of modern surgical practice, 15th edn. W.B. Saunders, Philadelphia, PA, p 1638 (italian version)
Sabiston DC (1997c) Subclavian artery aneurysms. In: Sabiston JC (ed) Textbook of surgery: the biological basis of modern surgical practice, 15th edn. W.B. Saunders, Philadelphia, PA, p 1662 (italian version)
Sabiston DC (1997d) Takayasu's arteritis. In: Sabiston JC (ed) Textbook of surgery: the biological basis of modern surgical practice, 15th edn. W.B. Saunders, Philadelphia, PA, pp 1679–1681 (italian version)
Sabiston DC (1997a) Leriche's syndrome. In: Sabiston JC (ed) Textbook of surgery: the biological basis of modern surgical practice, 15th edn. W.B. Saunders, Philadelphia, PA, pp 1689–1691 (italian version)
Shadman R, Criqui MH, Bundens WP, et al. (2004) Subclavian artery stenosis: Prevalence, risk factors, and association with cardiovascular diseases. J Am Coll Cardiol 44:618–623
Slocum MM, Silver D (1997) Alterations in upper limbs circulation. In: Sabiston JC (ed) Textbook of surgery: the biological basis of modern surgical practice, 15th edn. W.B. Saunders, Philadelphia, PA, pp 1747–1749 (italian version)
Tatli S, Lipton MJ, Davison BD, et al. (2003) MR imaging of aortic and peripheral vascular disease. Radiographics 23:S59–S78
Thomsen HS (2006) Nephrogenic systemic fibrosis: a serious late adverse reaction to gadodiamide. Eur Radiol 16:2619–2621
Tordoir JH, Mickley V (2003) European guidelines for vascular access: clinical algorithms on vascular access for hemodialysis. Edtna Erca J 29:131–136
Turnipseed WD (2002) Popliteal entrapment syndrome. J Vasc Surg 35:910–915
Unger EC, Schilling JD, Awad AN, et al. (1995) MR angiography of the foot and ankle. J Magn Reson Imaging 5:1–5
Utsunomiya D, Sawamura T (2007) Popliteal artery entrapment syndrome: Non-invasive diagnosis by MDCT and MRI. Australas Radiol 51:B101–B103
Van Grimberge F, Dymarkowski S, Budts W, et al. (2000) Role of magnetic resonance in the diagnosis of subclavian steal syndrome. J Magn Reson Imaging 12:339–342
Vavrik J, Rohrmoser G, Madani B, et al. (2004) Comparison of MR angiography versus digital subtraction angiography as a basis for planning treatment of lower limb occlusive disease. J Endovasc Ther 11:294–301
Visser K, Hunink MGM (2000) Peripheral arterial disease: gadolinium-enhanced MR angiography versus color-guided duplex US-A meta-analysis. Radiology 216:67–77
Vogt FM, Herborn CU, Parsons EC, et al. (2007) Diagnostic performance of contrast-enhanced MR angiography of the aortoiliac arteries with the blood pool agent Vasovist: initial results in comparison to intraarterial DSA. Rofo 179(4):412–420
Wasser MN (2003) MRA of peripheral arteries. In: Higgins CB, de Roos A (eds) Cardiovascular MRI and MRA. Lippincot Williams & Wilkins, Philadelphia, PA, pp 415–431
Willinek WA, von Falkenhausen M, Born M, et al. (2005) Noninvasive detection of steno-occlusive disease of the supra-aortic arteries with three-dimensional contrast-enhanced magnetic resonance angiography. Stroke 36:38–43
Winterer JT, Schleffer K, Paul G, et al. (2000) Optimization of contrast-enhanced MR angiography of the hands with a timing bolus and elliptically reordered 3D pulse sequences. J Comput Assist Tomogr 24(6):903–908
Winterer JT, Schaefer O, Uhrmeister P, et al. (2002) Contrast enhanced MR angiography in the assessment of relevant stenoses in occlusive disease of the pelvic and lower limb arteries: diagnostic value of a two-step examination protocol in comparison to conventional DSA. Eur J Radiol 41:153–160
Woo EY, Fairman RM, Velazquez OC, et al. (2006) Endovascular therapy of symptomatic innominate-subclavian arterial occlusive lesions. Vasc Endovasc Surg 40(1):27–33
Wyttenbach R, Gianella S, Alerci M, et al. (2003) Prospective blinded evaluation of Gd-DOTA- versus Gd-BOPTA-enhanced peripheral MR angiography, as compared with digital subtraction angiography. Radiology 227:261–269
Yamada I, Nakagawa T, Himeno Y, et al. (2000) Takayasu arteritis: diagnosis with breath-hold contrast-enhanced three-dimensional MR angiography. J Magn Reson Imaging 11:481–487
Yamada I, Numano F, Suzuki S (1993) Takayasu arteritis: evaluation with MR imaging. Radiology 188:89–94

Special Topics: Transplants

MRA in Liver and in Orthotopic Liver Transplants

12

Piero Boraschi, Francescamaria Donati, Simonetta Salemi, and Francesca Turini

CONTENTS

12.1 MRA in Liver *145*
12.1.1 Introduction *145*
12.1.2 Vascular Anatomy *146*
12.1.3 Vascular Imaging *147*
12.1.4 MR Angiography *148*
12.1.4.1 MRA Study Technique *149*
12.1.4.2 MRA Postprocessing *150*

12.2 MRA in Orthotopic Liver Transplants *150*
12.2.1 Introduction *150*
12.2.2 Surgery in Orthotopic Liver Transplants *150*
12.2.3 Complications After Orthotopic Liver Transplants *150*
12.2.3.1 Vascular Complications *151*
12.2.3.2 Vascular Complications: Imaging *153*
12.2.4 MR Angiography in OLT *153*

12.3 Conclusions *154*

References *154*

Piero Boraschi, MD
Francescamaria Donati, MD
Simonetta Salemi, MD
2nd Unit of Radiology, Department of Oncologic and Radiological Sciences, Pisa University Hospital, Via Paradisa 2, 56124 Pisa, Italy
Francesca Turini, MD
Diagnostic and Interventional Radiology, Department of Oncology, Transplants and Advanced Technologies in Medicine, University of Pisa, Via Roma 67, 56126 Pisa, Italy

ABSTRACT

MR imaging provides the necessary information about variations in the vascular and biliary anatomy and allows evaluation of the hepatic parenchyma for diffuse or focal abnormalities.

MR angiography is performed in very specific cases and in particular for getting a detailed knowledge of the hepatic angio-architecture, a prerequisite for successful, uncomplicated liver surgery.

Actually, 3D MR angiography can be a useful, noninvasive technique for evaluating vascular anatomy in preoperative imaging and an appropriate substitute for repetitive follow-up examinations in liver transplant recepients for early and accurate diagnosis of vascular complications.

12.1 MRA in Liver

12.1.1 Introduction

Currently, dynamic, contrast material-enhanced magnetic resonance imaging is performed in very specific cases, particularly for getting a detailed knowledge of the hepatic angio-architecture, a prerequisite for successful, uncomplicated liver surgery.

Indeed, a critical factor for a variety of liver surgeries including chemoembolization, transplantation, tumor resection, and laparoscopic hepatobiliary surgery is the presurgical planning of vascular anastomosis (Sahani et al. 2004).

The advantage of this technique is that it can be used as a single imaging test for comprehensive noninvasive presurgical evaluation. In fact MR imaging provides the necessary information about variations in the vascular and biliary anatomy and allows evaluation of the hepatic parenchyma for diffuse or focal abnormalities.

12.1.2 Vascular Anatomy

Preoperative imaging, especially in the orthotopic liver transplantation (OLT) candidates, has three main objectives:

1. Assessment of the arterial, portal and systemic venous anatomy and patency, in order to detect vascular variants or abnormalities and mismatches in donor and recipient vascular size.
2. Assessment of collateral circulation as a result of portal hypertension. This is needed because many patients who are candidates for liver transplantation are cirrhotic and the checking of their portal system is crucial (*Caramella and Perri 2008*).
3. Assessment of the presence or absence of malignancy (Boevea et al. 2001).

Knowledge of the exact vascularization of the liver is of utmost importance because variations in the arterial and venous anatomy occur in a large number of patients.

The classic hepatic arterial anatomy, with the proper hepatic artery (PHA) dividing into right and left hepatic arteries, is observed in approximately 55% of the population (Fig. 12.1).

In the standard, or classic, visceral anatomy, the celiac axis gives rise to three branches. The first branch is the left gastric artery (LGA), after which the vessel divides into the splenic artery and the common hepatic artery (CHA). The CHA then bifurcates into the gastro-duodenal artery (GDA) and PHA, and the PHA bifurcates into the right hepatic artery (RHA) and the left hepatic artery (LHA) (Covey et al. 2002).

It is important to recognize cases which present an anatomic variation, in order to plan the surgery (Figs. 12.2 and 12.3) (Sahani et al. 2004).

The Michel classification of hepatic arterial anatomy with the approximate frequency of occurrence of each type of variant in the general population is shown in Table 12.1 (Michel 1955).

In the classic hepatic venous anatomy, three main hepatic veins drain into the inferior vena cava (IVC). The left hepatic vein drains segments II and III, the middle hepatic vein drains segments IV, V and VIII, and the right hepatic vein drains segments V–VII. In approximately 60% of the population, the middle and left hepatic veins join to form a common trunk, which drains separately into the IVC (Soyer et al. 1996; Lerut et al. 1997).

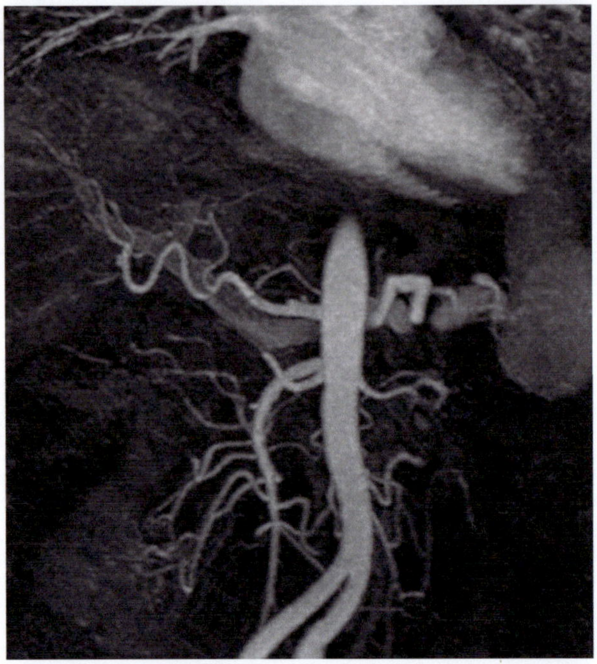

Fig. 12.1. Normal course and branching pattern of the hepatic artery. Coronal oblique MIP image from arterial phase of MR angiography shows the common hepatic artery (CHA) and the proper hepatic artery divides into the left hepatic artery and right hepatic artery (RHA)

Information relevant for the surgeon concerns the pattern of venous drainage into the IVC, the presence and number of accessory veins, as well as their size and distance from the main hepatic venous drainage site along the IVC. When a parenchymal dissection plane is planned, even small hepatic venous branches need to be left intact or reconstructed in order to avoid venous congestion, segmental necrosis and atrophy (Catalano et al. 2008).

The normal portal venous anatomy consists of the main portal trunk branching, at the porta hepatis, into the right and left portal veins, with the right portal vein subsequently dividing into anterior and posterior branches (Lerut et al. 1997), but portal vein trifurcation can be found in 10%–16% of patients (Fig. 12.4). Although this variant does not represent a contraindication to surgery, it needs to be known before the operation because extra surgical steps are necessary for isolation of portal vein branches (Catalano et al. 2008).

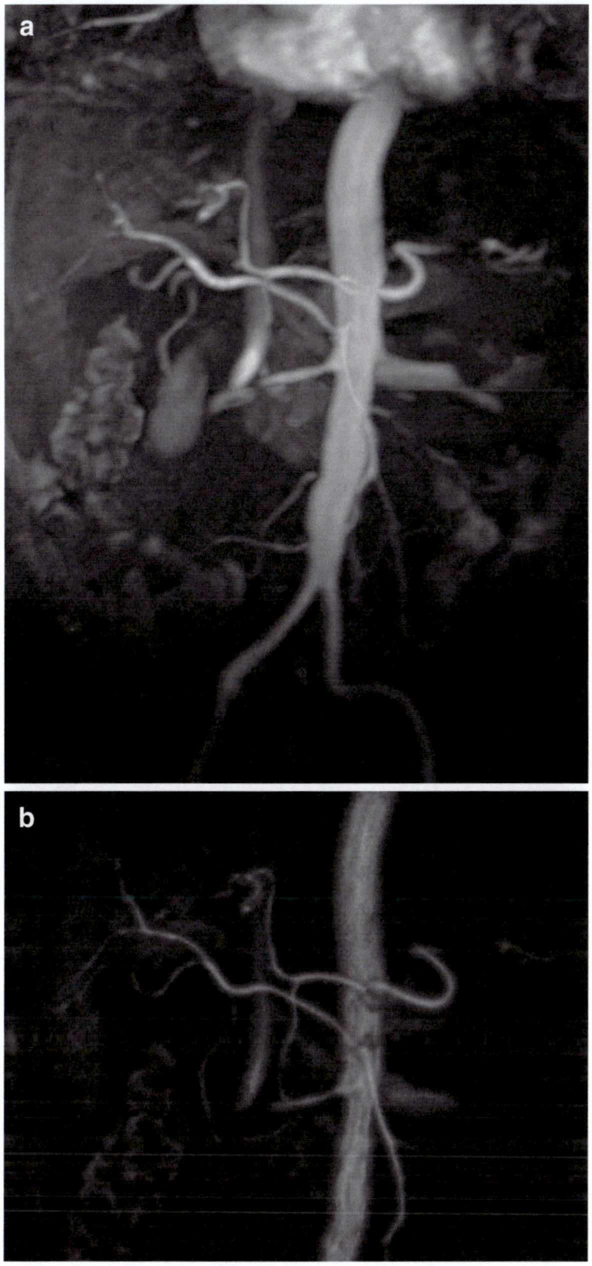

Fig. 12.2. Replaced RHA. Coronal oblique MIP (**a**) and VR (**b**) reconstructions from arterial phase of MRA demonstrate a replaced RHA from the superior mesenteric artery

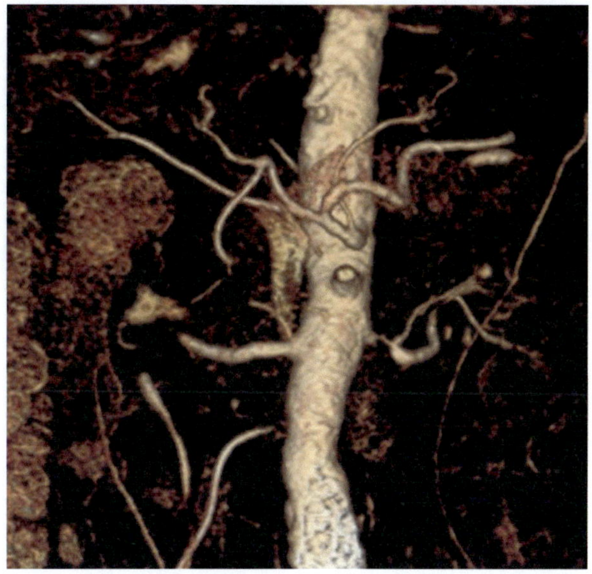

Fig. 12.3. Replaced RHA. Coronal oblique VR image from arterial phase of MRA reveals the RHA arising from CHA

Table 12.1. Michel classification of the hepatic arterial anatomy

I	55%	Hepatic trunk arise from common hepatic artery
II	10%	Replaced LHA arises from left gastric artery
III	11%	Replaced RHA arises from superior mesenteric artery
IV	1%	Replaced RHA and LHA
V	8%	Accessory LHA arises from left gastric artery
VI	7%	Accessory RHA arises from superior mesenteric artery
VII	1%	Accessory RHA and LHA
VIII	4%	Replaced RHA and accessory LHA or replaced LHA and accessory RHA
IX	4.5%	Entire hepatic trunk arises from superior mesenteric artery
X	0.5%	Entire hepatic trunk arises from left gastric artery

12.1.3 Vascular Imaging

Different approaches can be used for assessing the vascularisation.

Digital subtraction angiography (DSA) has long been considered the standard of reference for evaluation of the hepatic arterial anatomy. However, the morbidity and mortality associated with DSA, and its limitations in demonstrating the hepatic venous anatomy, have stimulated the development of non invasive methods of displaying the vascular anatomy (Sahani et al. 2004).

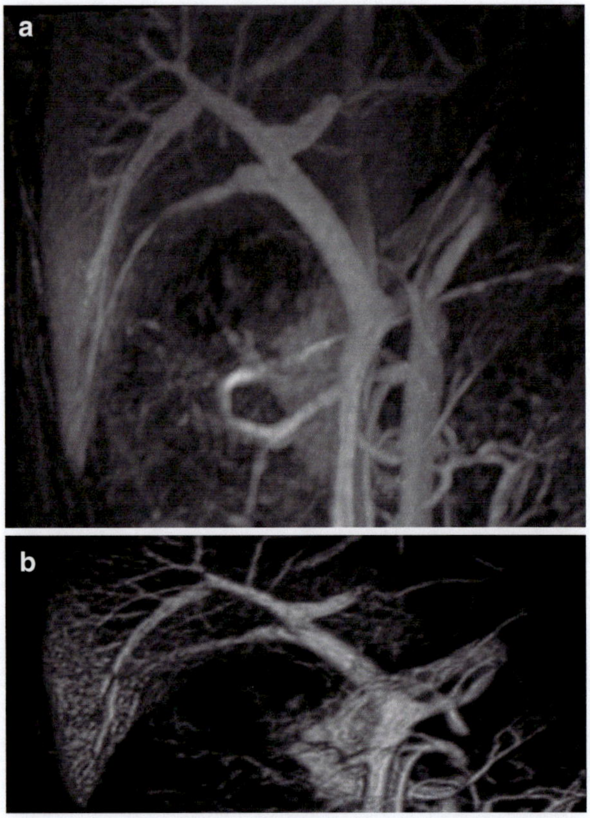

Fig. 12.4. Portal vein trifurcation. Coronal oblique MIP (**a**) and VR (**b**) images from portal phase of MRA show trifurcation of the portal vein into the right anterior, right posterior, and left portal veins

The current tendency is to avoid invasive imaging techniques and to rely instead upon non invasive techniques, such as ultrasound (US), multi-detector computed tomography (MDCT) and magnetic resonance (MR) imaging (Catalano et al. 2008).

The use of US evaluation in the pretransplantation phase requires Doppler examination of the portal and hepatic veins and real-time B-mode imaging of hepatic parenchyma and bile ducts (Leung et al. 1999; Pannu et al. 2001).

Moreover, Doppler US (DUS) allows the investigation of flow dynamics through the hepatic veins, the portal vein and its branches, and the intra- and extrahepatic parts of the hepatic artery, representing a sensitive tool for the detection of portal vein thrombosis as well as of arterial abnormalities. It also enables an accurate evaluation of the localization and flow of varices and porto-systemic shunts (Caramella and Perri 2008).

However, the limitations of the US examination are in the lack of an overall anatomic view and most importantly in its quality, which is strongly operator dependent (Boevea et al. 2001).

Abdominal computed tomography is a highly accurate method for the evaluation of the hepatic vessels and the liver parenchyma alike.

In particular, MDCT-angiography, used according to a three/four phase acquisition by means of a double-arterial phase followed by a portal-vein dominant phase, provides a complete description of the hepatic arterial and venous anatomy, as well as an accurate evaluation of liver parenchyma.

Several studies show that the use of multi-detector row scanners is effective in angiographic applications and, especially in transplantation candidates, multi-detector row CT angiography (CTA) provides results comparable to those from conventional angiography in the evaluation of the hepatic artery anatomy and in the detection of arterial abnormalities such as severe celiac artery stenosis and splenic or hepatic artery aneurysm that may lead to liver transplantation.

Furthermore, MDCT offers a precise evaluation of splenic, superior mesenteric and portal veins, and is useful in the assessment of complications due to portal hypertension such as the presence of retroperitoneal, spleno-renal, gastro-esophageal or paraumbilical varices, splenomegaly and ascites (Caramella and Perri 2008).

12.1.4 MR Angiography

MR angiography (MRA) is also becoming more and more used in preoperative planning for the assessment of the hepatic vascular anatomy because it combines the advantages of minimal invasiveness with both the assessment of the hepatic parenchymal morphology and the detailed analysis of the vascular anatomy.

However, we have to consider also the limitations of MRA. The spatial resolution of MRA is inferior to that of CTA and DSA.

Smaller segmental vessels that are readily seen at catheter and CTA may not be consistently visualized with MRA.

In addition, patients with pacemakers, metallic hardware or claustrophobia may not be suitable candidates for MR imaging. Furthermore, the longer breath hold required for MRA may introduce unwanted motion artifacts (Sahani et al. 2004).

Despite these limitations of MRA when compared to CTA and DSA, MRA provides a road map for planning angiographic interventions to minimize the dose of iodinated contrast material and catheter manipulation of conventional angiography (STAFFORD-JOHNSON et al. 1998).

Contrast-enhanced MR angiography (CEMRA) has anyway been established as a safe, accurate, noninvasive method of evaluating the hepatic vasculature in a single breath hold.

This modality exploits the transient shortening in blood T1 following the administration of gadolinium based contrast material.

The entire arterial system of importance to the liver blood supply can be evaluated with high diagnostic confidence; the only diagnostic problems are identified in arteries with small diameter. The arterial stenoses are seen in MRA, but a hemodynamic grading of celiac artery stenoses is hardly achieved because of the limited temporal resolution. Furthermore, CEMRA has a tendency to overestimate the stenosis. The causes of this are the limited spatial resolution, artifacts caused by bad timing of the contrast bolus (blurring and ringing) and an intravoxel dephasing component that also affects CEMRA (GLOCKNER et al. 2000).

Anyway, improvements in gradient performance and coil design, and refinements in MRA techniques permit faster imaging with improved spatial resolution (SAHANI et al. 2004).

Faster three-dimensional (3D) acquisition of MR imaging data, together with multiplanar imaging capability, allows excellent depiction of hepatic vessels with no saturation or turbulence-related artifact (SAHANI et al. 2004).

The detection of aneurysms of the visceral arteries with CEMRA is relatively affected by dephasing phenomena induced by turbulence. Usually the aneurysmatic sac is completely filled by contrast agent, with a presumed diagnostic gain.

Furthermore, the image quality and diagnostic confidence associated with CEMRA, in the evaluation of the hepatic and the portal venous systems and the delineation of the intrahepatic veins from the surrounding tissue, have been considered more convenient than those associated with CTA. This has been attributed to the minimal albumin binding of the administered MR contrast agent, thereby leading to an extended intravascular phase and a higher intravascular signal intensity (SCHROEDER et al. 2005) that allows the evaluation of veins patency and the identification of thrombosis.

12.1.4.1
MRA Study Technique

Until recently, MR imaging has been limited to two-dimensional GRE imaging with relatively thick sections (8–10 mm) and intersection gaps. The interpolated 3D MR imaging approach has the advantage of providing volumetric imaging of the entire liver with near-isotropic pixel size of 2.3 mm or less, thereby offering the possibility of improved detection and characterization of small lesions. Moreover, with the 3D sequence, the isotropic pixels pemit valuable angiographic reconstruction of the arterial phase data sets (LAVELLE et al. 2001).

MR imaging protocol usually involves a 1.5-T magnet with a phased-array multi-channel coil. First, axial breath-hold T1-weighted in-phase and opposed-phase gradient-echo images and T2-weighted images of the liver have to be obtained. Next, an axial breath-hold 3D interpolated spoiled gradient-echo sequence is performed after intravenous administration of 0.1 mL/kg of body weight of gadolinium chelate, injected with a power injector at a rate of 2–3 mL/s. Hepatic arterial phase and portal venous phase images are obtained after delays of 15–18 s and 70 s, respectively. Breath-hold images are obtained at end inspiration and usually require less than 25–30 s.

However, numerous improvements in the hardware (faster gradients, higher magnetic field strengths, multiple parallel receiver channels and dedicated coils), the sequence techniques (volume-interpolated breath-hold-sequences, such as VIBE, LAVA, THRIVE), the contrast agents (agents with higher relaxivity, such as Gd-BOPTA and Gadobutrol) and the post-processing software (curved planar reformats (CPR), volume-rendering and image fusion) have become available during the past few years (MICHAELY 2006; HEILMAIER et al. 2007).

In particular, these latter sequences (VIBE, LAVA, THRIVE) are basically very similar to 3D-GRE sequences as they are used for abdominal MRA; yet they sample only one part of the k-space and perform a zero-filling in the 3D (partition) phase-encode direction to complete the k-space. As fast 3D sequences, they are mainly applied for multi-phase studies of the liver, kidneys and pancreas. They are heavily T1-weighted and can provide substantial angiographic information when acquired immediately postcontrast agent administration; they also offer morphologic information (LEE et al. 2000).

Another new type of sequence which is becoming frequently used in abdominal MRA is time-resolved imaging (TRICKS, TREAT). One advantage of time-resolved MRA is that time-intensive bolus timing is avoided, with a temporal resolution of 2 s or less (FINK et al. 2005).

In addition to these above-mentioned technical improvements, alternative k-space trajectories have led to an increased quality of abdominal MRA; in fact, in centric or elliptic-centric MRA, the contrast defining k-space center is acquired first and the venous signal is suppressed as by the time the contrast agent reaches the veins the periphery of k-space is read-out (MICHAELY 2006).

12.1.4.2 MRA Postprocessing

With the introduction of high-performance workstations, source images can be processed on commercially available workstations for multiplanar reformation as well as 3D reconstruction with maximum intensity projection (MIP) CPR and volume rendering (VR). Three-dimensional reformatted images of the hepatic vessels are rendered using MIP, shaded surface display and VR. Source images are most helpful for defining vascular anatomy and 3D images are useful in understanding complex anatomy.

According to Catalano et al. the hepatic veins are best displayed in the axial plane, with the MHV determining the hepatectomy plane; and the portal venous anatomy is best displayed in the coronal plane (CATALANO et al. 2008).

12.2 MRA in Orthotopic Liver Transplants

12.2.1 Introduction

OLT has become an accepted therapy for acute and chronic end-stage liver disease (STANGE et al. 2003).

Today, liver transplant patients have a 5-year survival rate of approximately 75%. The improvement in survival can be attributed to better patient selection and preparation, advances in organ preservation, improved immunosuppressive therapy agents and refinement of surgical techniques (HUSSAIN and NGHIEM 2002).

Patients who are candidates for liver transplantation are those with acute liver failure (fulminant hepatic failure) or chronic liver disease for which conventional treatment is unavailable or has ceased to be effective. Fulminant hepatic failure, a disorder that ordinarily affects young people, causes a large portion of liver tissue to be destroyed in a short time, leading to failure of liver function, changes in mental status (hepatic encephalopathy), infections and kidney failure. Chronic liver diseases that may be treated by transplantation include chronic hepatitis B and C, autoimmune chronic hepatitis, primary biliary cirrhosis, primary sclerosing cholangitis, alcoholic liver disease and drug induced liver injury. Hereditary disorders of metal metabolism including hemochromatosis and Wilson disease (involving disordered iron and copper metabolism) also may require transplantation. Pediatric diseases requiring liver transplantation include biliary atresia and neonatal hepatitis (BORASCHI and DONATI 2004).

12.2.2 Surgery in Orthotopic Liver Transplants

Generally OLT is performed with standard surgical technique using a veno-venous by-pass:

the retro-hepatic vena cava is resected and an end-to-end interposition of the donor vena cava is made (GLANEMANN et al. 2002). Reperfusion of the graft is obtained after completion of both the portal vein and arterial anastomoses. Arterial anastomosis is usually performed between the donor hepatic artery with aortic patch or celiac trunk and the CHA of the recipient at the insertion of the GDA (Figs. 12.5 and 12.6).

In case of stenosis of the celiac trunk or small diameter of the hepatic artery of the recipient, arterial reconstruction is performed by the anastomosis between the donor hepatic artery and the recipient aorta using an (supraceliac or infrarenal) iliac artery interposition graft or connecting the donor hepatic artery and aortic patch with the splenic artery of the recipient (STANGE et al. 2003).

12.2.3 Complications After Orthotopic Liver Transplants

Despite the improvement in survival due to advances in organ preservation, improved immunosuppressive therapy agents and refinement of surgical techniques, there are still significant complications following liver transplantation. These complications mainly include biliary strictures, stones and leakage; arterial and venous stenosis and thrombosis; lymphoprolifeative disorders; recurrent tumors; hepatitis virus C infection; liver abscesses; right adrenal gland hemorrhage; fluid collections and hematomas. The diagnosis of

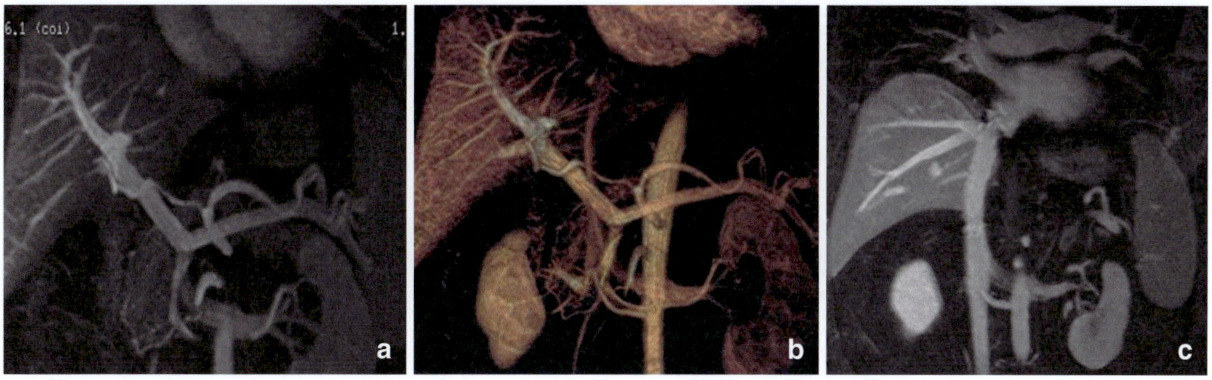

Fig. 12.5. Normal transplanted vascular anastomoses. Coronal oblique MIP (a) and VR (b) reconstructions from gadolinium-enhanced MRA demonstrate both patent portal vein at anastomotic site and normal transplanted hepatic artery. Coronal MIP (c) image from venous phase shows the veno-venous by-pass at the level of the retro-hepatic vena cava obtained with an end-to-end interposition of the donor vena cava

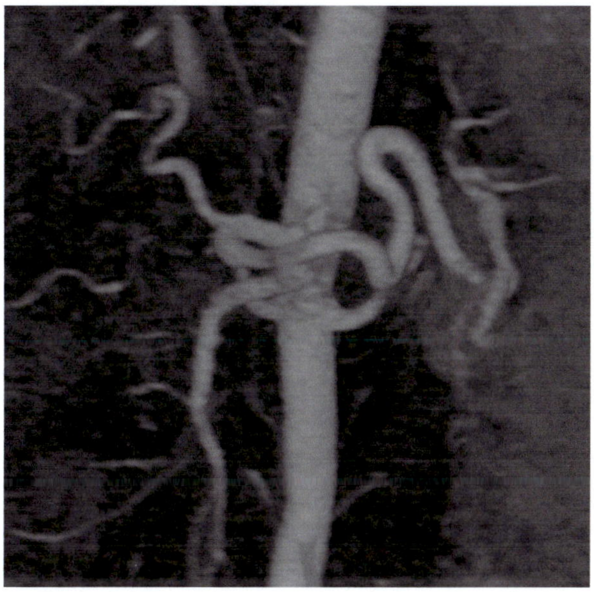

Fig. 12.6. Normal transplanted hepatic artery. Coronal oblique VR image from arterial phase of MRA shows normal transplanted hepatic artery

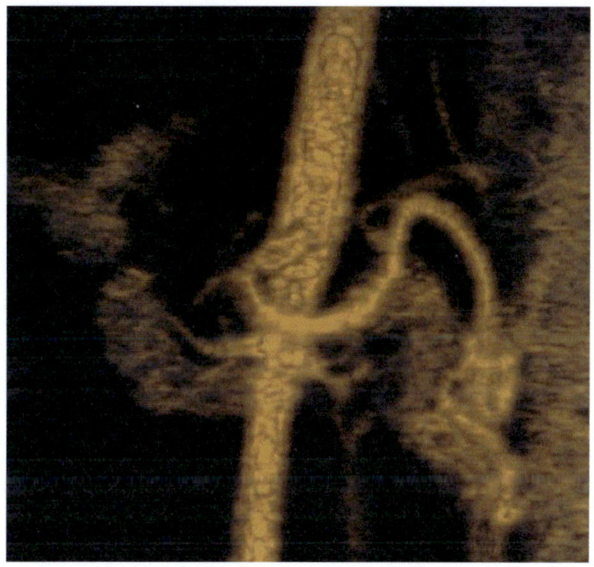

Fig. 12.7. Hepatic artery thrombosis. Coronal oblique VR image from arterial phase of MRA shows abrupt cut-off of hepatic artery

acute rejection, one of the most serious complications following liver transplantation, is established with graft biopsy and histologic study. The role of imaging methods consists of excluding the other complications, which can have clinical signs and symptoms similar to those of acute rejection (Boraschi and Donati 2004; Pandharipande et al. 2001)

12.2.3.1
Vascular Complications

Advances in surgical techniques have contributed to the improved results of living related liver transplantation, which, however, has a high risk of vascular complications because of complex reconstruction of the hepatic artery and portal vein. Thus, a significant percentage of liver grafts are still lost because of vascular complications after living related liver transplantation (Kim et al. 2003).

Vascular complications have been reported in approximately 9% of patients with liver transplants (Boraschi and Donati 2004).

Hepatic artery stenosis and thrombosis are the most common posttransplant vascular complications, and their incidence varies between 6% and 11% (Brancatelli et al. 2002) (Figs. 12.7–12.9).

Early diagnosis of these complications may allow successful treatment, with urgent surgical revascularization of the graft or with percutaneous

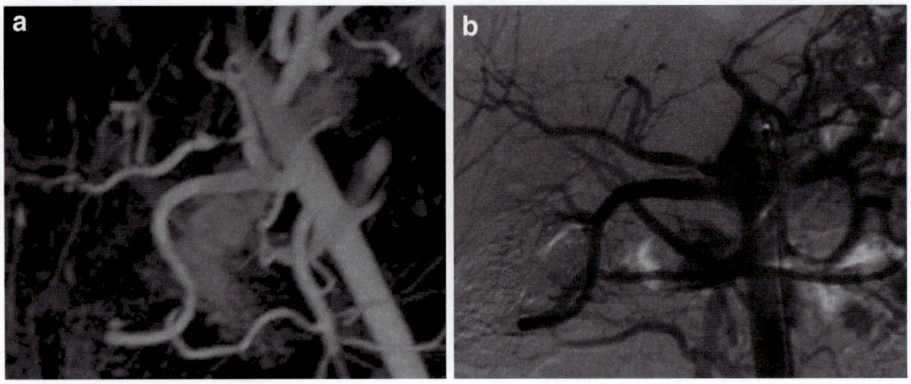

Fig. 12.8. Hepatic artery stenosis after liver transplantation. Coronal oblique VR (**a**) image from arterial phase of MRA shows severe hepatic artery stenosis at the anastomotic site, confirmed at DSA (**b**)

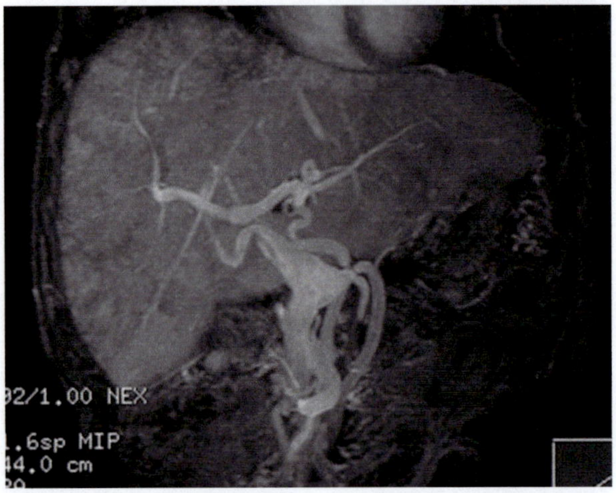

Fig. 12.9. Hepatic artery stenosis in hepatic-aortic anastomosis after liver transplantation. Coronal oblique MIP image from gadolinium-enhanced MRA shows arterial reconstruction, performed by the anastomosis between the donor hepatic artery and aortic patch with the splenic artery of the recipient, and demonstrates hepatic artery stenosis at the anastomotic site

angioplasty, and avoid re-transplantation before the development of severe hepatic failure or overwhelming sepsis (Cavallari et al. 2000).

Hepatic artery pseudoaneurysm is an uncommon complication of liver transplantation, which causes hematobilia, hemoperitoneum and gastrointestinal bleeding that may be life threatening. The most common site for pseudoaneurysm formation is the arterial anastomosis; it is generally accessible by endovascular techniques for transcatheter embolization (Leelaudomlipi et al. 2003).

Clinical presentations of vascular complications vary from mild elevation of hepatic function tests to fulminant hepatic failure. Because the clinical presentation is so varied, their early detection and adequate management are important elements for ensuring graft and patient survival after OLT. For this reason, imaging studies can be particularly helpful in finding or eliminating vascular complications with the differential diagnosis (Boraschi and Donati 2004).

Venous complications are less common and include stenosis and thrombosis of the portal vein or IVC. Portal vein stenosis and thrombosis are relatively rare complications of hepatic transplantation, occurring in fewer than 3% of transplant recipients, most commonly at the anastomotic site (Katyal et al. 2000; Marujo et al. 1991).

Portal vein complications may be asymptomatic or may cause symptoms of portal hypertension (Ito et al. 2000).

Portal vein thrombosis usually requires surgical thrombectomy, venous grafting or re-transplantation, although several studies have reported good results with trans-hepatic balloon dilation for the treatment of portal vein stenosis.

Stenosis of the IVC is a rare complication of hepatic transplantation (0.8–2.8%) (Glanemann et al. 2002).

Supra-hepatic caval stenosis causes hepatic venous obstruction and may lead to Budd-Chiari syndrome. Caval stenosis may occur in the early postoperative phase or several years after transplantation. Its consequences are extremely serious; graft outflow obstruction leads to persistent ascites, edema of the lower limbs, protein-losing enteropathy and progressive deterioration of the graft (Cavallari et al. 2000).

IVC thrombosis is rare, occurring in fewer than 3% of patients. It is caused by technical problems or compression of vessels by a fluid collection (Bowen et al. 1996).

However, if left unrecognized, caval thrombosis will develop in many patients with obstruction of the IVC and may extend into the hepatic veins (Pfammatter et al. 1997).

12.2.3.2
Vascular Complications: Imaging

Clinical presentations of vascular complications vary from mild elevation of hepatic function tests to fulminant hepatic failure. Because the clinical presentation is so varied, their early detection and adequate management are important elements for ensuring graft and patient survival after OLT. For this reason, imaging studies can be particularly helpful in finding or eliminating vascular complications with the differential diagnosis (Boraschi and Donati 2004).

In general OLTs are followed up routinely at 6 and 12 months and yearly after OLT, by techniques including DUS. In case of suspicious findings, further diagnostic investigations with abdominal CT scan, MRI, endoscopic retrograde cholangiography (ERC), or DSA are performed (Stange et al. 2003).

DUS is the primary modality used to evaluate the hepatic artery in the posttransplantation period.

The reported sensitivities of DUS for the detection of hepatic artery stenosis and thrombosis are 80–85% and 60–80%, respectively (Abbasoglu et al. 1997).

However, occasionally, US is technically limited by extensive bowel gas or ascites and can show false-negative results because of the development of extensive arterial collaterals that produce an arterial signal on DUS, leading to the diagnosis of a patent artery.

DSA is the gold standard for diagnosing hepatic artery complications and allows a concomitant angioplasty procedure under some circumstances; however, it is not an ideal screening test because of the associated high cost and invasive nature of the procedure with the associated risks and potential complications (Brancatelli et al. 2002).

CTA, particularly with the development of a multidetector row CT scanner with reconstruction algorithms, has been shown to accurately depict hepatic vasculature and its complications when compared with DSA (Katyal et al. 2000).

Furthermore with CT, higher injection rates and contrast material volumes are needed for optimized angiographic evaluations compared with those for gadolinium-enhanced MR imaging.

12.2.4
MR Angiography in OLT

Early and accurate diagnosis of vascular complications after liver transplantation is critical to the survival of the graft. MRA can demonstrate the entire vascular tree with a single injection of contrast material.

Three dimension CEMRA has recently become an alternative means of accurate, non invasive and rapid evaluation for hepatic vascular disease because image quality can be improved with thinner sections and shorter TR, resulting in a faster scanning time and fewer motion artifacts from incomplete breath-holding. Improved temporal resolution and repeated sequences after contrast administration allow optimal and separate enhancement of the arteries and veins (Watanabe et al. 2000).

A growing body of literature supports the accuracy of 3D CEMRA compared with that of DSA and in particular there has been good agreement between these techniques for the depiction of arterial and venous abnormalities, despite some limitations in small intrahepatic arteries (Hany et al. 1997; Ernst et al. 2000).

Glockner et al. reported high sensitivity of MRA in detecting vascular complications in liver-transplant recipients (86%) (Glockner et al. 2000).

Kim et al. reported a sensitivity of 100% for the detection of hepatic artery stenosis; however, the authors obtained a specificity of 74%, with 7 out of 23 false-positive cases, indicating that CEMRA tends to overestimate hepatic artery stenosis (Kim et al. 2003).

This fact might be explained by the following factors. First, the susceptibility artifact from the surgical clips adjacent to the hepatic arterial anastomotic site creates focal signal loss of the hepatic artery, mimicking stenosis. Second, low-flow states can play an important role in creating false-positive findings of arterial disease. Low-flow states are caused by the inflow of small amounts of contrast material into the intrahepatic artery because of elevated end-organ resistance (i.e., recently transplanted edematous liver, graft rejection) or a decrease of systemic (i.e., systemic hypotension) or localized (i.e., splenic artery steal) blood volume (Pandharipande et al. 2001).

Third, because of the small size of the lobar hepatic artery in living related liver transplantation grafts, subtle vascular signals cannot be distinguished from background signals in a small vessel, thereby producing an overestimation of the amount of stenosis .

Nevertheless, MRA has a high sensitivity and an excellent negative predictive value for the detection of clinically significant vascular stenosis (Kim et al. 2003).

In addition MRA shows a higher sensitivity and accuracy than DSA for the detection of thrombosis or the assessment of vessel patency in any part of the portal venous system (Mori et al. 1992).

MRA, may be also performed to evaluate the extent and degree of the porto-systemic collateral vessels resulting from portal hypertension. It is important that a diagnosis of portal vein stenosis correlates with both the clinical findings and DUS (most cases of portal vein stenosis are accompanied by symptoms of portal hypertension, such as varices, upper gastrointestinal hemorrhage, splenomegaly, massive ascites, hepatic failure and intestinal swelling).

Besides, CEMRA can be used to evaluate the IVC. IVC thrombus is depicted as an intraluminal defect. Coronal scanning is useful for determining the extent of IVC thrombus (Ito et al. 2000).

Furthermore, high-performance work-station and multiplanar reformations have entered the clinical routine. Most published studies on the topic state that volume rendered images are superior to MIP images for the grading of stenoses. Multiplanar reformations also allow for a superior conspicuity of the vascular structures, which is most valuable in case of a complex anatomy (Mallouhi et al. 2002; Baskaran et al. 2002).

12.3 Conclusions

The advances of hardware, software and contrast agents have effectively increased the quality of abdominal MRA. The high image quality and the small number of side effects have made MRA of the abdominal vessels a first line imaging tool for various pathological entities. Actually, 3D MRA can be a useful, noninvasive technique for evaluating vascular anatomy in preoperative imaging and an appropriate substitute for repetitive follow-up examinations for early and accurate diagnosis of vascular compromise in patients post-OLT. Besides, MR imaging can provide a comprehensive assessment of parenchymal, biliary and vascular findings both in presurgical and postsurgical evaluation, restricting the use of invasive procedures to patients for whom therapeutic procedures are advocated or whose MR studies are equivocal (Boraschi et al. 2008).

References

Abbasoglu O, et al. (1997) Hepatic artery stenosis after liver transplantation: incidence, presentation, treatment, and long-term outcome. Transplantation 63:250–255

Baskaran V, et al. (2002) Gadolinium-enhanced 3D MR angiography of renal artery stenosis: a pilot comparison of maximum intensity projection, multi planar reformatting, and 3D volume-rendering postprocessing algorithms. Acad Radiol 9:50–59

Boeve WJ, et al. (2001) Superior diagnostic strength of combined contrast enhanced MR-angiography and MR-imaging compared to intra-arterial DSA in liver transplantation candidates. Magn Reson Imaging 19:609–622

Boraschi P and Donati F (2004) Complications of orthotopic liver transplantation: imaging findings. Abdom Imaging 2:189–202

Boraschi P, et al. (2008) Complications after liver transplantation: evaluation with magnetic resonance imaging, magnetic resonance cholangiography and 3-dimensional contrast-enhanced magnetic resonance angiography in a single session. Can Assoc Radiol J 59:259–263

Bowen A, et al. (1996) Imaging in liver transplantation. Radiol Clin North Am 34:757–778

Brancatelli G, et al. (2002) Three-dimensional multislice helical computed tomography with the volume rendering techniques in the detection of vascular complications after liver transplantation. Transplantation 73:237–242

Caramella D, Perri M (2008) Transplantation Liver. In: Beart AL (ed) Encyclopedia of diagnostic imaging, vol 2. Springer, Berlin, pp 1842–1846

Cavallari A, et al. (2000) Complications of liver transplantation. In: Bücheler E, Nicolas V, Rogiers X, Krupski G (eds) Diagnostic and interventional radiology in liver transplantation. Springer, Berlin, pp 167–177

Catalano O, et al. (2008) Vascular and biliary variants in the liver: implications for liver surgery. Radiographics 28:359–378

Covey AM, et al. (2002) Variant hepatic arterial anatomy revisited: digital subtraction angiography performed in 600 patients. Radiology 224:542–547

Ernst O, et al. (2000) Comparing contrast-enhanced breath-hold MR angiography and conventional angiography in the evaluation of mesenteric circulation. AJR 174:433–439

Fink C, Ley S, Kroeker R, et al (2005) Time-resolved contrast-enhanced threedimensional magnetic resonance angiography of the chest: combination of parallel imaging with view sharing (TREAT). Invest Radiol 40:40–48

Glanemann M, et al. (2002) Results of end-to-end cavocavostomy in adult liver transplantation. World J Surg 26:342–347

Glockner JF, et al. (2000) Three-dimensional gadolinium-enhanced MR angiography of vascular complication after liver transplantation. AJR 174:1447–1453

Hany TF, et al. (1997) Evaluation of the aortoiliac and renal arteries: comparison of breath-hold, contrast-enhanced, three-dimensional MR angiography with conventional catheter angiography. Radiology 204:357–362

Heilmaier C, et al. (2007) Mapping of hepatic vascular anatomy: dynamic contrast-enhanced parallel MR imaging compared with 64-detector row CT. Radiology 245:872–880

Hussain HK and Nghiem HV (2002) Imaging of hepatic transplantation. Clin Liver Dis 6:247–270

Ito K, et al. (2000) MR imaging of complications after liver transplantation. AJR 175:1145–1149

Katyal S, et al. (2000) Detection of vascular complications after liver transplantation: early experience in multislice CT angiography with volume rendering. AJR 175:1735–1739

Kim BS, et al. (2003) Vascular complications after living related liver transplantation: evaluation with gadolinium-enhanced three-dimensional MR angiography. AJR 181:467–474

Lavelle MT, et al. (2001) Dynamic contrast-enhanced three-dimensional mr imaging of liver parenchyma: source images and angiographic reconstructions to define hepatic arterial anatomy. Radiology 218:389–394

Lee VS, et al. (2000) Hepatic MR imaging with adynamic contrast-enhanced isotropic volumetric interpolated breath-hold examination: feasibility, reproducibility, and technical quality. Radiology 215:365–372

Leelaudomlipi S, et al. (2003) Hepatic artery aneurysm in adult liver transplantation. Transplant Int 16:257–261

Lerut JP, et al. (1997) Adult liver transplantation and abnormalities of splanchnic veins: experience in 53 patients. Transplant Int 10:125–132

Leung DA, et al. (1999) Diagnostic imaging in liver transplantation. In: Bartolozzi C, Lencioni R (eds) Liver malignancies. Springer, Berlin, pp 423–442

Mallouhi A, et al. (2002) 3D MR angiography of renal arteries: comparison of volume rendering and maximum intensity projection algorithms. Radiology 223:509–516

Marujo WC, et al. (1991) Vascular complications following orthotopic liver transplantation: outcome and role of urgent revascularization. Transplant Proc 23:1484–1486

Michaely HJ, et al. (2006) MRA of abdominal vessels: technical advances. Eur Radiol 16:1637–1650

Michel NA (1955). Blood supply and anatomy of the upper abdominal organs with a descriptive atlas. Lippincott, Philadelphia, pp 64–69

Mori K, et al. (1992) The introduction of microvascular surgery to hepatic artery reconstruction in living-donor liver transplantation: its surgical advantages compared with conventional procedures. Transplantation 54:263–267

Pandharipande PV, et al. (2001) Vascular and extravascular complications of liver transplantation: comprehensive evaluation with three-dimensional contrast-enhanced volumetric MR imaging and MR cholangiopancreatography. AJR 177:1101–1107

Pannu HK, et al. (2001) Liver transplantation: preoperative CT evaluation. Radiographics 21:133–146

Pfammatter T, et al. (1997) Suprahepatic caval anastomotic stenosis complicating orthotopic liver transplantation: treatment with percutaneous transluminal angioplasty, Wallstent placement, or both. AJR 168:477–480

Sahani D, et al. (2004) Preoperative hepatic vascular evaluation with CT and MR angiography: implications for surgery. Radiographics 24:1367–1380

Schroeder T, et al. (2005) "All-in-one" imaging protocols for the evaluation of potential living liver donors: comparison of magnetic resonance imaging and multidetector computed tomography. Liver Transpl 11:776–787

Soyer P, et al. (1996) Three-dimensional helical CT of intrahepatic venous structures: comparison of three rendering techniques. J Comput Assist Tomogr 20:122–127

Stafford-Johnson DB, et al. (1998) Vascular complications of liver transplantation: evaluation with gadolinium enhanced MR angiography. Radiology 207:153–160

Stange BJ, et al. (2003) Hepatic artery thrombosis after adult liver transplantation. Liver Transpl 9:612–620

Watanabe Y, et al. (2000) Dynamic subtraction contrast-enhanced MR angiography: technique, clinical applications and pitfalls. Radiographics 20:135–152

MRA in Transplanted Pancreas and Kidneys

13

Piero Boraschi, Francescamaria Donati, and Valentina Battaglia

CONTENTS

13.1 **Introduction** *157*

13.2 **Vascular Anatomy in Transplants** *158*
13.2.1 Pancreas *158*
13.2.2 Kidney *158*

13.3 **Transplants and Diagnostic Imaging** *158*

13.4 **MR Angiography** *159*
13.4.1 Study Technique *159*

13.5 **MR Angiography and Vascular Complications of Pancreatic Transplant** *160*

13.6 **MR Angiography and Vascular Complications of Kidney Transplant** *163*

13.7 **Conclusion** *167*

References *167*

ABSTRACT

Thanks to the technical and technological advancements, MRA has been more and more applied in the pre and postoperative evaluation of transplants, being used both in cases of presurgical analysis of kidney living donors, as well as in the follow-up of transplanted pancreas and kidneys in patients, especially in cases of suspected vascular complications.

Actually CEMRA studies show a high diagnostic accuracy in the evaluation of vascular complications of transplanted pancreas and kidneys. CEMRA can therefore become an appropriate method of diagnostic angiography in the detection of transplant vessel stenosis and/or thrombosis, and can help to rapidly and noninvasively select patients for therapeutic interventional angiography, surgical correction or medical management.

13.1 Introduction

Pancreatic and renal transplants represent the definitive treatment for patients affected by diabetes mellitus of type I and chronic renal failure, respectively (Morris et al. 2004).

In these last few years, thanks to the improvements and refinement of surgical techniques and to the efficacy of posttransplant medical care, transplants have been quite frequently performed, showing high rates of peri- and postoperative survival, as well as a decreased peri-operative mortality or organ failure episodes (Hagspiel et al. 2007).

Anyway, transplantation still remains a technically challenging procedure, often affected by serious

Piero Boraschi, MD
Francescamaria Donati, MD
Second Unit of Radiology, Department of Oncologic and Radiological Sciences, Pisa University Hospital, Via Paradisa 2, 56124, Pisa, Italy
Valentina Battaglia MD
Diagnostic and Interventional Radiology, Department of Oncology, Transplants and Advanced Technologiesin Medicine, University of Pisa Via Roma 67, 56126 Pisa, Italy

complications, as for example, sepsis, rejection and, even more frequently, by graft-vessel diseases, which frequently result in transplant necrosis and necessitate graft removal (Krebs et al. 1995).

Since its introduction, magnetic resonance angiography (MRA) has stirred up an increasing interest among the multiple noninvasive exploring techniques, especially because of the widespread field of application. Recently, thanks to the technical and technological advancements, MRA has been more and more applied also in the pre and postoperative evaluation of transplants, being used both in cases of presurgical analysis of kidney living donors, as well as in the follow-up of transplanted patients, especially in cases of suspected vascular complications (Vosshenrich and Fischer 2002).

Nowadays, vascular complications represent in fact the most common cause of early transplant failure, both in cases of renal and pancreatic allograft, while, thanks to the introduction of powerful immunosuppressive drugs, acute or chronic allograft rejection has become only the second cause of graft loss (Hagspiel et al. 2007).

Among all possible diseases, graft vascular thrombosis is the most common cause of vascular pancreatic transplant dysfunction (Grewal et al. 1993; Douzdjian et al. 1993) with an incidence ranging between 2 and 19% (Hagspiel et al. 2007) and it often necessitates pancreatectomy, while in cases of renal transplant, arterial stenosis is the most common cause that usually occurs in a percentage ranging between 1 and 25% (Fervenza et al. 1998).

The importance of obtaining an accurate and early diagnosis of a suspected vascular complication after a pancreatic transplant finds its rationale in the successful treatment of a number of complications, by performing a surgical or endovascular approach, as reported in some published studies (Orsenigo et al. 2003).

13.2 Vascular Anatomy in Transplants

13.2.1 Pancreas

Apart from the different exocrine drainage (enteric or vescical), two types of vascular drainage can be applied when performing a pancreatic transplantation. The first introduced technique foresees the intraperitoneal placement of the allograft in the pelvis and the subsequent anastomosis between the donor splenic and superior mesenteric artery (SMA) and the recipient's iliac arteries by way of a donor Y-graft, with pancreatic venous outflow typically collecting into the common or external iliac vein and thus achieves the systemic circulation (Hagspiel et al. 2007).

A further variant of pancreatic transplantation, reintroduced in 1995, is the commonly referred "porto-enterical" variant. In this technique, the pancreatic allograft is placed intraperitoneally but higher in the recipient's abdomen. The pancreatic arteries, splenic artery and SMAs are anastomosed to the iliac arteries or aorta as well because of the higher position in the abdomen by way of a longer Y-graft formed from the donor's common, internal, and external iliac arteries. Pancreatic venous outflow is in this case achieved by anastomosing the transplant portal vein with the recipient's portal venous system, typically the superior mesenteric vein and less commonly the splenic vein (Heyneman et al. 1999).

13.2.2 Kidney

Different from pancreatic transplantation techniques, the method adopted in cases of renal transplant is simpler and contemplates a direct anastomosis between donor and recipient vessels without the need of a Y-graft. Transplanted kidney is preferentially placed extraperitoneally in the right iliac fossa. Generally the harvest kidney is the left one because of the longer renal vein with respect to that of controlateral. The donor artery is usually anastomosed to the ipsilateral external or common iliac artery, even if in some cases the anastomosis is realized to the recipient hypogastric vessel. The ureter is anastomosed to the bladder (Hohenwalter et al. 2001).

13.3 Transplants and Diagnostic Imaging

As clinical and laboratory analysis can often be insensitive and unspecific in detecting the first signs of allograft rejection, imaging modalities are nowadays considered as the reference methods in order to highlight a suspected graft dysfunction.

Diagnostic imaging, including computed tomography (CT), color-Doppler ultrasonography (US) and digital subtraction angiography (DSA), have demonstrated to be sensitive in detecting pancreatic dysfunction but are relatively unspecific in differentiation between vascular and parenchymal causes of organ failure.

Anyway, among all the different imaging techniques able to visualize and analyse transplanted organs and relative transplanted vessels, in these last few years MR, and particularly MRA, has stirred up an increasing interest thanks to the intrinsic characteristics of the method.

13.4 MR Angiography

MR angiography is actually considered as a valid and accurate alternative of DSA in the evaluation and analysis of suspected pathologies of vessels (TAN et al. 2002). The strong contribution of MRA in the diagnostic imaging of both on list and transplanted patients has markedly risen up during these last few years, mainly because of the characteristics of the technique, as well as of the remarkable improvements of technology.

As well known, the avoidance of ionizing radiations and the possibility to obtain images on multiple spatial planes are two fundamental characteristics of MR imaging that allow the performance of the technique even in patients with chronic renal failure, often permitting tissue characterization without the use of contrast medium. By acquiring conventional MR images of the abdomen before contrast administration, the same MR examination allows to evaluate not only the presence of incidental pathologies but also to accurately evaluate the graft parenchymal status, especially in cases of suspicion of acute or chronic organ failure (CHOYKE et al. 1997).

Moreover, differently from conventional angiography, MRA requires only intravenous administration of gadolinium-based contrast material, thereby averting the catheter-related risks.

Anyway, despite the multiple advantages in performing MRA, the technique shows some disadvantages mainly related to artefacts associated with respiratory and bulk motion, and frequent inadvertent venous enhancement.

Recently, the use of 2D time-of-flight and phase contrast MRA has been clinically abandoned for the study of abdominal vessels, because of the advent of newer contrast-enhanced MRA (CEMRA) techniques that enable the evaluation of the vasculature in less time, with greater detail and showing a greater stability (HAGSPIEL et al. 2007).

Faster 3D gradient-echo MR imaging pulse sequences can be in fact used within the time frame of a single breath hold without loss of spatial resolution, which help in minimizing motion artefacts, usually affecting the previously applied flow-based sequences.

Moreover, the decreased echo times of the newer sequences can help to reduce the signal intensity loss associated with spin de-phasing and therefore may help to minimize the overestimation of stenoses that, to varying degrees, first plagued all bright-blood MRA techniques (BOSMANS et al. 2001).

13.4.1 Study Technique

The evaluation of abdominal vessels with CEMRA must be performed with high-field strength magnets (1.5–3 T), by applying a dedicated phased-array coil (HAGSPIEL et al. 2007). Various improvements in the hardware (faster gradients, higher magnetic field strengths, multiple parallel receiver channels and dedicated coils), the sequence techniques (volume-interpolated breath-hold-sequences), the contrast agents (agents with higher relaxivity, such as Gd-BOPTA and Gadobutrol) and the postprocessing software have become available during the past few years (MICHAELY et al. 2006).

Baseline MR study usually includes axial T1-weighted fast low-angle shot sequences and axial T2-weighted turbo spin-echo sequences with fat suppression, mainly applied in order to identify peri-graft fluid collections and to evaluate transplants parenchyma. In cases of evaluation of pancreatic transplants, it can be also useful to perform T1-weighted images in order to evaluate normal pancreatic signal intensity (WINSTON et al. 1995) and to highlight peri-pancreatic hematoma or hemorrhagic necrosis; moreover, the use of fat suppression can aid in evaluating the signal intensity of the pancreas, which is normally hyperintense relative to the liver (DOBOS et al. 2005).

More recently, CEMRA study includes a breath-hold 3D interpolated spoiled gradient-echo sequence (VIBE, LAVA, THRIVE) in the axial or more frequently in the coronal plane; these latter sequences are basically very similar to 3D-GRE sequences as they are used for abdominal MRA, yet they only sample one part of the k-space and perform a zero-filling in the 3D (partition) phase-encode direction to complete the k-space. As fast 3D sequences, they are mainly applied for multi-phasic studies of the liver, kidneys and pancreas. They are heavily T1-weighted and can provide substantial angiographic information when acquired immediately postcontrast agent administration, but they also offer morphologic information (Lee et al. 2000).

The average amount of contrast media injected to perform a study of the abdominal vessels has been reported to be 0.01–0.02 mmol/kg (15–30 mL), administrated at a flow rate of 2–3 mL/s, with a minimum of two acquisitions performed with an inter-imaging delay of 10 s to depict both the arterial and the venous enhancement phases (Hagspiel et al. 2007).

Imaging acquisition is performed during both the arterial and venous phases of enhancement to evaluate the arterial and venous components of the transplant, respectively, followed by delayed axial fat-suppressed T1-weighted imaging (Dobos et al. 2005).

All examinations are acquired in breath-hold, sometimes with subtraction technique, both for the arterial and the venous phase. Imaging dataset is then analyzed at dedicated workstation, where it is possible to perform also 3D reconstructions such as multi planar volume reconstruction (MPVR), curved planar volume reconstruction (CPVR), maximum intensity projection (MIP), and volume rendering (VR). Even if these algorithms can furnish a paramount vision of the focused abdominal region, artefacts must be always be detected and considered; MIP reconstruction often results in an overestimation of the degree of stenosis and can also obscure small intraluminal filling defects such as dissection flaps, but it has proved to be very useful in the detection of small accessory vessels (Bakker et al. 1998).

13.5 MR Angiography and Vascular Complications of Pancreatic Transplant

The role of MR angiography in the detection of vascular complications of pancreatic transplants has been investigated till now in a limited number of studies (Hagspiel et al. 2007).

Multiphasic breath-hold gadolinium-enhanced 3D MR imaging is nowadays considered as an extraordinarily accurate diagnostic technique, which can provide information to make the specific diagnosis of vessel complications and to assess parenchymal viability after pancreas transplantation (Hagspiel et al. 2005). The technique allows the evaluation of the normal venous and arterial anatomy of pancreatic allograft (Fig. 13.1), showing the appearance of the major vascular complications.

The performance of diagnostic MRA, revealing a vascular complication amenable to intraluminal therapy, would in fact obviate the need for invasive procedure such as diagnostic conventional angiography. Instead, in case of a pathological finding, MRA furnishes anyway an accurate diagnosis, allowing the reduction of imaging frames and the amount of contrast medium when performing an eventual therapeutic DSA.

Vascular thrombosis is the most common cause of pancreatic allograft dysfunction, with a reported incidence of 2–19%, followed by rejection (Fernandez et al. 1991).

Transplanted pancreas is believed to be more vulnerable to thrombosis than renal graft because of the lower rate of blood flow received in transplanted pancreas (approximately the 1.3% instead of the 5% of cardiac output) (Krebs et al. 1995).

Graft thrombosis can be classified as early or late according to the time after surgery.

Early graft thrombosis is usually the result of vascular graft anastomotic error or microvascular damage from preservation injury. Late thrombosis (i.e., that which occurs more than a month after surgery), usually results from severe acute rejection with alloimmune arteritis and causes occlusion of small vessels, which leads to complete major vessel occlusion. Identification of total graft thrombosis and infarction is important, because the transplanted organ should be removed immediately to avoid the severe systemic effects of graft.

The evaluation with MR angiography has moreover the advantage to allow also the graft parenchymal status study. Initially in fact, MR study is performed as a standard baseline MR evaluation, in which T1- and T2-weighted images are performed in the axial plane more frequently than in the coronal plane, especially for the detection of areas of altered signal intensity with respect to normal parenchyma, consisting mainly in either partial or complete lack of enhancement of the graft parenchyma after the intravenous administration of Gadolinium-chelates, as well as the detection of fluid collections (Eubank et al. 2000). Then, contrast enhanced examination permits to highlight areas of altered parenchymal perfusion and moreover the presence of a graft vessel disease.

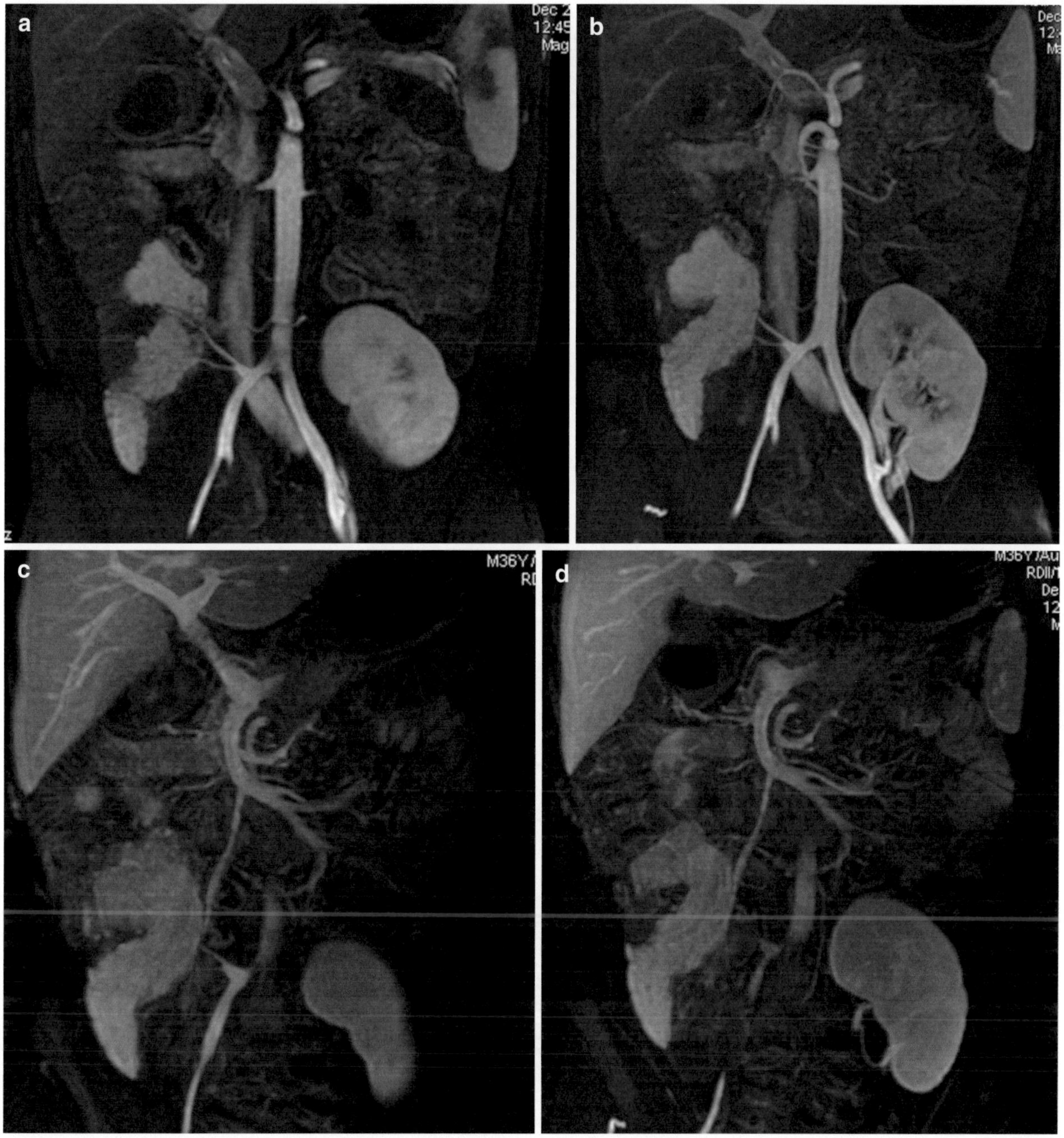

Fig. 13.1. MIP reformations (**a, b**) obtained from arterial phase of CEMRA show the vascularization of the pancreas in a case of combined kidney–pancreas transplantation with enteric-portal pancreatic drainage. Mesenteric and splenic branches of the arterial graft anastomosed with the right common iliac artery are depicted. MIP reconstructions (**c, d**) of the venous vessels from venous phase of CEMRA show the anastomotic junction between the donor's portal vein and the recipient's superior mesenteric vein which allows the insulin drainage in the portal vein

In the evaluation of a transplanted pancreas, the vessels usually analyzed are: (1) the distal aorta, the common, external and internal iliac arteries; the iliac Y-graft as well as the splenic artery (Fig. 13.2), SMA (Fig. 13.3) and the respective anastomoses; (2) the inferior cava vein; the common, external and internal iliac veins, the superior mesenteric vein, the portal vein and their respective anastomoses (Hagspiel et al. 2007).

The signs of vascular pathologies are anyway the same as those depicted in common DSA, as well as in the CT evaluation (Krebs et al. 1995).

The typical appearance of an acute thrombosis on contrast-enhanced MR angiography consists of a very

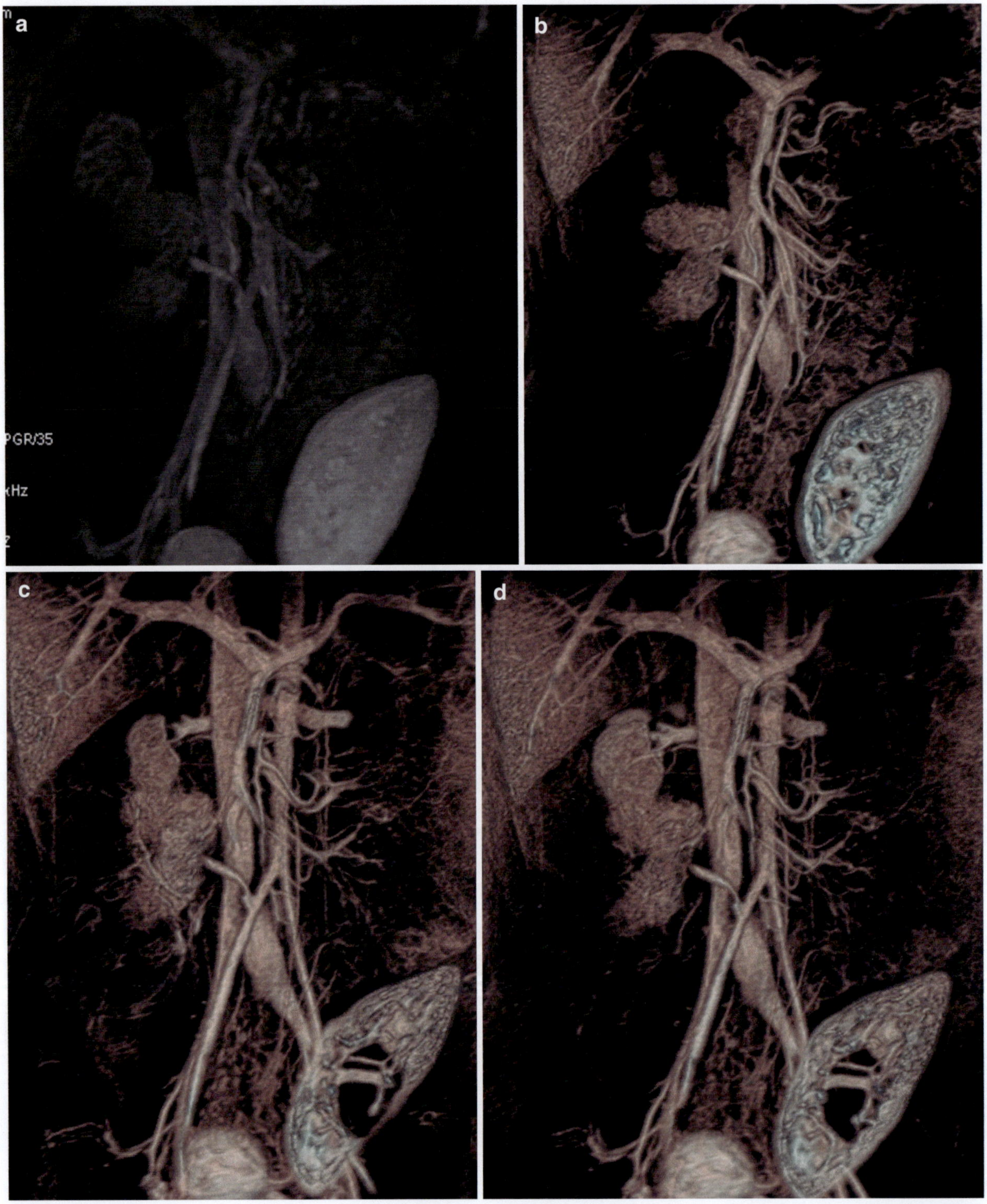

Fig. 13.2. MIP (**a**) and volume-rendered (**b**) images obtained from arterial phase of CEMRA, exhibit the complete thrombosis of the splenic branch of the pancreatic arterial graft in a case of combined kidney–pancreas transplantation with enteric-portal pancreatic drainage. Mesenteric branch of the arterial graft anastomosed with the common iliac artery is identified. Volume-rendered reconstructions (**c, d**) show regular patency of the venous vessels

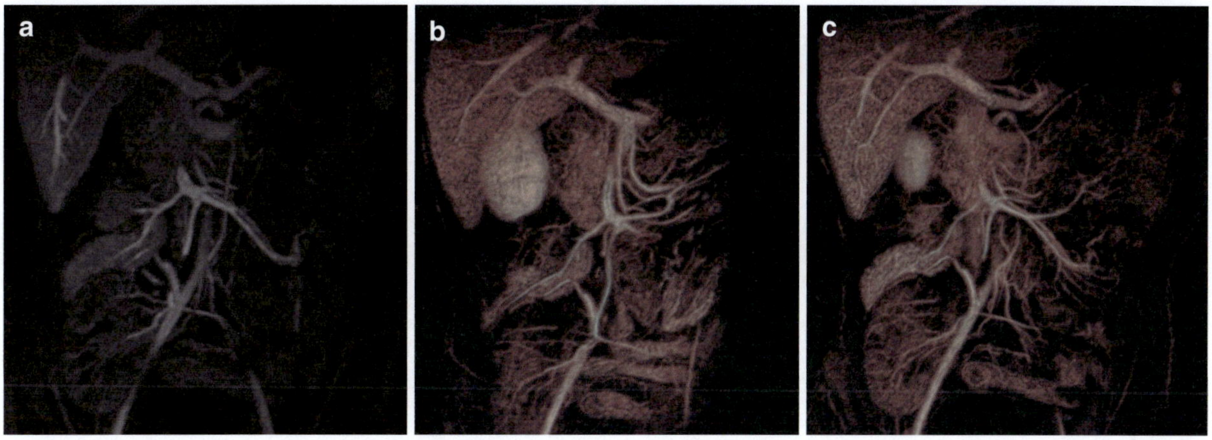

Fig. 13.3. MIP (**a**) and volume-rendered (**b, c**) reconstructions of CEMRA show the complete thrombosis of the mesenteric branch and the mild stenosis of the splenic branch of the pancreatic arterial graft anastomosed with the right common iliac artery in a case of isolated pancreas transplantation with enteric-portal drainage

dark hypointense filling defect occluding the arterial (Fig. 13.4) or the venous lumen (Fig. 13.5), with or without a trailing edge (Hagspiel et al. 2005). In cases of arterial thrombosis, parenchymal status can anyway appear as normal, because patency of only one allograft artery is sufficient to supply adequate perfusion to the entire gland (Hagspiel et al. 2005).

Apart from acute or chronic thromboses, other frequent complications that can occur are stenoses of anastomoses (Figs. 13.3 and 13.4) and kinking of the graft, especially in cases of porto-enteric drainage grafts, in which the longer Y-graft can undergo no such complication. In these cases, MR can furnish a detailed, non invasive, first diagnosis, permitting to accurately assess or exclude the presence of vascular complication and allowing then the performance of a therapeutic DSA only in true positive cases (Hagspiel et al. 2005).

13.6 MR Angiography and Vascular Complications of Kidney Transplant

Kidney transplantation can be single or double (Fig. 13.6).

Vascular thrombosis after renal transplantation usually occurs more frequently than in cases of pancreatic allografts (Krebs et al. 1995).

Vascular complications associated with renal transplantation merit urgent investigation as they are often correctable and timely intervention can help to keep the graft kidney safe. Renal transplantation is associated with a 1–15% incidence of vascular complications (Jain and Sawhney 2005).

The most important complication after renal transplantation is represented by arterial stenosis (TRAS), that can occur either in the native or in the donor artery as well as at the anastomosis (Fig. 13.7) and represents by itself the 75% of all posttransplant vascular complication (Hohenwalter et al. 2001).

It usually becomes apparent between 3 months and 2 years after renal transplantation, but it can present at any time. It frequently shows with worsening or refractory hypertension and/or graft dysfunction in the absence of rejection, ureteric obstruction, or infection.

It has been shown by some recent studies that CEMRA offers a diagnostic evaluation comparable to that of DSA, with the advantage of not exposing the patient to ionizing radiations (Bakker et al. 1998) and it can be applied as a screening method in suspected cases, with DSA reserved to selected patients who may be amenable to angioplasty (Hohenwalter et al. 2001). If the main renal artery shows a normal morphology and a normal intravascular blood flow at CEMRA, DSA can infact be avoided.

Stenosis at the peri-anastomotic site (Fig. 13.8) is the most common vascular complication of renal transplantation, with a 1–15% incidence; in the diagnosis, CEMRA has proved to be highly accurate, showing a sensitivity and specificity of 100 and 96%, respectively (Hohenwalter et al. 2001).

Donor renal artery stenosis is a rarer cause of TRAS: Fibromuscular dysplasia, vasculitis and extrinsic compressions due to fluid collections and masses can result in donor artery stenosis. Renal artery

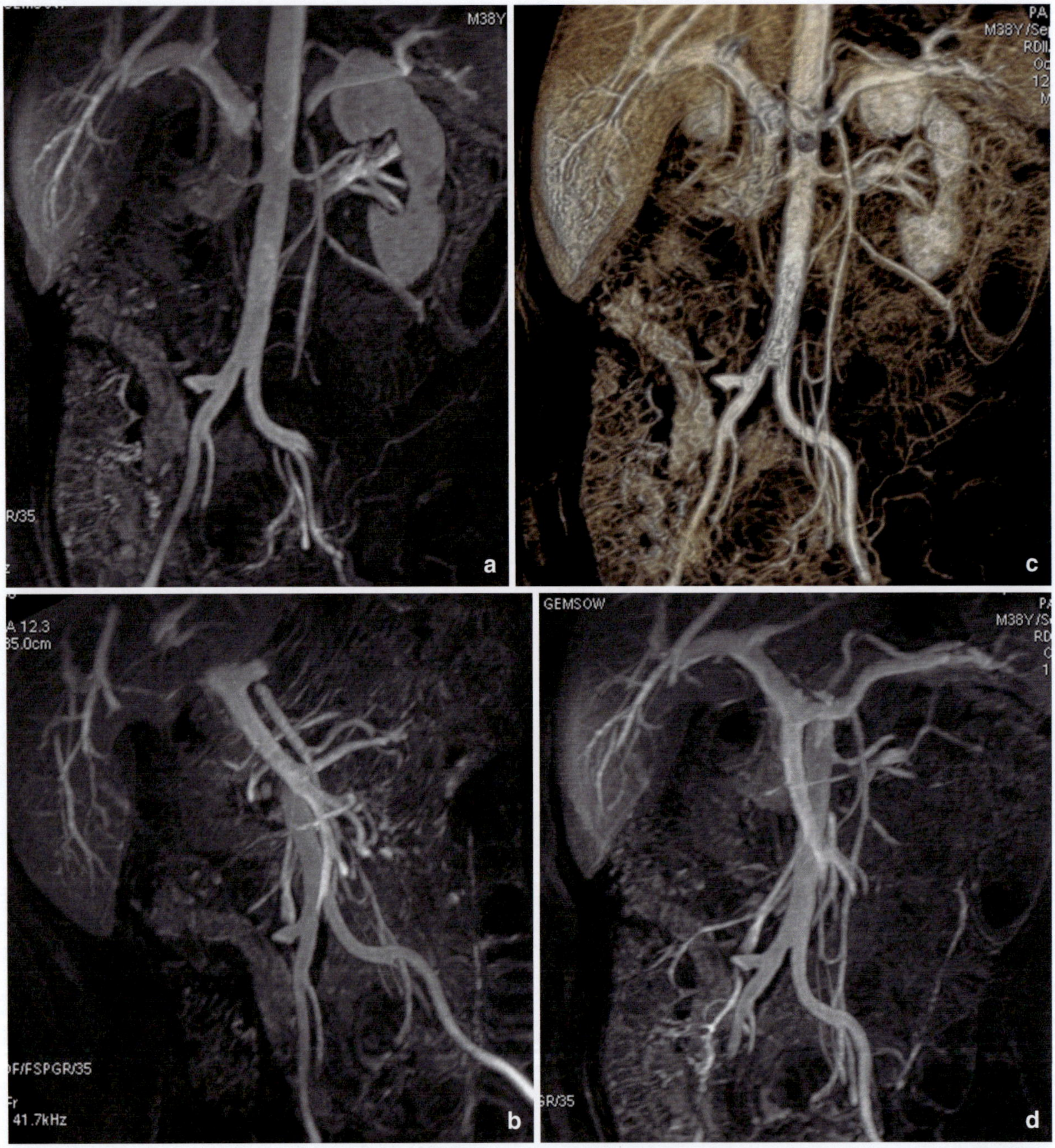

Fig. 13.4. MIP (**a**, **b**) and volume-rendered (**c**) images obtained from arterial phase of CEMRA demonstrate the complete thrombosis of both the splenic and the mesenteric branch of the pancreatic arterial graft anastomosed with the right common iliac artery in a case of isolated pancreas transplantation with enteric-portal drainage. A stenosis at the level of the anastomotic junction between the donor's portal vein and the recipient's superior mesenteric vein is showed on MIP reformation (**d**) obtained from venous phase of CEMRA in the same case

branch stenoses are also forms of donor artery stenosis.

On the contrary, older transplant recipients are prone to peripheral vascular disease and may have pseudo-TRAS because of atherosclerotic vascular disease involving their iliac system. Pretransplant assessment of the native arteries with CEMRA may be useful in high-risk diabetic and chronic hypertensive patients, in order to prevent this complication.

On imaging examination, the presence of TRAS can be indirectly assessed by many signs, as for example the arterial occlusion (visualized as a filling

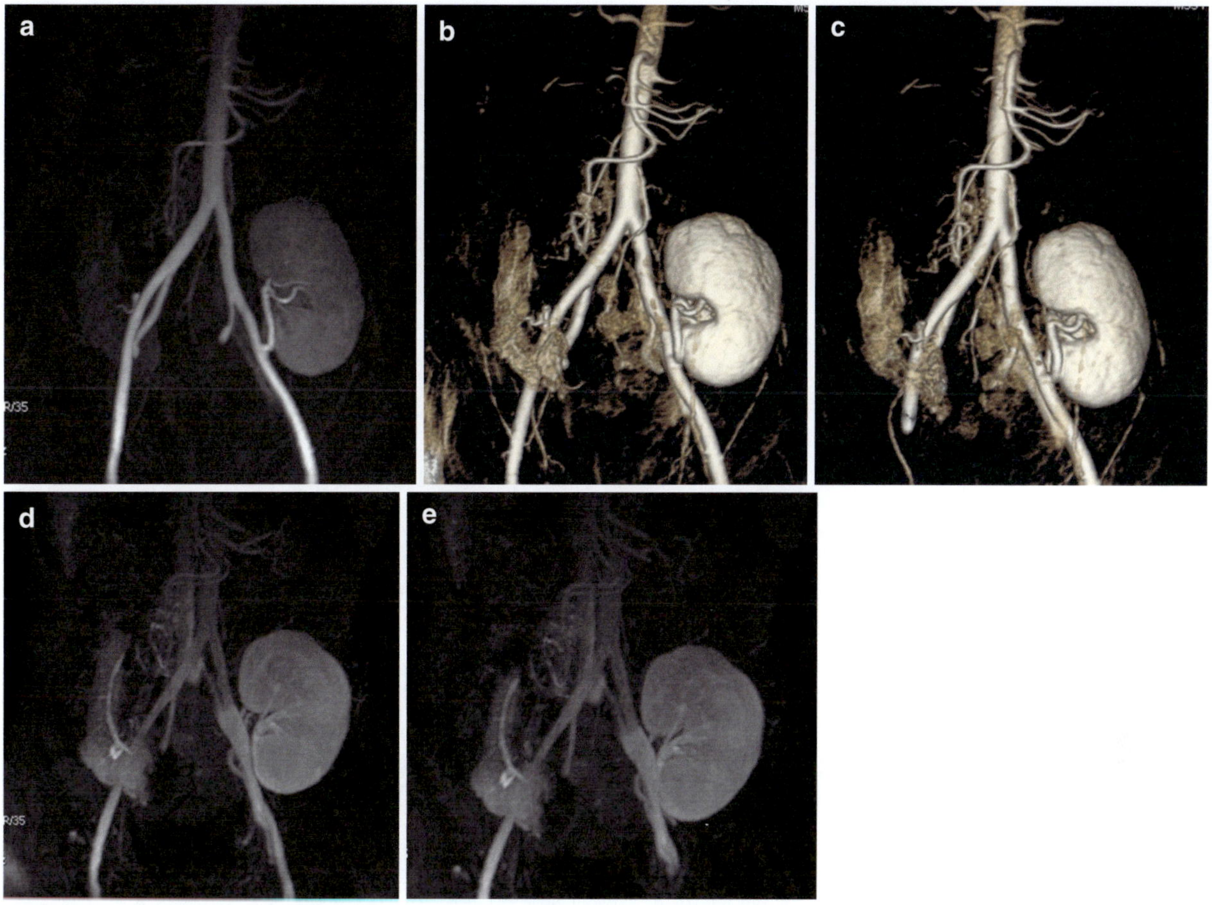

Fig. 13.5. MIP (**a**) and volume-rendered (**b, c**) images obtained from arterial phase of CEMRA highlight the pancreatic arterial graft anastomosed with the right external iliac artery (splenic branch is not visualized) and the renal graft anastomosed with the left external iliac artery in a case of combined kidney–pancreas transplantation performed by means of the systemic-bladder drainage technique. On MIP reconstructions (**d, e**) obtained from venous phase of CEMRA the splenic and the superior mesenteric veins are patent, while the anastomotic junction between the donor's portal vein (very thin) and the recipient's right iliac vein is completely obstructed. The venous graft of the kidney anastomosed with left external iliac vein is patent and normally represented

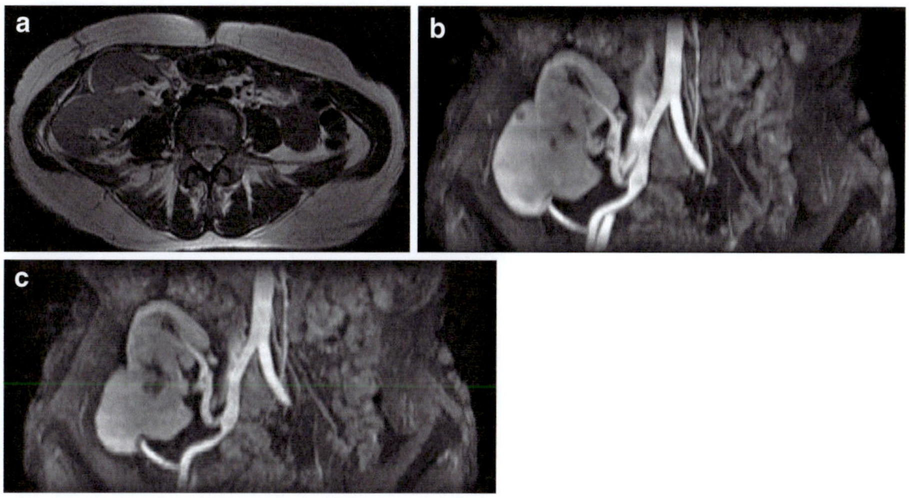

Fig. 13.6. Axial T2-weighted image (**a**) shows a double renal transplantation in the right iliac fossa. MIP reconstructions (**b, c**) show the renal grafts anastomosed with the right common iliac artery and with the right external iliac artery

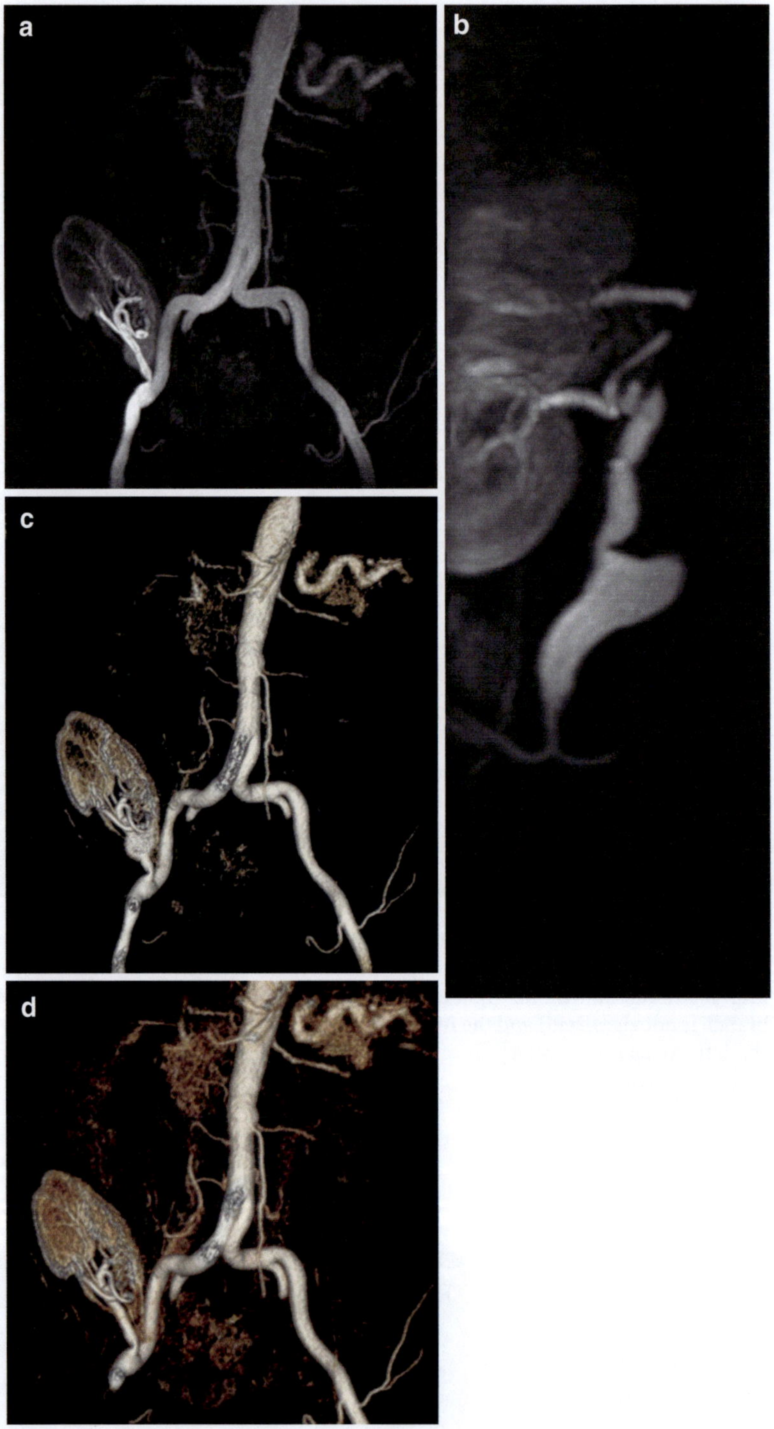

Fig. 13.7. MIP (**a**, **b**) and volume-rendered (**c**, **d**) images obtained from arterial phase of CEMRA demonstrate a stenosis at the anastomotic site between the donor's renal artery and the recipient's right external iliac artery in a case of isolated kidney transplantation in the right iliac fossa

defect of the vessels), a poststenotic dilation, as well as a delayed renal enhancement or a parenchymal mass reduction (cortical thinning and reduced renal length), especially in cases of chronic stenoses. Other signs of chronic arterial stenosis or occlusion of smaller branch arteries or thin accessory vessels are perfusion defects and infarcts, very well depicted as inhomogenously enhancing areas, especially in the arterial phase. MRA has revealed moreover an excellent method in identifying not only the main renal artery stenoses but also the possible kinking of the vessel that can occur in

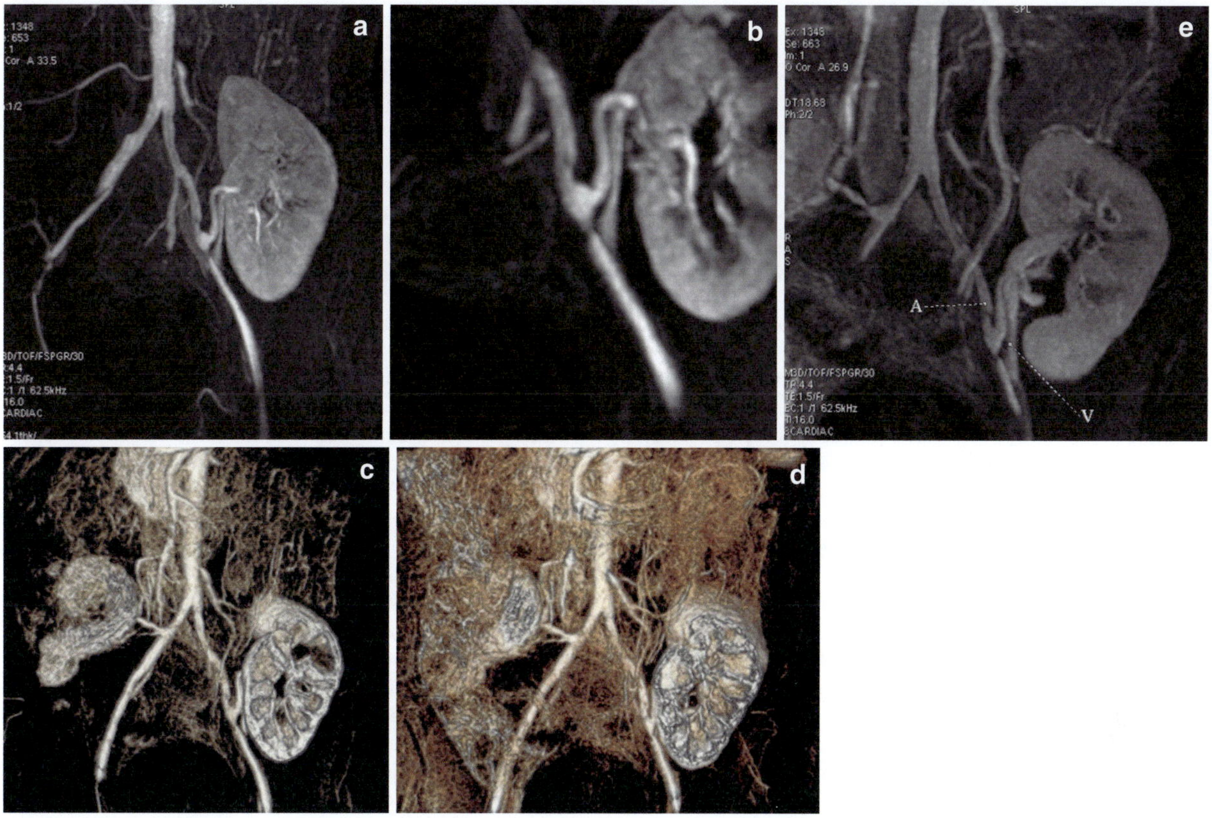

Fig. 13.8. MIP (**a**, **b**) and volume-rendered (**c**, **d**) images obtained from arterial phase of CEMRA well exhibit a stenosis immediately distal to the anastomotic site between the donor's renal artery and the recipient's left external iliac artery in a case of combined kidney–pancreas transplantation with enteric-portal pancreatic drainage. A prestenotic dilation of left external iliac artery is also appreciable. On MIP reconstruction (**e**) obtained from venous phase of CEMRA we can visualize both the renal vein (v) anastomosed with the left external iliac vein and the left external iliac artery (**a**)

cases of excessively long donor artery. In CEMRA the kinking can be both morphologically and functionally detected, as a turbulent blood flow (Amante and Kahan 1994).

Differently from pancreatic transplants, renal vein diseases are much more uncommon than arterial ones. Renal vein thrombosis is not so common and generally it occurs within the first week after transplantation, in up to 4% of patients. It is usually related to inadequate surgical technique, hypovolaemia or systemic venous thrombosis. Renal vein thrombosis often results in transplant infarction and graft loss. Even if the diagnosis is usually made by a not expensive, as well as not invasive, techniques (color-Doppler US), both contrast-enhanced CT and CEMRA can easily demonstrate the abnormality.

Renal vein stenosis as well as anastomotic stenosis or extrinsic renal vein compression caused by local fluid collections or local fibrosis represent even much rarer complications.

13.7 Conclusion

Actually CEMRA studies show a high diagnostic accuracy in the evaluation of vascular complications of transplanted pancreas and kidneys. CEMRA can therefore become an appropriate method of diagnostic angiography in the detection of transplant vessel stenosis and can help to rapidly and noninvasively select patients for therapeutic interventional angiography, surgical correction or medical management.

References

Amante AJM, Kahan BD (1994) Technical complications of renal transplantation. Surg Clin North Am 74:1117–113

Bakker J, Beek FJA, Beutler JJ, et al. (1998) Renal artery stenosis and accessory renal arteries: accuracy of detection and

visualization with Gadolinium-enhanced breath-hold MR angiography. Radiology 207:497–504
Bosmans H, Wilms G, Dymarkowski S, et al. (2001) Basic principles of MRA. Eur J Radiol 38:2–9
Choyke PL, Walther MM, Wagner JR, et al. (1997) Renal cancer: preoperative evaluation with dual phase three-dimensional MR angiography. Radiology 205:767–771
Dobos N, Roberts DA, Insko EK et al. (2005) Contrast-enhanced MR angiography for evaluation of vascular complications of the pancreatic transplant. Radiographics 25:687–695
Douzdjian V, Abecassis MM, Cooper JL, et al. (1993) Pancreas transplant salvage after acute venous thrombosis. Transplantation 56:222–223
Eubank WB, Schmiedl UP, AE Levy, et al. (2000) Venous thrombosis and occlusion after pancreas transplantation: evaluation with breath-hold Gadolinium-enhanced three-dimensional MR imaging. AJR Am J Roentgenol 175:381–385
Fernandez MP, Bernardino ME, Neylan JF, et al. (1991) Diagnosis of pancreatic transplant dysfunction: value of gadopentetate dimeglumine-enhanced MR imaging. AJR Am J Roentgenol 156:1171–1176
Fervenza FC, Lafayette RA, Alfrey EJ, et al. (1998) Renal artery stenosis in kidney transplants. Am J Kidney Dis 31:142–148
Grewal HP, Garland L, Novalk K, et al. (1993) Risk factors for postimplantation pancreatitis and pancreatic thrombosis in pancreas transplant recipients. Transplantation 56:609–612
Hagspiel KD, Nandalur K, Burkholder B, et al. (2005) Contrast-enhanced MR angiography after pancreas transplantation: normal appearance and vascular complications. AJR Am J Roentgenol 184:465–473
Hagspiel KD, Nandalur K, Pruett TL, et al. (2007) Evaluation of vascular complications of pancreas transplantation with high-spatial-resolution contrast enhanced MR angiography. Radiology 242:590–599
Heyneman LE, Keogan MT, Tuttle-Newhall JE, et al. (1999) Pancreatic transplantation using portal venous and enteric drainage: the postoperative appearance of a new surgical procedure. J Comput Assist Tomogr 23:283–290
Hohenwalter MD, Skowlund CJ, Erickson SJ, et al. (2001) Renal transplant evaluation with MR angiography and MR imaging. Radiographics 21:1505–1517
Jain R, Sawhney S. (2005) Contrast-enhanced MR angiography (CE-MRA) in the evaluation of vascular complications of renal transplantation. Clin Radiol 60:1171–1181
Krebs TL, Daly B, Wong JJ, et al. (1995) Vascular complications of pancreatic transplantation: MR evaluation. Radiology 196:793–798
Lee VS, Lavelle MT, Rofsky NM, et al. (2000) Hepatic MR imaging with a dynamic contrast-enhanced isotropic volumetric interpolated breath-hold examination: feasibility, reproducibility, and technical quality. Radiology 215:365–372
Michaely HJ, Dietrich O, Nael K, et al. (2006) MRA of abdominal vessels: technical advances. Eur Radiol 16:1637–1650
Morris MC, Santella RN, Aaronson ML, et al. (2004) Pancreas transplantation. SDJ Med 57(7):269–272
Orsenigo E, De Cobelli F, Salvioni M, et al. (2003) Successful endovascular treatment for gastroduodenal artery pseudoaneurysm with an arteriovenous fistula after pancreas transplantation. Transpl Int 16:647–696
Tan KT, Van Beek EJR, Brown PWG, et al. (2002) Magnetic resonance angiography for the diagnosis of renal artery stenosis: a meta-analysis. Clin Radiol 57:617–624
Vosshenrich R, Fischer U (2002) Contrast-enhanced MR angiography of abdominal vessels: Is there still a role for angiography? Eur Radiol 12:218–230
Winston CB, Mitchell DG, Outwater EK, et al. (1995) Pancreatic signal intensity on T1-weighted fat saturation MR images: clinical correlation. J Magn Reson Imaging 5:267–271

Subject Index

A

Abdomen and pelvic veins
- common iliac veins, 66
- external iliac vein, 65–66
- hepatic veins, 67
- hypogastric vein, 66
- inferior vena cava, 66–67
- lumbar veins, 67
- ovarian veins, 67
- portal system, 67–69
- renal veins, 67
- spermatic veins, 67

Abdominal aorta. *See also* Thoracic aorta
- aortic occlusion, 119
- arterial spin labeling (ASL), 117
- atherosclerotic and inflammatory aneurysms, 117–119
- balanced fast gradient echo, 117
- computed tomography angiography (CTA), 116
- contrast-enhanced MR angiography (CEMRA), 116
- dissection, 119
- Erdheim–Chester disease, 120–121
- infected aneurysm, 118
- Leriche's syndrome, 119
- multislice computed tomography (MSCT), 115–116
- perfusion technique, 117
- phase-contrast (PC) technique, 116–117
- retroperitoneal fibrosis, 120–121
- vascular stents and endoleak, 119–120

Aneurysm
- aortic aneurysm, 93–96
- CE-MRA, 131–132
- clinical manifestation, 131, 136
- femoral, 132
- lower limb, 131–132
- popliteal, 132
- subclavian, 136
- upper limb, 136

Aortic dissection
- CEMRA, 119
- clinical findings, 96
- histopathology, 97
- MR features, 97–98

Aortic occlusion, 119

Aortic prostheses, 119–120

Arterial anastomosis, 152

Arterial spin labeling (ASL), 117

Arteries. *See also* Thoracic aorta
- abdominal arteries, 53–57
- limb arteries, 57–60
- pulmonary arteries, 52–53
- thoracic aorta, 48–52

Artifacts
- classification, 28
- flow artifacts, 28–29
- geometric artifacts, 30
- hinge artifact, 29–30
- magnetic susceptibility artifacts, 30
- Maki artifacts, 31
- patient artifacts, 31–32
- postprocessing, 32–33
- radiofrequency artifacts, 28
- vascular blurring, 31, 32

Atherosclerotic disease
- advantages, 79
- classification error, 80
- contrast enhancement, 119
- diagnostic accuracy, 79
- histopathology, 118
- MIP coronal reconstruction, 118
- preoperative evaluation, 79–80
- radiological findings, 117–118
- stenosis overestimation, 80

B

Balanced fast field echo (BFFE) imaging, 117
- bright blood technique, 92
- dissection entry site, 98
- pulmonary vessel, 102

Balanced fast gradient echo, 117

Black blood technique, 91–92

Blood-pool imaging, 103

Breath-hold imaging, 106

Bright blood technique, 92

C

Carotid arteries
- atherosclerotic disease, 78–81
- cervical artery dissection (CAD), 82–83

Carotid endarterectomy (CEA), 78
Cervical artery dissection (CAD)
- aneurismal dilation, 82–83
- hematoma, 82
- predisposing factors, 82
Chemical exchange saturation transfer (CEST), 15
Computed tomography angiography (CTA), 116
Contrast agents
- chemical exchange saturation transfer (CEST), 15
- dose and administration, 90–91
- gadolinium, 90
- gadolinium-based particulate agents, 15
- high field, 15
- hyperpolarized contrast agents, 15
- magnetic properties, 7–8
- relaxivity, 8–9
- safety, 14–15
- susceptibility effect, 9
- vascular imaging, 9–14
- vasovist®, 129
Contrast-enhanced magnetic resonance angiography (CE-MRA)
-3D-GRE sequences, 19–21
- abdominal aorta, 118–120
- atherosclerotic disease (*see* Atherosclerotic disease)
- contrast administration, 19
- imaging and acquisition technique, 19–24
- k-space filling strategies, 23, 24
- liver, 148
- lower limb arterial disease assessment, 128–133
- neck vessel examination (*see* Neck vessel examination, CE-MRA)
- parallel imaging, 21
- principle, 18–19
- pulmonary vessels, 101–103
- renal arteries, 122–124
- steady-state imaging, 21, 23, 24
- thoracic aorta, 89–90
- upper limb arterial disease assessment, 133–139
Coronary magnetic resonance angiography (CMRA)
- advantages, 106–107
- arterial switch procedure, 111
- cardiac and respiratory motion suppression, 106
- clinical application, 110–111
- contrast agent, 106, 110
- coronary artery and bypass functional evaluation, 112–113
- flow measurement, 112
- guidelines, 110
- image quality, 110
- IMA grafts, 112–113
- intravascular contrast agent, 110
- Kawasaki disease, 110–111
- k-space segmentation, 106
- limitations, 105–106, 110
- parallel imaging technique, 109
- phase-array coil design, 109
- phase-contrast velocity quantification, 112
- stenting and bypass, 111–112
- technological requirement, 106–107
- three-dimensional image acquisition, 110
- whole heart CMRA, 107–108
Coronary stenting and bypass. *See* Stenting and bypass

D

Diethylenetriaminepentaacetic acid, 107
Digital subtraction angiography (DSA), 78–79
Dipyridamole, 113
Donor renal artery stenosis, 163–164

E

Endoleak, 120
Endovascular aortic repair (EVAR) procedure, 98–100
Erdheim–Chester disease, 121

F

Fast imaging employing steady-state acquisition (FIESTA) image, 108, 117, 121
Fast spin echo (FSE) pulse sequences, 91
Femoral aneurysm, 132
Flow-based MRA
- MR sequences, 3
- phase contrast (PC) MR angiography, 4û5
- T2-based MR angiography, 5
- time-of-flight MR angiography, 4
Flow quantification technique, 87–88
Free-breathing imaging, 106

G

Gadolinium 107
Gadolinium-based particulate agents, 15
Geometric artifacts, 30
Gradient echo (GRE) imaging, 92
Graft thrombosis, 160

H

Hinge artifact, 29–30
Hyperpolarized contrast agents, 15

I

IMA graft stenosis, 112–113
Infected aortic aneurysm, 118. *See also* Mycotic aneurysm

Inferior vena cava (IVC) stenosis, 152
Inflammatory aneurysms. *See* Atherosclerotic disease
Intramural hematoma
- CEMRA, 119
- clinical findings, 96
- histopathology, 97
- MR features, 97–98

K

Kawasaki disease, 110–111
K-space
- filling strategies, 23, 24
- segmentation, 106

L

Leriche's syndrome, 119, 131
Liver
- 3D-gradient-echo sequence, 149
- contrast enhanced MR angiography (CE-MRA), 149
- digital subtraction angiography (DSA), 147
- Doppler ultrasound (DUS), 148
- image postprocessing, 150
- MDCT angiography, 148
- Michel hepatic artery classification, 147
- MR angiography (MRA) vs. CTA and DSA, 148
- portal vein trifurcation, 148
- time-resolved MRA, 149–150
- vascular anatomy, 146–147
- vascular imaging, 147–148
- volumetric imaging, 149
Luminography, 122

M

Magnetic resonance sequences, 92–93. *See also* Pulse sequence imaging
Magnetic susceptibility artifacts, 30
Maki artifacts, 31
Maximum intensity projection (MIP), 37–39
- aortic aneurysm, 118
- popliteal artery aneurysm, 133
- pulmonary vessels, 101–102
- renal artery, 122
- renal artery stenosis (RAS), 120, 124
- steno-occlusive disease, 130
- thoracic aorta, 91, 94
Michel hepatic artery classification, 147
Multifocal fibrosclerosis, 121
Multiplanar reconstruction (MPR), 35–37
- aortic dissection, 119
- image analysis, 102
- stenotic branch vessel, 118
- thoracic vascular anomalies, 91
Multislice computed tomography (MSCT), 115–116
Mycotic aneurysm, 93, 95, 118

N

Neck vessel examination, CE-MRA
- acquisition timing, 74
- advantages, 74
- ALS-based MRA, 77
- artifacts, 74–75
- bolus geometry, 74, 75
- carotid arteries, 78–83
- consequences, 74
- contrast media and agent, 76
- nonenhanced MRA technique, 76–77
- parallel imaging technique, 75–76
- subclavian arteries, 77–78
- time resolution improvement, 75
- vertebral arteries, 83–84
Nephrogenic systemic fibrosis (NSF), 14, 90
Nitinol, 119

O

Orthotopic liver transplants (OLT)
- clinical manifestation, 150
- complications, 150–153
- IVC stenosis, 152
- MR angiography, 153–154
- surgery, 150, 151

P

Pancreatic transplants
- acute thrombosis, 161, 163–165
- arterial stenosis, 163
- contrast-enhanced MRA (CEMRA) techniques, 159
- contrast media, 159–160
- diagnostic imaging, 158–159
- diagnostic performance, 160
- evaluation, 161–163
- MR angiography study technique, 159–160
- multiphasic breath-hold gadolinium-enhanced 3D MR imaging, 160
- pre and postoperative evaluation, 157–158
- vascular anatomy, 158
- vascular and graft thrombosis, 160
- vascular complication, 160–163
Parallel imaging technique, 109
Para-sagittal imaging, 90
Patient artifacts, 31–32
Pelvic veins. *See* Abdomen and pelvic veins

Perfusion imaging
- abdominal aorta and renal arteries, 117, 123
- pulmonary vessels, 102
Periaortic edema, 118
Peripheral arterial occlusive disease (PAOD), 129
Peripheral vessels
- aneurysm, 131–132, 136
- contrast agents, 129
- digital subtraction angiography (DSA), 128
- duplex ultrasound (DUS), 128
- fast field-echo (FFE) T1-weighed 3D diagnostic imaging, 128
- imaging modality, 134–135
- inflammatory arteritis, 134
- moving table technology, 128–129
- popliteal artery entrapment syndrome, 132–134
- steady-state imaging, 129
- steno-occlusive disease, 129–131, 135–136
- Takayasu's arteritis, 136–139
- test-bolus technique, 128
- thoracic outlet syndrome, 138, 139
- three-dimensional contrast-enhanced MRA (3D-CEMRA), 128
- time-of-flight (TOF) angiography, 127–128
- vasospastic disease, 134
- vasovist®, 129
Phase-array coil design, 109
Phase contrast (PC) MR angiography
- 3D acquisition, 5
- abdominal aorta and renal artery, 116–117
- bipolar gradient, 4
- encoding gradient characteristics, 4–5
- limitations, 5
Popliteal aneurysm, 132
Popliteal artery entrapment syndrome, 132–134
Portal vein trifurcation, 148
Proton movement, 87–88
Pulmonary vessels
- anatomy, 100–101
- BFFE imaging, 102
- blood-pool imaging, 103
- MIP reconstruction, 101–102
- perfusion imaging and ventilation, 102
- thrombus imaging, 103
- time-resolved MRA, 102
Pulse sequence imaging
- 2D spin-echo, 108
- 3D gradient echo, 108–109
- black blood images, 107–108
- bright blood images, 107
- contrast agent, 107
- image quality, 110
- motion suppression, 107
- parallel imaging, 109
- self-navigated image reconstruction approach, 109
- SSFP sequence, 108–109
- variable sampling in time (VAST) technique, 109

R

Radiofrequency artifacts, 28
Renal artery stenosis (RAS)
- arterial spin labeling (ASL), 117
- atherosclerotic disease, 121–122
- balanced fast gradient echo, 117
- CEMRA, 122, 123
- clinical manifestation, 121
- computed tomography angiography (CTA), 116
- contrast-enhanced MR angiography (CEMRA), 116
- fibromuscular dysplasia, 122
- gamma-variate function, 123
- MR perfusion imaging, 123
- multislice computed tomography (MSCT), 115–116
- PC technique, 122
- perfusion technique, 117
- phase-contrast (PC) technique, 116–117
- renal blood flow measurement, 122–124
Renal transplants
- arterial stenosis, 163
- CEMRA, 163
- chronic stenosis, 166
- contrast-enhanced MRA (CEMRA) techniques, 159
- contrast media, 159–160
- diagnostic imaging, 158–159
- donor renal artery stenosis, 163–164
- imaging, 164, 166–167
- MR angiography study technique, 159–160
- pre and postoperative evaluation, 157–158
- single and double transplants, 163, 165
- vascular anatomy, 158
- vascular complication, 163–167
- vein diseases, 167
Retroperitoneal fibrosis, 96
- clinical manifestation, 120–121
- MRI imaging, 121
Retrospective gating approach, 106

S

Segmented k-space fast GRE imaging, 92
Shaded surface display (SSD), 39–40
Single-shot fast spin echo (SSFSE) sequences, 91–92
Small particles of iron oxide (SPIOs), 9
3D Spoiled gradient echo (SPGR), 90
Steady state free precession (SSFP) technique, 92, 108–109
Steno-occlusive arterial disease
- asymptomatic and symptomatic, 135
- bypass grafts, 131, 135
- CE-MRA, 129, 130
- DSA, 129
- Leriche's syndrome, 131
- PAOD, 129
- subclavian steal syndrome, 136
- TOF angiography, 129

Stenting and bypass, 110–111
Subclavian steal syndrome, 136

T

T2-based MR angiography, 5
Takayasu's arteritis
- clinical manifestation, 136–137
- diagnosis, 138–139
- symptoms, 137
Thoracic aorta
- anatomy, 88–89, 93
- aortic aneurysm, 93–96
- aortic dissection, 96–98
- arterial circle of Willis, 51
- ascending aorta, 48
- basilar artery, 51
- black blood techniques, 91–92
- bright blood techniques, 92
- common carotid arteries, 49
- contrast dose and administration, 90–91
- contrast-enhanced magnetic resonance angiography (CE-MRA), 89–90
- contrast injection synchronization, 90
- coronary artery, 48
- disease prevalence and spectrum, 88
- external carotid artery, 49
- fundamental sequences, 92–93
- image processing, 91
- innominate artery, 48–49
- internal carotid artery, 49–50
- internal mammary artery, 51–52
- intramural hematoma, 96–98
- parietal branches, 52
- patient positioning, 89
- posterior cerebral artery, 51
- post EVAR evaluation, 98–100
- subclavian artery, 50–51
- thoracic descending aorta, 52
- thyrocervical trunk, 51
- trauma, 98
- vascular thoracic outlet syndrome, 98
- vertebral artery, 51
- visceral branches, 52
Thoracic outlet syndrome (TOS), 78, 138, 139
Time-of-flight (TOF) angiography, 4
- peripheral vessels, 127–128
- thoracic aorta, 88
Time-resolved MRA
- CE-MRA, 21, 22
- pulmonary vessel, 102
- thoracic aorta, 97–98, 100

U

Ultrasmall particles of iron oxides (USPIOs), 9

V

Variable sampling in time (VAST) technique, 109
Vascular imaging
- arteries (*see* Arteries)
- paramagnetic gadolinium agents, 9–13
- superparamagnetic USPIOs, 13–14
- veins (*see* Veins)
Vascular stents 119
Vascular thoracic outlet syndrome, 98
Vascular thrombosis, 160
Vasospastic disease, 134
Vasovist®, 129
Veins, 62–63
- abdomen and pelvis, 65–69 (*see also* Abdomen and pelvic veins)
- heart, 65
- limbs, 69–71
- pulmonary veins, 65
- superior vena cava, 63–65
Vertebral arteries
- atherosclerosis, 83
- dissection, 83–84
Virtual endoscopy (VE), 41–43
Volume rendering (VR), 40–42, 91, 150, 160

W

Whole heart CMRA, 107–108

List of Contributors

MICHELE ANZIDEI, MD
Department of Radiology
University of Rome
"La Sapienza," Policlinico Umberto I,
viale Regina Elena
324, 00161 Roma
Italy

VALENTINA BATTAGLIA, MD
Diagnostic and Interventional Radiology,
Department of Oncology
Transplants and Advanced Technologies in Medicine
University of Pisa
Via Roma 67
56126 Pisa
Italy

ELENA BELLONI, MD
Radiology Department, Vita-Salute University
San Raffaele Scientific Institute
Via Olgettina 60
20132 Milan
Italy
Email: belloni.elena@hsr.it

FRANCESCA BIANCHI, MD
Diagnostic and Interventional Radiology
University of Pisa
Via Roma 67
56100 Pisa
Italy

PIERO BORASCHI, MD
Second Unit of Radiology
Department of Oncologic and Radiological Sciences
Pisa University Hospital
Via Paradisa 2
56124 Pisa
Italy
Email: p.boraschi@do.med.unipi.it

GABRIELE CAPRONI, MD
Diagnostic and Interventional Radiology
Department of Oncology
Transplants and Advanced Technologies in Medicine
University of Pisa
Via Roma 67
56100 Pisa
Italy

DAVIDE CARAMELLA, MD
Diagnostic and Interventional Radiology
Department of Oncology, Transplants,
and Advanced Technologies in Medicine
University of Pisa
Via Roma 67
56100 Pisa
Italy
Email: caramella@med.unipi.it

ALESSANDRO CARRIERO, MD
Diagnostic and Interventional Radiology Department
University del Piemonte Orientale "A. AVOGADRO"
Corso Mazzini 18
28100 Novara
Italy
Email: profcarriero@virgilio.it

CARLO CATALANO, MD
Department of Radiology
University of Rome
"La Sapienza," Policlinico Umberto I,
viale Regina Elena
324, 00161 Roma
Italy
Email: carlo.catalano@uniroma1.it

Mirco Cosottini, MD
Unit of Neuroradiology
Department of Neurosciences
University of Pisa
Via Roma 67
56100 Pisa
Italy
Email: mircocosottini@libero.it

Francesco De Cobelli, MD
Radiology Department, Vita-Salute University
San Raffaele Scientific Institute
Via Olgettina 60
20132 Milan
Italy
Email: francesco.decobelli@hsr.it

Alessandro Del Maschio, MD
Radiology Department
Vita-Salute University
San Raffaele Scientific Institute
Via Olgettina 60
20132 Milan
Italy

Marco Di Terlizzi, MD
Diagnostic and Interventional Radiology Department
Corso Mazzini 18
28100 Novara
Italy

Francescamaria Donati, MD
Second Unit of Radiology
Department of Oncologic and Radiological Sciences
Pisa University Hospital
Via Paradisa 2
56124 Pisa
Italy

Lorenzo Faggioni, MD
Diagnostic and Interventional Radiology
Department of Oncology, Transplants,
and Advanced Technologies in Medicine
University of Pisa
Radiodiagnostica 1 Universitaria
Ospedale Cisanello. Via Paradisa 2
56100 Pisa
Italy

Rita Fossaceca, MD
Diagnostic and Interventional Radiology Department
Corso Mazzini 18
28100 Novara
Italy

Marco Francone, MD, PhD
Department of Radiology
University of Rome "La Sapienza"
Policlinico Umberto I
viale Regina Elena
324, 00161 Roma
Italy

Ilaria Iacucci, MD
Department of Radiology
University of Rome "La Sapienza"
Policlinico Umberto I
viale Regina Elena
324, 00161 Roma
Italy

Andrea Laghi, MD
Department of Radiological Sciences
"Sapienza" – University of Rome
Polo Pontino, I.C.O.T. Hospital
Viale Franco Fagiana 34
04100 Latina
Italy
Email: andrea.laghi@uniroma1.it

Massimo Lombardi, MD
U.O.C. Cardiovascular RM
C.N.R./"G. Monasterio" Foundation
Via Moruzzi 1
56124 Pisa
Italy
Email: massimo.lombardi@ifc.cnr.it

Paola Lucchesi, MD
Department of Radiological Sciences
"Sapienza" - University of Rome
Polo Pontino
I.C.O.T. Hospital
Viale Franco Fagiana 34
04100 Latina
Italy
Email: paola.lucchesi@email.it

Matteo Milanesi, PhD
U.O.C. Cardiovascular RM
C.N.R./"G. Monasterio" Foundation
Via Moruzzi 1
56124 Pisa
Italy

Emanuele Neri, MD
Diagnostic and Interventional Radiology
Department of Oncology, Transplants,
and Advanced Technologies in Medicine
University of Pisa
Radiodiagnostica 1 Universitaria
Ospedale Cisanello. Via Paradisa 2
56100 Pisa
Italy
Email: neri@med.unipi.it

Pasquale Paolantonio, MD
Department of Radiological Sciences
"Sapienza" – University of Rome
Polo Pontino,
I.C.O.T. Hospital
Viale Franco Fagiana 34
04100 Latina
Italy
Email: paolantonipasquale@hotmail.com

Ilaria Pesaresi, MD
Diagnostic and Interventional Radiology
Department of Oncology, Transplants,
and Advanced Technologies in Medicine
University of Pisa
Via Roma 67
56100 Pisa
Italy

Marco Regno, MD
Department of Radiological Sciences,
"Sapienza" – University of Rome
Polo Pontino
I.C.O.T. Hospital
Viale Franco Fagiana 34
04100 Latina
Italy
Email: marco.rengo@gmail.com

Ambrosini Roberta, MD
Diagnostic and Interventional Radiology Department
Corso Mazzini 18
28100 Novara
Italy

Simonetta Salemi, MD
Second Unit of Radiology
Department of Oncologic and Radiological Sciences
Pisa University Hospital
Via Paradisa 2
56124 Pisa
Italy

Francesca Turini, MD
Diagnostic and Interventional Radiology
Department of Oncology, Transplants, and Advanced
Technologies in Medicine
University of Pisa
Via Roma 67
56100 Pisa
Italy

Valentina Vallini, MD
Diagnostic and Interventional Radiology
University of Pisa
Via Roma 67
56100 Pisa
Italy

Angelo Vanzulli, MD
Niguarda Cà Granda Hospital
Piazza Ospedale Maggiore 3
20162 Milano
Italy
Email: angelo.vanzulli@ospedaleniguarda.it; angelo.vanzulli@fastwebnet.it

Claudio Vignali, MD
Diagnostic and Interventional Radiology
University of Pisa
Via Roma 67
56100 Pisa
Italy
Email: cvignali@med.unipi.it

Francesco Vullo, MD
Department of Radiology
University of Rome "La Sapienza"
Policlinico Umberto I
viale Regina Elena
324, 00161 Roma
Italy

Virna Zampa, MD
Diagnostic and Interventional Radiology,
Department of Oncology, Transplants,
and Advanced Technologies in Medicine
University of Pisa
Via Roma 67
56100 Pisa
Italy

Email: v.zampa@med.unipi.it

Medical Radiology Diagnostic Imaging and Radiation Oncology

Titles in the series already published

Diagnostic Imaging

Radiological Imaging of Sports Injuries
Edited by C. Masciocchi

Modern Imaging of the Alimentary Tube
Edited by A. R. Margulis

Diagnosis and Therapy of Spinal Tumors
Edited by P. R. Algra, J. Valk and J. J. Heimans

Interventional Magnetic Resonance Imaging
Edited by J. F. Debatin and G. Adam

Abdominal and Pelvic MRI
Edited by A. Heuck and M. Reiser

Orthopedic Imaging
Techniques and Applications
Edited by A. M. Davies and H. Pettersson

Radiology of the Female Pelvic Organs
Edited by E. K. Lang

Magnetic Resonance of the Heart and Great Vessels
Clinical Applications
Edited by J. Bogaert, A. J. Duerinckx, and F. E. Rademakers

Modern Head and Neck Imaging
Edited by S. K. Mukherji and J. A. Castelijns

Radiological Imaging of Endocrine Diseases
Edited by J. N. Bruneton in collaboration with B. Padovani and M.-Y. Mourou

Radiology of the Pancreas
2nd Revised Edition
Edited by A. L. Baert. Co-edited by G. Delorme and L. Van Hoe

Trends in Contrast Media
Edited by H. S. Thomsen, R. N. Muller, and R. F. Mattrey

Functional MRI
Edited by C. T. W. Moonen and P. A. Bandettini

Emergency Pediatric Radiology
Edited by H. Carty

Liver Malignancies
Diagnostic and Interventional Radiology
Edited by C. Bartolozzi and R. Lencioni

Spiral CT of the Abdomen
Edited by F. Terrier, M. Grossholz, and C. D. Becker

Medical Imaging of the Spleen
Edited by A. M. De Schepper and F. Vanhoenacker

Radiology of Peripheral Vascular Diseases
Edited by E. Zeitler

Radiology of Blunt Trauma of the Chest
P. Schnyder and M. Wintermark

Portal Hypertension
Diagnostic Imaging and Imaging-Guided Therapy
Edited by P. Rossi.
Co-edited by P. Ricci and L. Broglia

Virtual Endoscopy and Related 3D Techniques
Edited by P. Rogalla, J. Terwisscha van Scheltinga and B. Hamm

Recent Advances in Diagnostic Neuroradiology
Edited by Ph. Demaerel

Transfontanellar Doppler Imaging in Neonates
A. Couture, C. Veyrac

Radiology of AIDS
A Practical Approach
Edited by J. W. A. J. Reeders and P. C. Goodman

CT of the Peritoneum
A. Rossi, G. Rossi

Magnetic Resonance Angiography
2nd Revised Edition
Edited by I. P. Arlart, G. M. Bongartz, and G. Marchal

Applications of Sonography in Head and Neck Pathology
Edited by J. N. Bruneton in collaboration with C. Raffaelli, O. Dassonville

3D Image Processing
Techniques and Clinical Applications
Edited by D. Caramella and C. Bartolozzi

Imaging of the Larynx
Edited by R. Hermans

Pediatric ENT Radiology
Edited by S. J. King and A. E. Boothroyd

Imaging of Orbital and Visual Pathway Pathology
Edited by W. S. Müller-Forell

Radiological Imaging of the Small Intestine
Edited by N. C. Gourtsoyiannis

Imaging of the Knee
Techniques and Applications
Edited by A. M. Davies and V. N. Cassar-Pullicino

Perinatal Imaging
From Ultrasound to MR Imaging
Edited by F. E. Avni

Diagnostic and Interventional Radiology in Liver Transplantation
Edited by E. Bücheler, V. Nicolas, C. E. Broelsch, X. Rogiers and G. Krupski

Imaging of the Pancreas
Cystic and Rare Tumors
Edited by C. Procacci and A. J. Megibow

Imaging of the Foot & Ankle
Techniques and Applications
Edited by A. M. Davies, R. W. Whitehouse and J. P. R. Jenkins

Radiological Imaging of the Ureter
Edited by F. Joffre, Ph. Otal and M. Soulie

Radiology of the Petrous Bone
Edited by M. Lemmerling and S. S. Kollias

Imaging of the Shoulder
Techniques and Applications
Edited by A. M. Davies and J. Hodler

Interventional Radiology in Cancer
Edited by A. Adam, R. F. Dondelinger, and P. R. Mueller

Imaging and Intervention in Abdominal Trauma
Edited by R. F. Dondelinger

Radiology of the Pharynx and the Esophagus
Edited by O. Ekberg

Radiological Imaging in Hematological Malignancies
Edited by A. Guermazi

Functional Imaging of the Chest
Edited by H.-U. Kauczor

Duplex and Color Doppler Imaging of the Venous System
Edited by G. H. Mostbeck

Multidetector-Row CT of the Thorax
Edited by U. J. Schoepf

Radiology and Imaging of the Colon
Edited by A. H. Chapman

Multidetector-Row CT Angiography
Edited by C. Catalano and R. Passariello

Focal Liver Lesions
Detection, Characterization, Ablation
Edited by R. Lencioni, D. Cioni, and C. Bartolozzi

Imaging in Treatment Planning for Sinonasal Diseases
Edited by R. Maroldi and P. Nicolai

Clinical Cardiac MRI
With Interactive CD-ROM
Edited by J. Bogaert, S. Dymarkowski, and A. M. Taylor

Dynamic Contrast-Enhanced Magnetic Resonance Imaging in Oncology
Edited by A. Jackson, D. L. Buckley, and G. J. M. Parker

Contrast Media in Ultrasonography
Basic Principles and Clinical Applications
Edited by E. Quaia

Paediatric Musculoskeletal Disease
With an Emphasis on Ultrasound
Edited by D. Wilson

MR Imaging in White Matter Diseases of the Brain and Spinal Cord
Edited by M. Filippi, N. De Stefano, V. Dousset, and J. C. McGowan

Imaging of the Hip & Bony Pelvis
Techniques and Applications
Edited by A. M. Davies, K. Johnson, and R. W. Whitehouse

Imaging of Kidney Cancer
Edited by A. Guermazi

Magnetic Resonance Imaging in Ischemic Stroke
Edited by R. von Kummer and T. Back

Diagnostic Nuclear Medicine
2nd Revised Edition
Edited by C. Schiepers

Imaging of Occupational and Environmental Disorders of the Chest
Edited by P. A. Gevenois and P. De Vuyst

Virtual Colonoscopy
A Practical Guide
Edited by P. Lefere and S. Gryspeerdt

Contrast Media
Safety Issues and ESUR Guidelines
Edited by H. S. Thomsen

Head and Neck Cancer Imaging
Edited by R. Hermans

Vascular Embolotherapy
A Comprehensive Approach
Volume 1: *General Principles, Chest, Abdomen, and Great Vessels*
Edited by J. Golzarian. Co-edited by S. Sun and M. J. Sharafuddin

Vascular Embolotherapy
A Comprehensive Approach
Volume 2: *Oncology, Trauma, Gene Therapy, Vascular Malformations, and Neck*
Edited by J. Golzarian.
Co-edited by S. Sun and M. J. Sharafuddin

Vascular Interventional Radiology
Current Evidence in Endovascular Surgery
Edited by M. G. Cowling

Ultrasound of the Gastrointestinal Tract
Edited by G. Maconi and G. Bianchi Porro

Parallel Imaging in Clinical MR Applications
Edited by S. O. Schoenberg, O. Dietrich, and M. F. Reiser

MRI and CT of the Female Pelvis
Edited by B. Hamm and R. Forstner

Imaging of Orthopedic Sports Injuries
Edited by F. M. Vanhoenacker, M. Maas and J. L. Gielen

Ultrasound of the Musculoskeletal System
S. Bianchi and C. Martinoli

Clinical Functional MRI
Presurgical Functional Neuroimaging
Edited by C. Stippich

Radiation Dose from Adult and Pediatric Multidetector Computed Tomography
Edited by D. Tack and P. A. Gevenois

Spinal Imaging
Diagnostic Imaging of the Spine and Spinal Cord
Edited by J. Van Goethem, L. van den Hauwe and P. M. Parizel

Computed Tomography of the Lung
A Pattern Approach
J. A. Verschakelen and W. De Wever

Imaging in Transplantation
Edited by A. Bankier

Radiological Imaging of the Neonatal Chest
2nd Revised Edition
Edited by V. Donoghue

Radiological Imaging of the Digestive Tract in Infants and Children
Edited by A. S. Devos and J. G. Blickman

Pediatric Chest Imaging
Chest Imaging in Infants and Children
2nd Revised Edition
Edited by J. Lucaya and J. L. Strife

Color Doppler US of the Penis
Edited by M. Bertolotto

Radiology of the Stomach and Duodenum
Edited by A. H. Freeman and E. Sala

Imaging in Pediatric Skeletal Trauma
Techniques and Applications
Edited by K. J. Johnson and E. Bache

Image Processing in Radiology
Current Applications
Edited by E. Neri, D. Caramella, C. Bartolozzi

Screening and Preventive Diagnosis with Radiological Imaging
Edited by M. F. Reiser, G. van Kaick, C. Fink, S. O. Schoenberg

Percutaneous Tumor Ablation in Medical Radiology
Edited by T. J. Vogl, T. K. Helmberger, M. G. Mack, M. F. Reiser

Liver Radioembolization with ^{90}Y Microspheres
Edited by J. I. Bilbao, M. F. Reiser

Pediatric Uroradiology
2nd Revised Edition
Edited by R. Fotter

Radiology of Osteoporosis
2nd Revised Edition
Edited by S. Grampp

Gastrointestinal Tract Sonography in Fetuses and Children
A. Couture, C. Baud, J. L. Ferran, M. Saguintaah and C. Veyrac

Intracranial Vascular Malformations and Aneurysms
2nd Revised Edition
Edited by M. Forsting and I. Wanke

High-Resolution Sonography of the Peripheral Nervous System
2nd Revised Edition
Edited by S. Peer and G. Bodner

Imaging Pelvic Floor Disorders
2nd Revised Edition
Edited by J. Stoker, S. A. Taylor, andJ. O. L. DeLancey

Coronary Radiology
2nd Revised Edition
Edited by M. Oudkerk and M. F. Reiser

Integrated Cardiothoracic Imaging with MDCT
Edited by M. Rémy-Jardin and J. Rémy

Multislice CT
3rd Revised Edition
Edited by M. F. Reiser, C. R. Becker, K. Nikolaou, G. Glazer

MRI of the Lung
Edited by H.-U. Kauczor

Imaging in Percutaneous Musculoskeletal Interventions
Edited by A. Gangi, S. Guth, and A. Guermazi

Contrast Media. Safety Issues and ESUR Guidelines
2nd Revised Edition
Edited by H. Thomsen, J.A.W. Webb

Inflammatory Diseases of the Brain
Edited by S. Hähnel

Imaging of Bone Tumors and Tumor-Like Lesions - Techniques and Applications
Edited by A.M. Davies, M. Sundaram, and S.J. James

MR Angiography of the body
Technique and Clinical Applications
Edited by E. Neri, M. Cosottini and D. Caramella

Virtual Colonoscopy
A Practical Guide
Edited by P. Lefere and S. Gryspeerdt

Diffusion-Weighted MR Imaging
Applications in the Body
Edited by D.-M. Koh and H.C. Thoeny

MRI of the Gastroinestinal Tract
Edited by J. Stoker

Digital Mammography
Edited by U. Bick and F. Diekmann

Springer

Medical Radiology Diagnostic Imaging and Radiation Oncology

Titles in the series already published

Radiation Oncology

Lung Cancer
Edited by C. W. Scarantino

Innovations in Radiation Oncology
Edited by H. R. Withers and L. J. Peters

Radiation Therapy of Head and Neck Cancer
Edited by G. E. Laramore

Gastrointestinal Cancer – Radiation Therapy
Edited by R. R. Dobelbower, Jr.

Radiation Exposure and Occupational Risks
Edited by E. Scherer, C. Streffer, and K.-R. Trott

Interventional Radiation
Therapy Techniques – Brachytherapy
Edited by R. Sauer

Radiopathology of Organs and Tissues
Edited by E. Scherer, C. Streffer, and K.-R. Trott

Concomitant Continuous Infusion
Chemotherapy and Radiation
Edited by M. Rotman and C. J. Rosenthal

Intraoperative Radiotherapy – Clinical Experiences and Results
Edited by F. A. Calvo, M. Santos, and L. W. Brady

Interstitial and Intracavitary Thermoradiotherapy
Edited by M. H. Seegenschmiedt and R. Sauer

Non-Disseminated Breast Cancer
Controversial Issues in Management
Edited by G. H. Fletcher and S. H. Levitt

Current Topics in Clinical Radiobiology of Tumors
Edited by H.-P. Beck-Bornholdt

Practical Approaches to Cancer Invasion and Metastases
A Compendium of Radiation Oncologists' Responses to 40 Histories
Edited by A. R. Kagan with the Assistance of R. J. Steckel

Radiation Therapy in Pediatric Oncology
Edited by J. R. Cassady

Radiation Therapy Physics
Edited by A. R. Smith

Late Sequelae in Oncology
Edited by J. Dunst, R. Sauer

Mediastinal Tumors. Update 1995
Edited by D. E. Wood, C. R. Thomas, Jr.

Thermoradiotherapy and Thermochemotherapy
Volume 1: *Biology, Physiology, and Physics*
Volume 2: *Clinical Applications*
Edited by M. H. Seegenschmiedt, P. Fessenden and C. C. Vernon

Carcinoma of the Prostate
Innovations in Management
Edited by Z. Petrovich, L. Baert, and L. W. Brady

Radiation Oncology of Gynecological Cancers
Edited by H. W. Vahrson

Carcinoma of the Bladder
Innovations in Management
Edited by Z. Petrovich, L. Baert, and L. W. Brady

Blood Perfusion and Microenvironment of Human Tumors
Implications for Clinical Radiooncology
Edited by M. Molls and P. Vaupel

Radiation Therapy of Benign Diseases
A Clinical Guide
2nd Revised Edition
S. E. Order and S. S. Donaldson

Carcinoma of the Kidney and Testis, and Rare Urologic Malignancies
Innovations in Management
Edited by Z. Petrovich, L. Baert, and L. W. Brady

Progress and Perspectives in the Treatment of Lung Cancer
Edited by P. Van Houtte, J. Klastersky, and P. Rocmans

Combined Modality Therapy of Central Nervous System Tumors
Edited by Z. Petrovich, L. W. Brady, M. L. Apuzzo, and M. Bamberg

Age-Related Macular Degeneration
Current Treatment Concepts
Edited by W. E. Alberti, G. Richard, and R. H. Sagerman

Radiotherapy of Intraocular and Orbital Tumors
2nd Revised Edition
Edited by R. H. Sagerman and W. E. Alberti

Modification of Radiation Response
Cytokines, Growth Factors, and Other Biolgical Targets
Edited by C. Nieder, L. Milas and K. K. Ang

Radiation Oncology for Cure and Palliation
R. G. Parker, N. A. Janjan and M. T. Selch

Clinical Target Volumes in Conformal and Intensity Modulated Radiation Therapy
A Clinical Guide to Cancer Treatment
Edited by V. Grégoire, P. Scalliet, and K. K. Ang

Advances in Radiation Oncology in Lung Cancer
Edited by B. Jeremi´c

New Technologies in Radiation Oncology
Edited by W. Schlegel, T. Bortfeld, and A.-L. Grosu

Multimodal Concepts for Integration of Cytotoxic Drugs and Radiation Therapy
Edited by J. M. Brown, M. P. Mehta, and C. Nieder

Technical Basis of Radiation Therapy
Practical Clinical Applications
4th Revised Edition
Edited by S. H. Levitt, J. A. Purdy, C. A. Perez, and S. Vijayakumar

CURED I • LENT
Late Effects of Cancer Treatment on Normal Tissues
Edited by P. Rubin, L. S. Constine, L. B. Marks, and P. Okunieff

Radiotherapy for Non-Malignant Disorders
Contemporary Concepts and Clinical Results
Edited by M. H. Seegenschmiedt, H.-B. Makoski, K.-R. Trott, and L. W. Brady

CURED II • LENT
Cancer Survivorship Research and Education
Late Effects on Normal Tissues
Edited by P. Rubin, L. S. Constine, L. B. Marks, and P. Okunieff

Radiation Oncology
An Evidence-Based Approach
Edited by J. J. Lu and L. W. Brady

Primary Optic Nerve Sheath Meningioma
Edited by B. Jeremi´c, and S. Pitz

Function Preservation and Quality Life in Head and Neck Radiotherapy
Edited by P.M. Harari, N.P. Connor, and C. Grau

Nasopharyngeal Cancer
Edited by J.J. Lu, J.S. Cooper and A.W.M. Lee

Springer

MIX
Papier aus verantwortungsvollen Quellen
Paper from responsible sources
FSC® C105338

If you have any concerns about our products,
you can contact us on
ProductSafety@springernature.com

In case Publisher is established outside the EU,
the EU authorized representative is:
Springer Nature Customer Service Center GmbH
Europaplatz 3, 69115 Heidelberg, Germany

Printed by Libri Plureos GmbH
in Hamburg, Germany